SATELLITE MICROWAVE REMOTE SENSING

ELLIS HORWOOD SERIES IN MARINE SCIENCE

Series Editor: T. D. ALLAN, Institute of Oceanographic Sciences, Wormley, Surrey

This programme of authoritative books will keep professional scientists and students up-to-date with new advances in the marine disciplines — physics, chemistry, biology, geology and geophysics. It will pay particular attention to advanced techniques and will include comprehensive studies of the role of remote sensing, amongst a wide range of important topics in the field.

SATELLITE MICROWAVE REMOTE SENSING
T. D. ALLAN, Institute of Oceanographic Sciences, Wormley, Surrey

PHYSICAL OCEANOGRAPHY OF COASTAL WATERS
K. F. BOWDEN, University of Liverpool

REMOTE SENSING IN METEOROLOGY, OCEANOGRAPHY AND HYDROLOGY
Edited by A. P. CRACKNELL, Carnegie Laboratory of Physics, University of Dundee

SATELLITE OCEANOGRAPHY
I. S. ROBINSON, University of Southampton

NEW PERSPECTIVES IN MARINE GEOLOGY
R. C. SEARLE and R. B. KIDD, Institute of Oceanographic Sciences, Surrey

MARINE CORROSION IN OFFSHORE STRUCTURES
Edited by J. R. MERCER, University of Aberdeen

SATELLITE MICROWAVE
REMOTE SENSING

Editor:

T. D. ALLAN, B.Sc., D.I.C., Ph.D.
Institute of Oceanographic Sciences
Wormley, Godalming, Surrey
and Chairman, SEASAT Users Research Group of Europe (SURGE)

ELLIS HORWOOD LIMITED
Publishers · Chichester

Halsted Press: a division of
JOHN WILEY & SONS
New York · Brisbane · Chichester · Toronto

First published in 1983 by

ELLIS HORWOOD LIMITED

Market Cross House, Cooper Street, Chichester, West Sussex, PO19 1EB, England

The publisher's colophon is reproduced from James Gillison's drawing of the ancient Market Cross, Chichester.

Distributors:

Australia, New Zealand, South-east Asia:
Jacaranda-Wiley Ltd., Jacaranda Press,
JOHN WILEY & SONS INC.,
G.P.O. Box 859, Brisbane, Queensland 40001, Australia

Canada:
JOHN WILEY & SONS CANADA LIMITED
22 Worcester Road, Rexdale, Ontario, Canada.

Europe, Africa:
JOHN WILEY & SONS LIMITED
Baffins Lane, Chichester, West Sussex, England.

North and South America and the rest of the world:
Halsted Press: a division of
JOHN WILEY & SONS
605 Third Avenue, New York, N.Y. 10016, U.S.A.

© 1983 T. D. Allan/Ellis Horwood Limited

Images from SEASAT's synthetic aperture radar which appear in this book were supplied through the European Space Agency and are published with their permission.

British Library Cataloguing in Publication Data
Satellite microwave remote sensing. —
(Ellis Horwood series on marine science)
1. Oceanography — Remote sensing — Congresses
2. Artificial satellites in oceanography — Congresses
I. Allan, Thomas D.
551.46'0028   GC10.4R4

Library of Congress Card No. 82-15774

ISBN 0-85312-494-9 (Ellis Horwood Limited, Publishers)
ISBN 0-470-27397-6 (Halsted Press)

Typeset in Press Roman by Ellis Horwood Ltd.
Printed in Great Britain by Butler & Tanner, Frome, Somerset

Table of Contents

Table of Contents

Foreword

It is well over ten years since I was actively involved in scientific space affairs, but the excitement and the scope of the field made an indelible impression on me. It came, therefore, as a special delight to me that in the fields of NERC activities space is assuming greater and greater importance, not least in oceanography. I am, therefore, very happy to have been asked to contribute the foreword to this volume.

SEASAT, the first satellite dedicated to studying the sea surface, was launched in June 1978 and operated for little more than 100 days before its mission was brought to an untimely end by a severe power failure. The spacecraft carried a suite of very high resolution microwave sensors which, unlike the majority of earth surveillance satellites that use optical sensors, provided all-weather, day/night capability.

A year prior to SEASAT's launch a group of European scientists, who became known as SURGE (SEASAT Users Research Group of Europe), submitted a joint proposal to NASA and NOAA for an oceanographic/geodetic, glaciological experimental programme based on the use of SEASAT data. The proposal was accepted by NASA who then cooperated with ESA (European Space Agency) in commissioning a receiving station at RAF Oakhanger in Southern England — the only one outside the North American continent.

Almost four years after SEASAT's launch a meeting was held at the Royal Society, London, to present the results of this European research together with invited contributions from Canada and the USA. This book presents their findings.

The SURGE proposal contained contributions from some 30 European laboratories. In the event none of the ice studies planned for the winter 1978–79 was carried out, owing to the early demise of the satellite in October 1978, but several of the other proposals were successfully completed — in particular the North Sea altimeter project in which satellite trackers, geodesists, and oceanographers from France, Germany, Netherlands, and the United Kingdom combined to produce a model to match the satellite's observation. Another element of the SURGE proposal to be successfully completed was the comparison of SEASAT's wind and wave sensors with surface data collected during JASIN. It was a happy coincidence that caused one of the largest joint air-sea interaction experiments ever conceived to take place for a period of almost two months within the satellite's three-month operation. The results of the comparison are fully reported in this book.

SEASAT succeeded in establishing the soundness of the notion that wind and wave conditions over the oceans can be measured from a satellite with an accuracy comparable to that achieved from surface platforms. This is a remarkable breakthrough offering the possibility of permanently closing what is perhaps the biggest gap in the data base for climatology (if climatic forecasting could be made useful, its importance needs no emphasis) and for forecasting weather over the oceans. The biggest problem in these fields is lack of observations — a problem that would be greatly reduced by a single polar-orbiting satellite and virtually overcome by an operational series of observational satellites.

But, apart from wind and wave measurements, SEASAT sensors detected other ocean features of fundamental interest to marine scientists. Imagery from the satellite's synthetic aperture radar revealed detailed patterns of coastal bathymetry and complex slick patterns produced by internal waves. The altimeter measured its own height above the sea surface to a remarkable 10 cm precision, thus revealing a future potential for monitoring the very small changes in ocean topography produced by circulation patterns.

Future satellite missions for ocean surveillance are planned by several space agencies. In particular, the first remote sensing satellite to be launched by the European Space Agency in 1987/88 will be primarily an ocean mission. Such programmes are largely based on the experience of SEASAT. This book produced between SEASAT's demise and the proposed launch of ERS-1 acts as a timely review of the performance of all four microwave sensors carried by the spacecraft, and thus of the scope for the future in this exciting scientific field.

Sir Hermann Bondi FRS
Chairman, Natural Environment
Research Council,
Swindon, England.

Preface

The SEASAT Users Research Group of Europe wishes to acknowledge the encouragement and enthusiastic support given by JPL/NASA at Pasadena especially in the period before and immediately after the satellite's launch. The co-operation of many individuals at the European Space Agency and at the Royal Aircraft Establishment, Farnborough proved very helpful.

Some financial support for the meeting on which this book is based was provided by ESA and by the European Association of Remote Sensing Laboratories. Production of the colour plates in the first chapter was funded by the Natural Environment Research Council.

All speakers without exception submitted their manuscripts in a timely fashion which made the editor's job comparatively straightforward.

Much of the proof-reading was done by Mrs. N. Dodd who also carried out all the secretarial work. Finally, it is a pleasure to acknowledge the co-operation received from the Director and colleagues at the Institute of Oceanographic Sciences.

PART 1

Introduction

Chapter 1

A review of SEASAT

T. D. ALLAN
Institute of Oceanographic Sciences, Wormley, Surrey, England, GU8 5UB

1.1 INTRODUCTION

In many repects SEASAT was a satellite ahead of its time. The first satellite dedicated to studying the ocean surface SEASAT carried a suite of microwave sensors giving it an all-weather, day/night capability. Its orbital height was just under 800 km, its speed over the ground 6.7 km/s, and it completed $14\frac{1}{3}$ revolutions of the earth each day. Launched on 27 June 1978 (European time) its 'proof-of-concept' mission was brought to an untimely end by a power failure after little more than 100 days of operation.

As an experimental mission there was no replacement for SEASAT following its early failure. The marine community, having glimpsed the potential value of modern microwave sensors for monitoring the sea surface, was brought to realise that almost a decade would pass before a similar payload was placed in orbit. Such a long gap does provide the opportunity, however, for a thorough analysis of the three-months data record so that future ocean missions can now be planned around the SEASAT experience. The succeeding chapters of this book describe in some detail the current state of this analysis with particular emphasis on experiments carried out in Europe.

SEASAT brought together a suite of four microwave sensors not previously mounted on a single space platform. They were a radar altimeter (ALT), a synthetic aperture radar (SAR), a wind scatterometer (SASS), and a scanning multi-channel microwave radiometer (SMMR). These microwave sensors were supported by a visible and infrared radiometer (VIRR).

The swaths swept out over the ground by the individual sensors are shown in Fig. 1.1, while the experimental objectives for each sensor are listed in Table 1.1. The major mission of the satellite was the accurate measurement of sea surface winds and waves; however, several other phenomena were observed at the surface — not all of which have found unambiguous explanations. The synoptic view of the sea surface provided by a fast-moving imaging radar is altogether different from the view from a research vessel at sea. The novelty of the technique may make interpretation difficult, but it is already clear that a full understanding of physical processes at the sea surface must be capable of offering an explanation of the often strange and unexpected imagery produced by SEASAT.

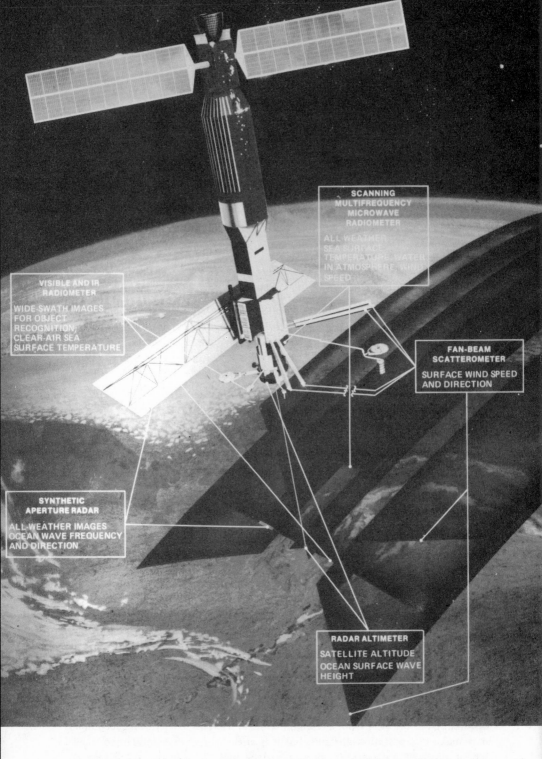

Fig. 1.1 – The sensors carried on SEASAT showing the individual swaths swept out on the ground.

Table 1.1
SEASAT objectives

ALT
- To measure very precisely ($\sim$ 10 cm) the satellite altitude above the sea surface
- To measure the significant wave height of the ocean surface at the subsatellite point
- To utilise the altitude measurement, confirmed with precision OD, to extract oceanographic and marine geoid information

SAR
- To obtain radar imagery of ocean wave patterns in deep oceans
- To obtain ocean wave patterns and water-land interaction data in coastal regions
- To obtain radar imagery of sea- and fresh-water ice and snow cover

SASS
- To deduce local wind vector information from ocean radar scattering coefficient information
- To obtain synoptic ocean radar scattering coefficient measurements over a wide variety of sea and weather conditions and instrument parameters
- To obtain radar scattering-wind vector interaction data over the ocean

SMMR
- To measure ocean surface temperatures
- To measure ocean surface wind speeds
- To measure liquid and vapour water in the atmosphere
- To measure ice coverage and ice characteristics
- To provided propagation corrections for other experiments

VIRR
- To provide image feature identification (land, clouds, etc.) in support of other experiments
- To obtain thermal images of the ocean for various oceanographic purposes

---

1.2 REVIEW OF SENSORS' PERFORMANCE

1.2.1 Radar altimeter

The SEASAT altimeter was a third-generation instrument, previous models having been flown on SKYLAB and GEOS–3. It was designed to measure altitude to 10 cm r.m.s. in order to detect currents, tides, storm surges, and to refine the geoid. It operated at a frequency of 13.5 GHz, its data rate was 10 Kbits/s, and it covered a footprint of 2–12 km depending on sea state. By measuring the

slope of the leading edge of the return pulse and averaging over a sufficient
number of pulses it also provided an estimate of sea surface roughness which could
be translated to a measurement of significant wave height accurate to 0.5 m or
10% in the range 1–20 m.

A precise altimeter is a necessary but not sufficient prerequisite for measuring
the speed of moderate surface currents from a satellite. A current of 10 cm/s, at
latitude 30°, will produce a surface elevation of no more than 7 cm over a
horizontal distance of 100 km; at that level of precision a number of other
effects contribute to the altimeter's signal. Before altimetry can reveal information on surface currents it is necessary to know

 (i) the absolute height of the spacecraft

 (ii) the shape of the marine geoid

(iii) the effect of atmospheric and sea surface elevations on the signal.

Considering the last of these first, the separate atmospheric and sea surface
effects are listed in Table 1.2, and the scale of the corrections is shown in Table
1.3. Although the ionospheric and atmospheric effects can be modelled, an
accuarcy of 2–3 cm for the water vapour contribution can only be achieved by
simultaneous direct measurement. On SEASAT this measurement was made by
SMMR.

Ocean surface topography responds not only to the effect of currents but
also to tides, waves, and to variations in atmospheric pressure (where approximately 1 millibar = 1 cm sea elevation). Deep sea tidal amplitudes can reach
1 metre, and so their variation across an ocean basin must be taken into account.
Of course altimetry may be employed to extract information on these tides
(as discussed by Cartwright in Chapter 19 of this book and in Gower (1981)) but
if the objective is the measurment of surface currents then tides must be considered as part of the 'noise' in the signals, and appropriate corrections made.

Table 1.2
Altimeter path length
Contributions from atmosphere and ocean surface

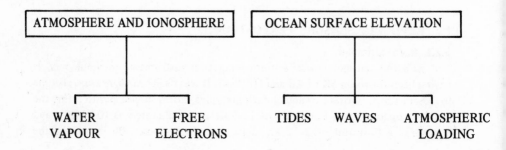

Table. 1.3
Scale of corrections

	Amplitude of correction	Estimated error	Approximate space scale
Sea surface			
Tides	100 cm	10 cm	Basinwide
Waves	5% H_s	10–15 cm	200–2000 km
Atmos. loading	1 mb = 1 cm	2–3 cm	200–2000 km
Atmosphere			
Water vapour			
Dry component	230 cm	1–2 cm	100–2000 km
Wet component	5–25 cm	3–6 cm	100–2000 km
Ionosphere			
Free electrons	1–30 cm	3–5 cm	Global

The effect of waves is to bias the mean reflecting surface of the sea towards the troughs. The extent of this effect has not been determined directly but an empirical treatment suggests a correction of 5% $H_{1/3}$ (Born *et al.* 1982).

Uncertainties in reconstructing the satellite's orbit also add to the altimeter's error budget. SEASAT was tracked by Doppler and laser ranging stations which reduced the r.m.s. error to about 2 m. Subsequent adjustments to minimise the discrepancies at track intersections have reduced the error still further to an r.m.s. value of 70 cm worldwide and to as little as 12 cm on a regional scale (Marsh *et al.* 1982).

The greatest single source of error in the estimation of surface currents from satellite altimetry remains the uncertainty in the marine geoid – the gravity equipotential surface corresponding to the mean level of the sea which would obtain over the oceans in the absence of winds, tides, and currents. Its variation over the surface of the globe from a smooth reference ellipsoid is about ± 100 m; over the Atlantic the geoid slopes from a high of + 65 m west of Ireland to a value of − 50 m off Florida – a gradient of 2 m/100 km. It is against this time-invariant geoidal variation that changes of a few centimetres due to water movement must be detected.

Before the advent of SEASAT models of the geoid had been largely constructed from some 20 years of observations of satellite orbit perturbations, combined with regional marine gravity surveys. These models had been successively improved until they incorporated harmonic terms up to degree and order 36. In some areas, notably the western Atlantic and North Sea, the error in the geoid was estimated to be no greater than 2 m. The data provided by SEASAT's altimeter over the few weeks of its operation have produced a series of updated models containing much greater detail of geoidal variations at higher frequencies.

Orbits

Measurement of the absolute value of the surface current requires the geoid to be known to a higher accuracy than that part of the signal due to the current, that is, to better than 5 cm or so. Nowhere is the geoid known to such an accuracy. It is possible, however, to monitor small *changes* in surface current by arranging for the satellite to repeat its orbit at regular intervals so that the time-invariant geoid can be averaged out. For example, a geoid uncertainty of 1 m would be reduced to 10 cm in less than a year of measurements at a sampling interval of 3 days.

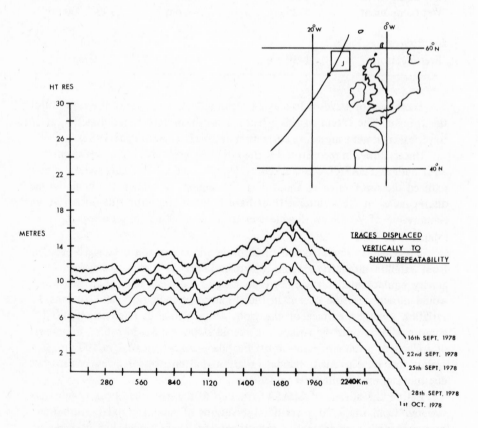

Fig. 1.2 – Residuals of the altimeter's measurement of its own height above the sea surface after removal of a constant value. Five profiles taken at 3-day intervals over the JASIN area are shown.

During the 3 months of SEASAT's operation a near-repeat of 17 days was the pattern followed in the period 26 June–25 Aug. After 16 days of slight adjustments the satellite entered into an exact 3-day repeat period, and during the latter few weeks of its operation achieved a series of 8 repeat orbits; it was those 3-day interval orbits which were subsequently analysed to reveal variations in boundary currents and mesoscale eddies. (Cheyney *et al.* 1981).

A series of 5 altimeter profiles through the JASIN area is shown in Fig. 1.2. Elevations of the geoid at the sub-metre level can be clearly traced from one profile to another.

Wave measurements
The attraction of the altimeter as an instrument for monitoring sea surface conditions lies in its versatility; for not only does it provide an accurate profile of sea surface topography, but analysis of the shape of the return echo can yield data on significant wave height and surface wind speed. The altimeter appears to be capable of measuring significant wave height to an accuracy comparable to that of the surface instruments against which it was compared, as described by Webb in Chapter 25.

Global charts of waves and winds have been constructed from altimeter data by Chelton *et al.* (1981). In its brief life SEASAT provided more wave observations in the southern hemisphere than had been collected by surface vessels and buoys up to then.

1.2.2 Synthetic Aperture Radar (SAR)
Unlike the altimeter the synthetic aperture radar brought no space heritage to the SEASAT mission. Its main purpose was to obtain high-resolution imagery of the sea surface and sea ice; in particular it was not known with any certainty how well, and under what conditions, the directional wavenumber spectra of ocean waves could be imaged from a satellite.

SAR operated at L-band (1.4 GHz, $\lambda \sim 22$ cm) and was designed to image a 100 km swath offset some 250 km to the right of the satellite's sub-orbital track, (i.e. from 250–350 km). The high-resolution goal of 25 m required such a high data rate that no on-board recording was possible, and all SAR data were transmitted in real time to a ground station. Four such stations operated on the North American continent – three in the USA in Alaska, California, and Florida; and one in Newfoundland, Canada. The Oakhanger station, commissioned by ESA and operated by the Royal Aircraft Establishment, Farnborough, was the only SAR receiving station outside North America (see Chapter 2). Thus, no SAR imagery of the southern hemisphere is available. The total SAR coverage over Europe was approximately 10 million km^2 (see Fig. 1.3) yet the *total duration* of the SAR transmissions received at Oakhanger was less than 4½ hours.

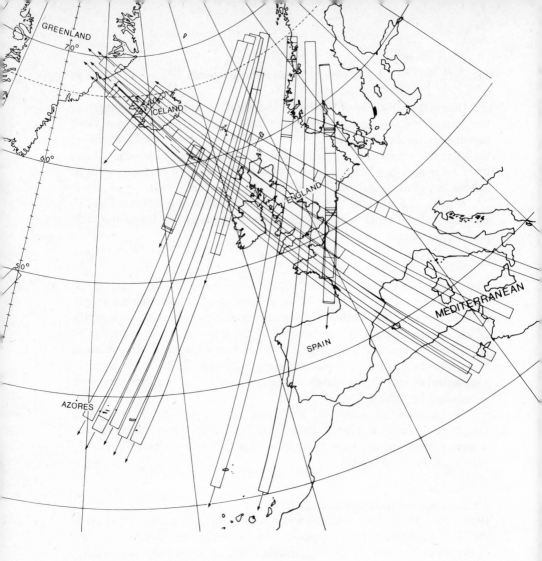

Fig. 1.3 – The total coverage achieved by SEASAT's SAR over Europe. Data received at Oakhanger, England, over a total period of less than 4½ hours. Area covered is 10^7 km². Note the dense coverage over JASIN.

Principles of operation
The scattering of electromagnetic waves from the sea surface depends on:

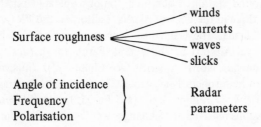

Surface roughness — winds
— currents
— waves
— slicks

Angle of incidence ⎫
Frequency ⎬ Radar parameters
Polarisation ⎭

Although speculation remains on certain aspects of SAR's interaction with a moving sea surface having random and organised components, it is generally accepted that the radar's pulses are Bragg scattered by short gravity waves of wavelength $\dfrac{\lambda_{SAR}}{2\sin\theta} \sim 30$ cm (for L-band SAR looking down at an angle of $23°$ from the vertical). These small ripples are modulated by the longer gravity waves, so that the tilting of the small-wave specular points reflects the wave pattern of the larger waves on the SAR image. The other features which affect surface roughness — winds, currents, and slicks — have all been clearly identified on SEASAT SAR imagery.

SAR achieves fine resolution in the range (or cross-track) direction by transmitting a very short pulse. In SEASAT the pulse was compressed to about 50 ns so that the resolution in the look direction was $\delta r = \dfrac{ct}{2} = 7.5$ m which became $7.5/\sin 23° = 25$ m ground resolution. The time interval between successive pulses was $600\,\mu s$ for SEASAT, corresponding to a spacing of 200 km.

High resolution in the azimuthal (along-track) direction relied on the synthetic aperture technique. This operates on the principle of coherently summing the echoes received from a target while the radar travels a distance L to provide in effect the signal that would have been received from a real array antenna of length L. In other words a long antenna is synthesised through the forward movement of the spacecraft.

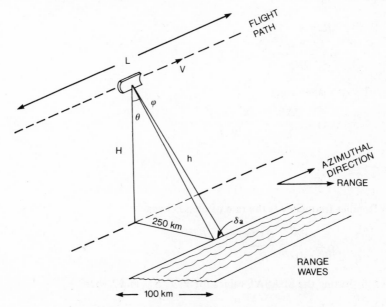

Fig. 1.4 – Geometry of SAR operation.

The angular resolution of an antenna (ϕ in Fig. 1.4) is inversely proportional to the aperture width expressed in radar wavelengths. For 2-way transmission then

$$\varphi = \frac{\lambda}{2L}$$

where L is the length of the synthetic aperture.

Thus,

$$\frac{\delta a}{h} = \frac{\lambda}{2L}$$

If V = speed of satellite, and T is the total time a target remains in the beam, then $L = VT$. For SEASAT, $V = 7.5$ km/s; $T = 2$s; $\lambda = 0.22$ m; $h = 850$ km Thus, $\delta a = 6$ m.

To reduce speckle, four azimuthal resolution cells are summed incoherently so that the final 'four-look' azimuthal resolution is made equal to the range resolution of 25 m. It should be noted that azimuthal resolution depends only on the width of the real antenna aperture and is independent of radar wavelength and range to target.

To image a point target at the correct position in the image plane the SAR processor makes use of the perceived phase history of the point during the 2 seconds it remains in the beam. For stationary targets this can be derived as follows (see Fig. 1.5):

$$\Delta R = R - R_0$$
$$\sim \frac{(x - x_0)^2}{2R_0}$$

For a 2-way phase change,

$$\varphi(x) = \frac{2\Delta R}{\lambda} = \frac{(x - x_0)^2}{R_0 \lambda}$$

$$= \frac{V^2 t^2}{R_0 \lambda}$$

The Doppler frequency is the rate of phase change

$$f_D = \frac{d\varphi}{dt} = \frac{2V^2 t}{R_0 \lambda}$$

and substituting the SEASAT values for V, T, R_0, and λ then

$$f_D \sim \pm 600 \text{ Hz for} -1 \leqslant t \leqslant 1 \ .$$

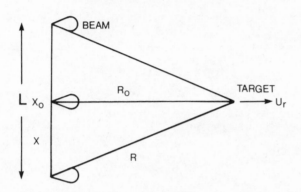

Fig. 1.5 – Geometry of Doppler shift from a SAR target.

Now if the target moves with a radial velocity U_r during the time it remains in the beam then the effect will be to shift the Doppler frequency by $-2\dfrac{U_r}{\lambda}$. By equating the two Doppler terms the displacement in the azimuth direction can be seen to be

$$\Delta x = \frac{R_0}{V} \cdot U_r$$

and for SEASAT $R_0/V \sim 100$. Thus a radial velocity 1 m/s would produce a displacement of 100 m.

The azimuthal shift of targets moving with a constant radial velocity during the imaging time can be clearly detected in images where ships are displaced relative to their wakes. Fig. 1.6 shows the effect for four ships in the English Channel (rev 762). The measured displacements lay in the range 280–370 m which, taking account of a 23° radar look angle, corresponds to speeds of 7.6–9.4 m/s (15–19 knots).

Imaging of ocean waves
Three mechanisms have been proposed for imaging of ocean waves by synthetic aperture radar. These are (i) tilting of the small gravity waves by the longer waves, (ii) a hydrodynamic effect causing the short waves to be concentrated on the forward slope of the longer waves, (iii) the orbital velocity of the long gravity waves causing individual water particles to move in and out of phase with the radar look direction. The separate mechanisms are illustrated in Fig. 1.7, and their effects are discussed in considerable detail in the chapters on SAR operation. We shall limit ourselves here to a simplified description of the degradation in resolution that may be introduced into SAR imagery through the orbital motion of waves.

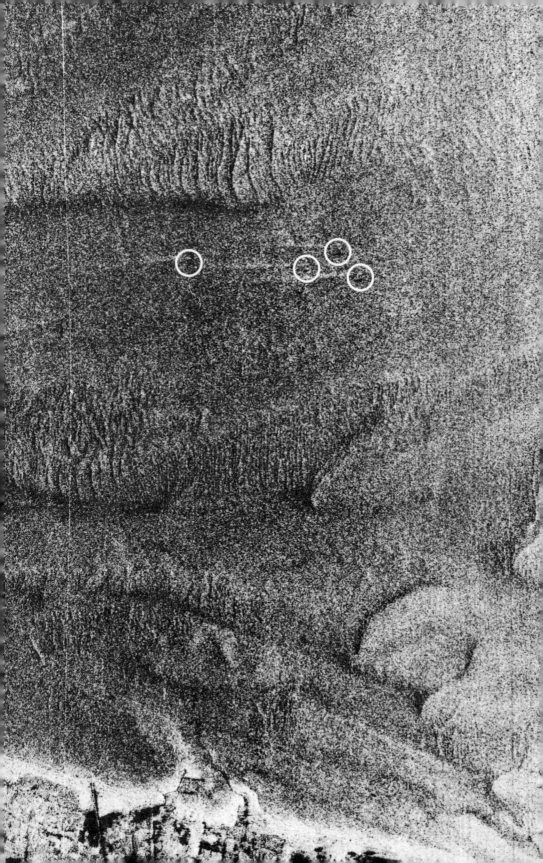

In the previous section we saw that a radial velocity of U_r will displace a target in the image plane by $\dfrac{R}{V_0}$ U_r. If we consider an average swell wave with a 10-second period and an amplitude, a, of 2 m (peak-to-trough) then its orbital velocity is $\dfrac{\pi a}{T} \sim 0.63$ m/s. Thus from Fig. 1.7 c the crest of a wave may be moved 63 m in one direction and the trough 63 m in the opposite direction — that is, a 150 m long wave could suffer a relative displacement in the image plane of 126 m. It was this argument, presented before the launch of SEASAT, that convinced several oceanographers that a SAR deployed from a satellite would be incapable of imaging waves. But waves were clearly imaged by SEASAT SAR, as can be seen in Figs 1.8 and 1.9 and in other chapters. There remains, however, the important question of the degree of distortion in the image. Are some wavelengths imaged selectively and other suppressed?

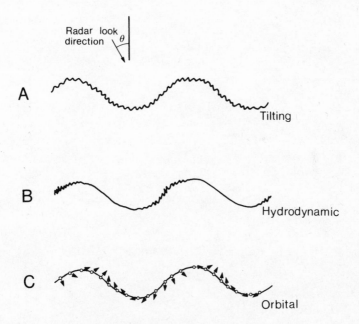

Fig. 1.7 — Three effects which are believed to play a part in the interaction of SAR with a moving sea surface: (a) tilting; (b) hydrodynamic effect; (c) orbital velocity.

Facing page. Fig. 1.6 — SAR image of English Channel. The uneven texture reflects the bottom topography. Note the displacement of the four ships (sailing in a cross-track direction) from their wakes.

Fig. 1.8 – An example of range (cross-track) travelling swell waves together with some internal wave trains. Rev. 1049, JASIN area.

Facing page. Fig. 1.9 – A comparatively rare example of quasi-azimuthal (along-track) travelling swell. Rev. 1087, JASIN area.

For range-travelling waves the effect of the azimuthal shift would be a displacement along the direction of crests and troughs; in that case the dominant wavelengths may be retained in the image. Conversely, azimuthal waves should suffer significant distortion since the direction of displacement is at right angles to the wave front. However, where the displacement equals $\dfrac{n\lambda}{4}$, then enhancement of those wavelengths should occur while other wavelengths are smeared out.

From the JASIN record both range waves and waves travelling within 20° of azimuth were clearly imaged. However, it does appear that the imagery of azimuthal waves displays longer dominant wavelengths than similar imagery of range waves. Range-travelling waves also appear to have been more prevalent over the JASIN area but this may have more to do with the prevailing wave direction and the ratio of ascending to descending passes over which SAR was operated. Where waves were detected, range waves were imaged on ascending passes, near-azimuthal waves on descending passes. Examples of optically-processed imagery displaying range and azimuthal waves over JASIN are shown in Figs 1.8 and 1.9.

Fourier transforms of SAR wave imagery
Digitally processed SAR imagery allows fast two-dimensional discrete Fourier transforms to be calculated on the spatial dimension of an image. The CCTs of SAR imagery provided by ESA contain a maximum of 4376 azimuthal lines each containing 3960 pixels 12.5×12.5 m^2 in area. Thus the maximum area of a single scene available on tape is about 50×56 km^2. Such a scene can be subdivided and the separate parts Fourier transformed to reveal any changes in the wave-number spectra across an image. Col. Fig. 1.10 illustrates the technique. A 3×3 array of Fourier transforms, each 512×512 pixels corresponding to an area 6.4×6.4 km^2 over a scene from rev 762, is shown. The waves are range-travelling, indicated by the N–S orientation of the transform. Dominant wavelengths (measured from the red-tinted highligths on the transforms) lay in the range 200–300 m across the 9 samples, but the individual transforms indicate significant energy across a wide distribution of wavelengths in the range 100–350 m. One ship R/V *John Murray,* was in the region of the SAR swath, but, although her wave recorder was operational, she was steaming in waves measured at 5 m signficant height, and her recordings of wave period are probably not reliable. Two wave buoys in the JASIN area some 300 km to the west measured the period of wave spectral maximum as 11.7 s, which corresponds to a dominant wavelength around 210 m, in reasonable agreement with the transform values.

A similar example of 3×3 Fourier transforms equally spaced across a digitally processed image is shown in Col. Fig. 1.11 (rev. 1044). Here the dominant direction of the transforms shows the waves to be travelling within 30° of azimuth. Again, areas of highest intensity equate to dominant wavelengths and

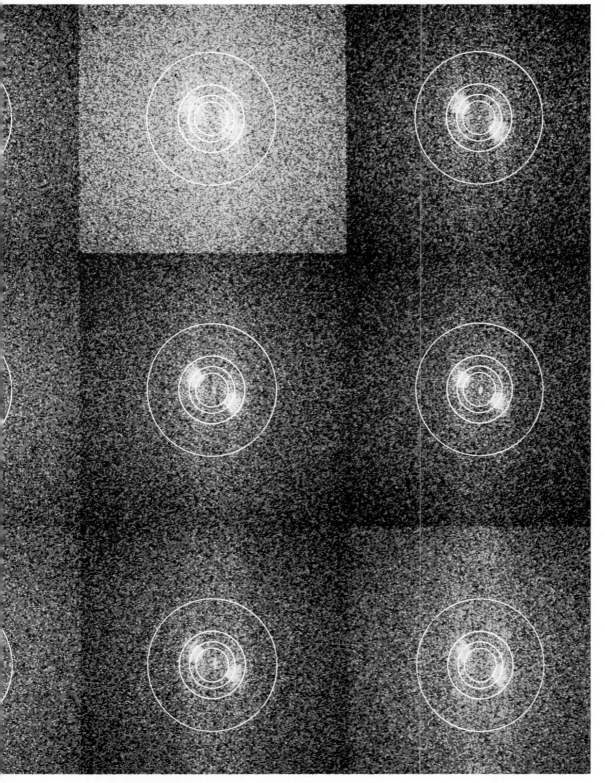

Col. Fig. 1.11 – (*see over for legend*)

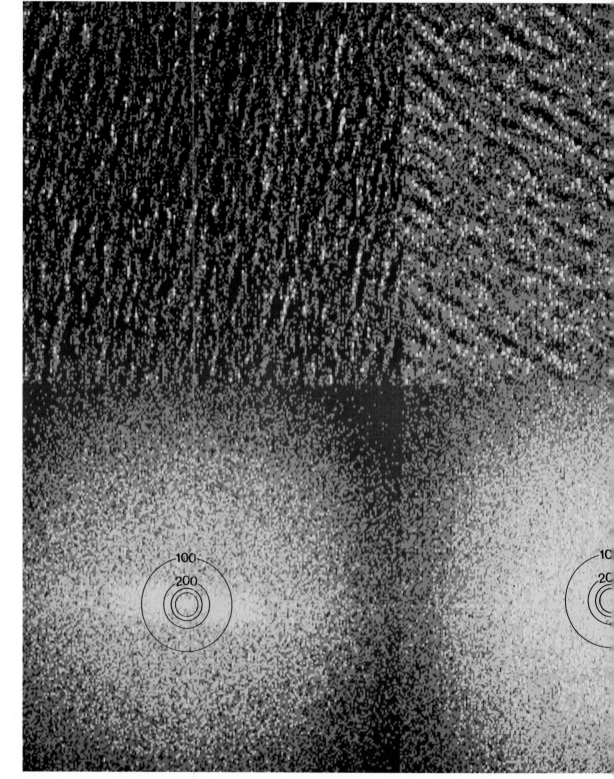

Col. Fig. 1.12 – (*see opposite for legend*)

Col. Fig. 1.10 — A 3 × 3 array of digital transforms carried out on rev 762 over JASIN. Very little difference in wavenumber and direction is revealed across the area of the image which is approximately 40 km on the side.

Col. 1.11 — A similar 2-D Fourier transform made on an image from rev 1044 showing a wave direction within 30° of azimuthal.

Col. Fig. 1.12 — A comparison of digitally-processed wave imagery and their Fourier transforms on successive ascending and descending revs over JASIN. The areas imaged do not overlap, and there is an 8-hour interval. There appears to be a significant difference in the character of the wave field.

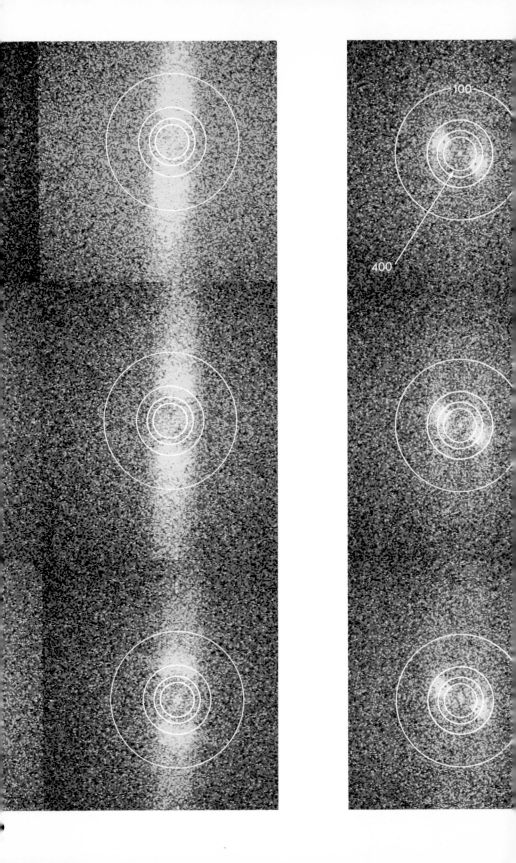

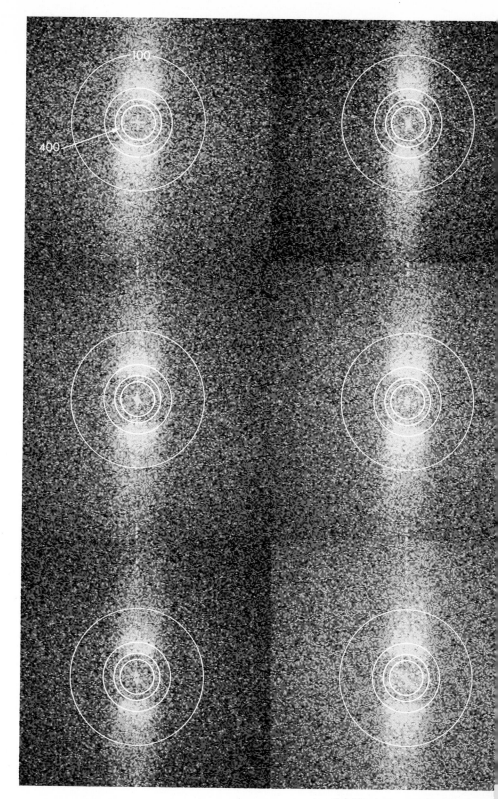

Col. Fig. 1.10 — (*see over for legend*)

are coloured red on the transforms. Over the 3 × 3 arrays these wavelengths lay in the range 300–330 m and show no significant change in dominant direction.

One method of checking differences in the imaged wave field of range and azimuth-travelling waves would be to examine the same scene on consecutive descending and ascending passes. Unfortunately, over the JASIN area the time interval was over 8 hours (5 revs at 100 minutes per rev), which is long enough for the character of the waves to change. Notwithstanding this drawback a portion of the image produced on rev 1044 (descending) was compared with an image from rev 1049 of an area close to (but not overlapping) that of 1044. The two images and their Fourier transforms are shown in Col. Fig. 1.12. It is reasonably clear from the wave imagery alone that 1044 displays longer dominant wavelengths than 1049, and this is confirmed in the Fourier transforms where the dominant wavelengths over the areas shown are 320 m and 240 m respectively.

RRS *Discovery* was on station at position 59° 16′ N 13° 06′W during the SAR overflights. From her ship-borne wave recorder the periods of the wave spectral maximum were calculated as 12.7 s (rev 1044) and 11.2 s (rev 1049), which equate to dominant wavelengths of 240 m and 185 m respectively. Thus, although the absolute values do not agree particularly well with the Fourier transform values of 320 and 240 m it is significant that the wavelengths decrease at about the same ratio over the 8-hour interval. Such a significant change with time will be reflected in corresponding spatial changes in the wave field, so it is not improbable that *Discovery*, on station some 50–100 kilometres from the SAR imagery, would record a different wave field.

SAR imagery of internal waves, slicks and streaks

Trains of internal waves had been identified and studied in many different parts of the oceans many years before the launch of SEASAT. Nevertheless they were not generally considered to be as ubiquitous as the SAR record eventually proved them to be. Because of their longer wavelengths and their dispersive nature they stand out more clearly on the average SAR record than the fine 'finger-printing' caused by surface gravity waves. A good example is shown in Fig. 1.13 which is a 140 × 100 km image of the continental shelf off the coast of Portugal showing intersecting trains of internal waves propagating towards the coast. Surface swell is also present over most of the image, running parallel to the internal waves over part of the area and crossing them at right angles over other parts. The dominant wavelength of the swell (measured directly from the image) is 250–270 m decreasing to 150 m nearshore.

The coastline shown in the image runs between C. Mondego in the South to a point north of Aveiro. The whole of the coast is sandy, and it is reported in the *British Pilot* of the area (Hydrographic Dept. 1972) that "due to the influence of wind and currents, new sand banks, *on which the sea breaks heavily* are continually forming and disappearing". Breaking waves can be clearly identified in the image as a white band − indicating increased backscatter −

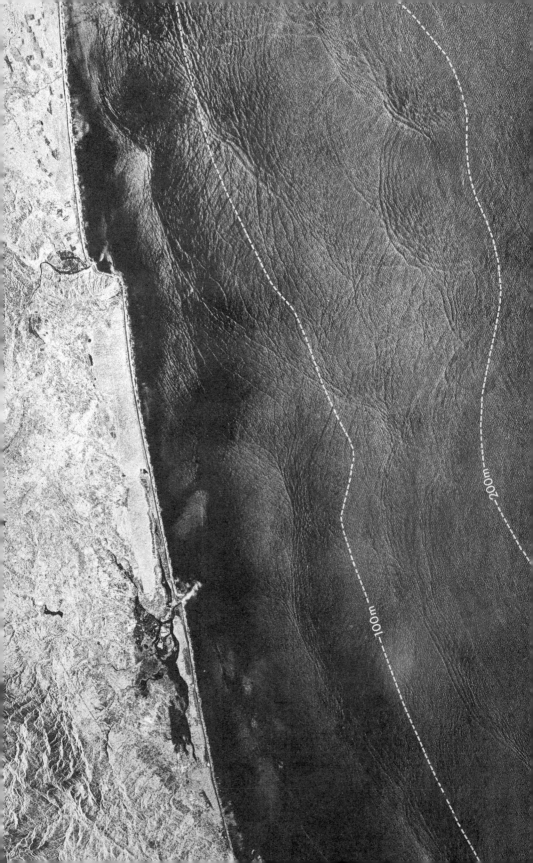

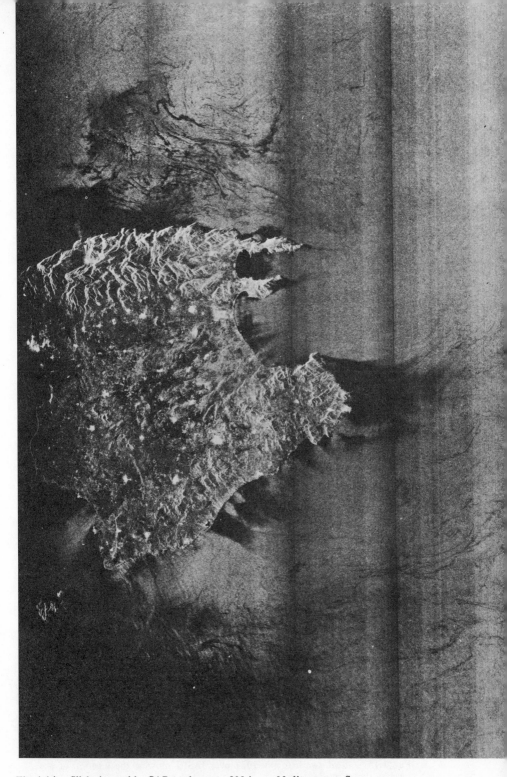

Fig. 1.14 – Slicks imaged by SAR to the east of Majorca, Mediterranean Sea

Facing page. Fig. 1.13 – A proliferation of internal wave trains west of Portugal. Swell waves were also clearly imaged and can be seen breaking against the sandy beach.

running along the shore. The wavelength of the internal waves varies in the range 500 m to over 1000 m. At the western edge of the image the water depth reaches over 1000 m and wave trains are less pronounced; the greater part over which the internal waves are propagating represents the continental shelf shoaling from 200 m as shown.

Slicks more randomly patterned than internal wave trains were also imaged by SAR, and an example off Majorca is shown in Fig. 1.14. The L-Band employed in SEASAT SAR appears to have been particularly sensitive to small changes in surface roughness.

As distinct from slicks, several SEASAT images have revealed SAR streaks. Over JASIN they occur on rev 714, 1006, and 1087 when winds were in the range 16–20 m/s. SAR streaks are quite distinct from surface swell and internal waves. They appear to be spaced 1–2 km apart, and their appearance on a SAR image is accompanied by the absence, and on one occasion the replacement of imaged waves. The two images shown in Fig. 1.16 are from rev 1087, a descending pass on 11 Sept. 1978, and they are separated by 400 km (Fig. 1.15). The top image shows quasi-azimuthal travelling waves clearly imaged; to the south the wind appeared to increase from 15 m/s to 20 m/s (according to the scatterometer) and the streaks replace the wave imagery. A possible explanation is that at a critical wind speed breaking waves create enough foam over the whole of the wave profile to cause the difference in backscatter from crest and trough to become indistinguishable (Allan & Guymer, in press). Further research into their effect is required.

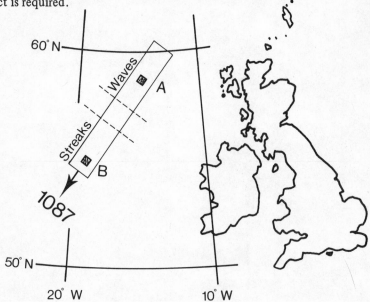

Fig. 1.15 – SAR swath of rev 1087 over JASIN showing the location of the imagery shown in Fig. 1.16.

Fig. 1.16 — The top image shows swell recorded over the northerly part of the swath. In the bottom image the swell has given way to streaks as the wind strength increased.

1.2.3 Wind scatterometer (SASS)

The scatterometer was designed to measure ocean surface wind speed from 4 m/s to > 26 m/s with an accuracy of 2 m/s or 10%. It operated at Ku-Band (14.6 GHz, $\lambda \sim 2$ cm). Four antenna beams arranged at $45°$ to the spacecraft's track were deployed to estimate wind direction. SASS was designed to sweep out two 500 km swaths separated by 400 km, together with a narrower swath around the sub-satellite track (see Fig. 1.1). Its resolution cell was about 50 X 100 km.

SASS was another sensor that had flown previously on SKYLAB. Research into radar interaction with the sea surface sprang from the initial investigations carried out in World War II into the causes of radar sea clutter. The physical principle behind the operation of a wind scatterometer is that the strength of the radar backscatter is proportional to the capillary wave amplitude which, in turn, is in equilibrium with the surface wind stress. However, it is not entirely clear why the capillary wave spectrum does not saturate at low to moderate wind speeds. Other mechanisms may play a role. Certainly, there is a large degree of empiricism relating $\sigma°$ to surface wind speed based on a comprehensive set of field experiments.

The algorithms that were developed prior to SEASAT's launch were tested against the 'in situ' recordings of calibrated anemometers during the specially-commissioned Gulf of Alaska SEASAT Experiment (GOASEX) and, subsequently, against similarly well-calibrated surface measurements in JASIN.

SASS could transmit and receive at both vertical and horizontal polarisation. The switching of this facility between the four antennas gave the instrument a total of nine operating modes which were selected according to the nature of the experiment and the quality of the available surface data. On several of those modes coverage was restricted to one side of the spacecraft.

The performance of SASS is described in some detail by Pierson (Chapter 4) and Guymer (Chapter 5) of this book. The instrument appears to have met its objectives. The r.m.s. difference in wind speed between SASS and JASIN observations over some 23 passes was less than the ± 2 m/s goal and wind directions agreed to better than $020°$. Reliable synoptic wind velocity data over the surface of the global oceans are a prospect which can now be seriously contemplated with an operational system of a few satellites. This demonstration may be regarded as one of SEASAT's greatest achievements.

1.2.4 Scanning Multi-channel Microwave Radiometer (SMMR)

SMMR, the only passive sensor of SEASAT's suite of four microwave instruments, covered a 600 km swath to the right of the spacecraft. It received both vertically and horizontally polarised radiation at 5 frequencies – 6.6, 10.7, 18.0, 21.0, and 37.0 GHz. From this matrix of 10 channels of information the main goals for the instrument were the extraction of:

(i) sea surface temperature to $2°K$ absolute and $1°K$ relative over resolution cells approximately $80 × 150$ km;

(ii) ocean surface wind speed (but not direction) to 2 m/s;

(iii) integrated liquid water content and water vapour.

(iv) rain rate;

(v) ice age, concentration, and dynamics.

SMMR is the only one of SEASAT's sensors now functioning on another satellite, NIMBUS-7, so that the complex algorithms and their subsequent refinement during validation experiments can be applied to an operational instrument.

We have seen that to improve the accuracy of the altimeter's measurement of sea surface topography a simultaneous monitoring of atmospheric water vapour is required, and on SEASAT this information was supplied primarily from the 18 and 21 GHz channels.

SMMR's all-weather capability gives it a considerable advantage over infra-red devices for monitoring sea-surface temperature. If it could reliably monitor changes in heat flux over the oceans it would be an invaluable tool in future climate studies. Its spatial resolution of about 150 km may be poor for studying localised features such as fronts or eddies, but would be acceptable for studies at ocean basin or global scales provided it could achieve high absolute accuracy and this is where doubt remains over the present generation of microwave radiometers. SMMR's estimates of SST usually fell within $1.5°K$ of surface observations but undoubtedly part of this discrepancy stems from errors in the surface measurements.

It appears that in its present state SMMR would require additional support in measuring sea-surface temperature to make a significant impact on climate studies. However, its ability to measure integrated atmospheric water vapour to the same accuracy as radiosondes is a significant breakthrough; equally, its role as a back-up instrument to the altimeter is now confirmed.

Later chapters in the book include a comprehensive review of the performance of SEASSAT's SMMR (Chapter 29), together with a description of how the geophysical algorithms were developed (Chapter 30).

1.3 JASIN

A major proposal included in the SURGE submission to NASA was a comparison of wind and wave fields between SEASAT sensors and JASIN ships. It was a piece of good luck that one of the largest air-sea interaction multi-ship experiments ever to take place should coincide with the first two months of the spacecraft's mission — the more so because of its early failure. The JASIN experiment was carried out over an area in the North East Atlantic roughly halfway between Northern Scotland and Iceland (Royal Society 1979). A total of 14 ships participated from 5 countries as well as 3 meteorological research

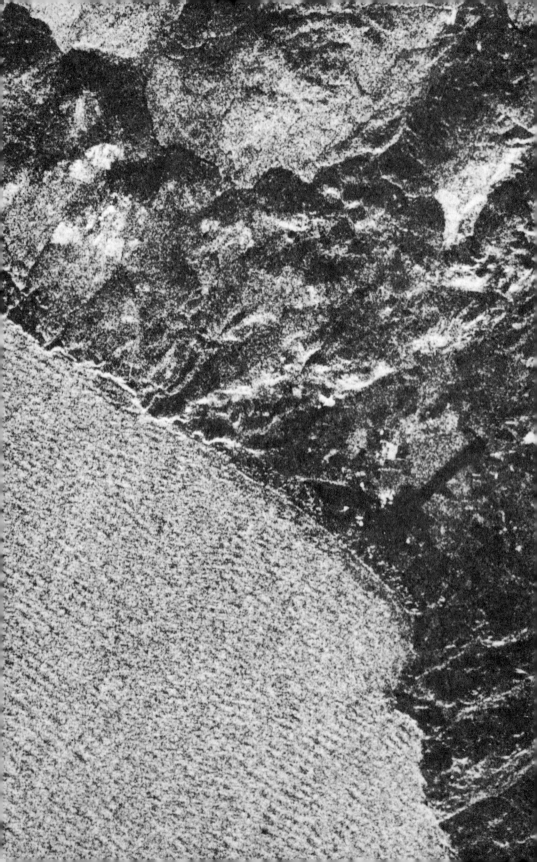

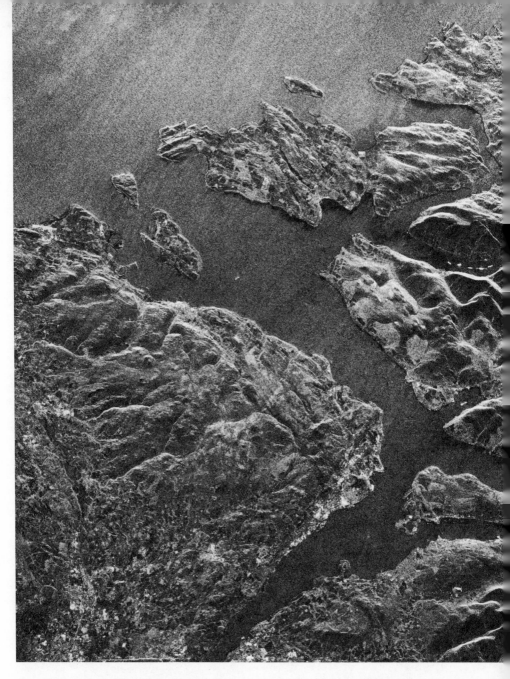

Fig. 1.18 — A SAR image of the Firth of Clyde used by JASIN ships for a mid-cruise break. Note how clearly ships at anchor in a sea loch are imaged.

Facing page. Fig. 1.17 — A SAR image of Machrihanish, Kintyre, Scotland, from which the the meteorological research aircraft flew to the JASIN area and from which SEASAT operations were coordinated.

aircraft operating from RAF Machrihanish on the Mull of Kintyre; from there ships' observations were coordinated with SEASAT overflights of the area. A SAR image of the airfield is shown in Fig. 1.17 and Fig. 1.18 is an image of the area of the Firth of Clyde — used by the JASIN ships for their mid-term breaks. The clarity with which ships, even when stationary, can be detected is evidenced by the image of several vessels at anchor in the sea loch north-west of the Clyde.

Table 1.4
Analysis of SEASAT sensors performance over JASIN

SENSORS	OBJECTIVE	JASIN COMPARISON
(1) SASS	$\pm$ 2 m/s, 020° for 4 $<$ V $<$ 26 m/s	0 $\pm$ 1.6 m/s, 018° for 3 $<$ V $<$ 16 m/s
(2) SMMR (wind speed only)	$\pm$ 2 m/s for 7 $<$ V $<$ 50 m/s	Before rev 900 SMMR surface 2.5 $\pm$ 1.8 m/s After rev 900 SMMR surface 0.6 $\pm$ 0.9 m/s for 3 $<$ V $<$ 16 m/s
(3) Altimeter H_s Wind speed	0.5 m or 10% $\pm$ 3 m/s	$\pm$ 10 cm 0 $<$ H_s $<$ 4 m $\pm$ 2 m/s 0 $<$ V $<$ 10 m/s Begins to saturate above 10 m/s
(4) SAR a) Directional Wave spectra		Dominant wavelength to $\pm$ 15% Dominant direction to $\pm$ 10° (when waves clearly visible)
b) Surface features : Surface streaks		$\lambda \sim$ 2-3 Boundary Layer Depth. (When observed wind speed $>$ 15 m/s).
: Internal waves		In vicinity of seamounts

JASIN was well-covered by SEASAT. The importance attached to the area as a source of reliable validation data is clear from the plot of SAR coverage over Europe shown in Fig. 1.3 in which JASIN is seen as the pivot for many ascending and descending passes. Such a coverage eventually allowed SAR mosaics of Iceland and the United Kingdom (4 sheets) to be prepared (Huntings 1980).

Succeding chapter of this book describe in some detail the comparisons made between JASIN surface observations and SEASAT estimates of scatterometer winds, significant wave height, and atmospheric water vapour. In each case the sensors exceeded their accuracy objectives. A summary of JASIN comparisons is shown in Table 1.4.

1.4 CONCLUSIONS

SEASAT succeeded in proving that modern microwave sensors deployed from a polar-orbiting satellite can provide very accurate data on sea surface conditions — useful both in the short-term for weather forecasting, ship routing, and offshore operations, and in the longer term for air-sea interaction and climate studies.

In future operational missions the data on wind and wave fields should initally provide a valuable complement to existing observational networks at sea; eventually satellite data may represent the major input to meterological centres.

The longer-term prediction of sea conditions is an important consideration in the design of rigs and platforms, especially in more remote areas where surface observations may be sparse. At present a single satellite equipped with a radar altimeter could provide reasonably frequent observations of significant wave height over a fixed sampling grid at the expense of a wide spacing between the grid lines. As an example wave height could be measured twice every 3 days (on ascending and descending passes) if the equatorial spacing between adjacent tracks were 950 km. This may leave unacceptably large areas unsampled from a single satellite. The alternative is to increase the sampling interval (in SEASAT the intervals were 3 and 17 days) or employ more than one satellite. The trade-off between sampling frequency and spatial coverage for satellites in SEASAT's orbit is shown in Fig. 1.19.

Perhaps the major revelation of primarily scientific interest to emerge from the SEASAT mission was the potential value of very accurate monitoring of sea surface topography provided by a precise radar altimeter. When adjustments are made to the calculated orbits in order to minimise the discrepancies at the many track intersections, the details of the geoid that are revealed are quite astonishing as Marsh *et al.* (1982) have shown. An example of their geoid plotted at 1 m interval over the North East Pacific highlighting the major fracture zones is shown in Fig. 1.20.

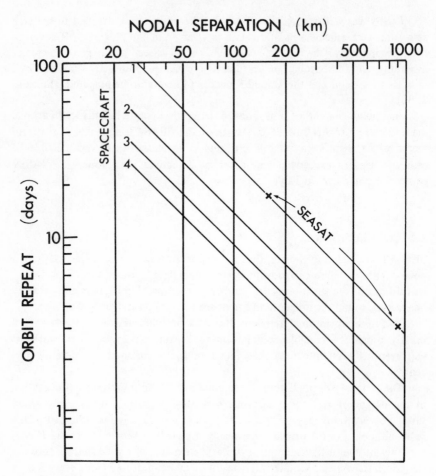

Fig. 1.19 – Plot of distance between adjacent orbits across the equator as a function
of sampling interval over one location, and number of satellites in orbit.

Already the SEASAT altimeter has demonstrated its ability to detect western
boundary currents and mesoscale eddies. Future missions such as NASA's
proposed TOPEX satellite, which would be dedicated to measuring only sea
surface topography, and Europe's ERS-1, may help to demonstrate that satellites
represent the only feasible method of measuring ocean circulation patterns on
global or ocean-basin scales.

SEASAT more than justified its promise. In the history of marine science
and exploration its brief mission may yet come to be considered as one of the
most important events since *Challenger* left port over a century ago to sail
round the world on a 3½ years voyage and lay the foundations for the systematic
study of the sea.

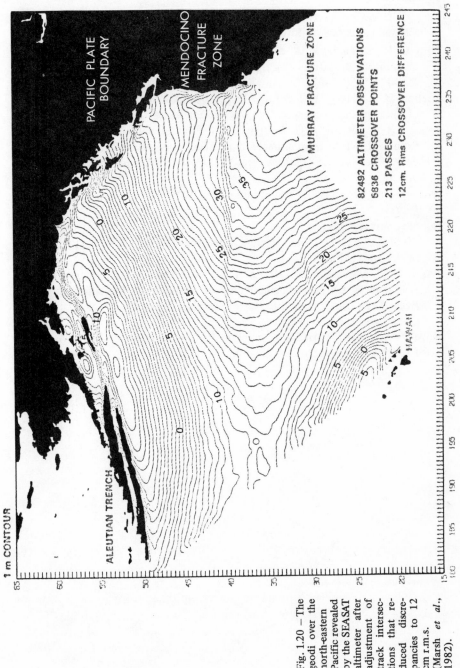

Fig. 1.20 – The geodi over the north-eastern Pacific revealed by the SEASAT altimeter after adjustment of track intersections that reduced discrepancies to 12 cm r.m.s. (Marsh *et al.*, 1982).

REFERENCES

Allan, T. D. & Guymer, (in press). SEASAT measurements of wind and waves on selected passes over JASIN. *Int. J. Rem. Sensing.*

Born, G. H., Richards, M. A. & Rosborough, G. W. (1982). The empirical determination of the effects of sea state bias on SEASAT altimetery, *J. Geophys. Res.* **87** C5, 3221-3227.

Chelton, D. B., Hussey, K. J. & Parke, M. E. (1981). Global satellite measurements of water vapour, wind speed and wave height, *Nature* **294** (5841) 529-532.

Cheney, R. E., Marsh, J. G. & Grano, V. (1981) Global mesoscale variability from SEASAT collinear altimeter data. *EOS Trans. AGU* **62** (17) 298.

Gower, J. F. (editor) (1981) *Oceanography from space,* Plenum Press, New York, 987 pp.

Hydrographic Dept. (1972) 5th edition of the *West Coast of Spain and Portugal Pilots.*

Huntings Surveys 1980
    (i) SEASAT Radar Mosaic, United Kingdom, 4 X 1/500,000
    (ii) Iceland, 1/500,000

Marsh, J. G., Cheney, R. E., Martin, T. V., McCarthy, J. J. & Brenner, A. C. (1982) Mean sea surface computations based upon satellite altimeter data. *Proc. Third Int. Symp. on Use of Artifical Satellites for Geodesy and Geodynamics,* Ermioni, Greece.

Royal Society (1979) Air-Sea Interaction Project *Summary of the 1978 field Experiment,* Royal Society, London.

Chapter 2

SEASAT–data acquisition and processing by the Royal Aircraft Establishment

D. HARDY
Royal Aircraft Establishment, Farnborough

2.1 INTRODUCTION

Towards the end of 1976 the member states of the European Space Agency expressed interest in acquiring and disseminating among them the data from experimental remote-sensing spacecraft which were to be launched by NASA in 1978. The task would be coordinated by the Agency's own remote sensing network, known as EARTHNET, making use of facilities offered by the member states.

In January 1977 the UK delegation to ESA offered the use of the ground station at Oakhanger and the data processing facilities at the Space Department, Royal Aircraft Establishment, Farnborough. Although the offer was not confined to SEASAT, this mission was identified as the one which was of greatest scientific and technological interest to the UK. This paper describes the preparation for, and execution of, the data acquistion and processing operations.

2.2 BACKGROUND

The UK interest in SEASAT arose from the national preoccupation with things maritime and from a desire to explore the potential of microwave sensors for all-weather monitoring of the sea and land surface. There was also the techno-logical challenge presented by the requirement to acquire and process the data from the synthetic aperture radar (SAR) on board SEASAT.

The UK offered a favourable coverage of the North Atlantic and European waters from a station which, ideally, would have been sited in the north-west of the country. Establishment of a new station on such a site was not favoured, partly because of cost but also because the spacecraft was due to be launched in May 1978 and there was barely sufficient time for such a venture. Consequently, in December 1976, the Department of Industry requested Space Department RAE to search for an existing station which could be used for SEASAT data acquisition.

The requirements were essentially that it could operate at S-band (2200–2300 MHz), that it had a large aperture ($\geqslant 10$ m), and that it be capable of tracking SEASAT which would be in a low orbit (800 km orbital height). There were relatively few large S-band stations in the UK, and none of them had all the characteristics required. The most promising was the Tracking and Command Station (TCS) at Oakhanger (Fig. 2.1) which was then operated by the Royal Air Force. Because of a temporary decline in its workload, it would have spare capacity during the SEASAT mission. The station radio horizon (Figs 2.2 and 2.3) was a good match to the SEASAT coverage requirements and it had an antenna of diameter 18.3 metres, which would enable a good signal-to-noise ratio to be achieved in the reception of the telemetered SAR signals. Its proximity to Farnborough, and to Heathrow Airport, was to prove extremely valuable in achieving operational readiness by the time of launch.

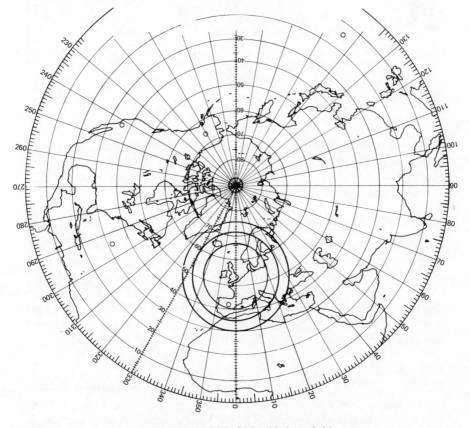

Fig. 2.2 – Coverage of SEASAT orbit from Oakhanger

Facing page: Fig. 2.1 – Telemetry and command station – RAE Oakhanger.

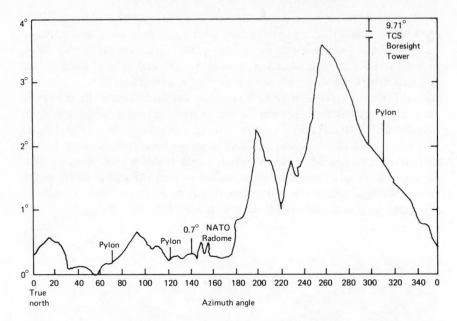

Fig. 2.3 – The radio horizon of TCS Oakhanger

It was established that the station could be made available for civil use on a basis of sharing with military requirements, and that SEASAT could be given priority in allocation of station time during the life of the mission. Accordingly Oakhanger was selected and offered to ESA.

2.3 MODIFICATION AND PREPARATION

The offer was made less than 18 months before the planned launch of SEASAT. Other competing offers had been made, and it was inevitable that the process of technical evaluation and selection would take some time. It was therefore decided to assume that Oakhanger and Farnborough would be used for the most technically demanding of the experimental missions, which was SEASAT, and to put in hand the design of the modifications needed at the station and the additional data-processing facilities at RAE up to and including the order of long lead items.

2.3.1 TCS Modifications

TCS Oakhanger was a conventional station for supporting geostationary satellites in their transfer and final orbits. It had an altazimuth wheel and track antenna mount with a slew rate in azimuth of $2°$ sec^{-1} and acceleration of $2°$ sec^{-1} sec^{-1}. This, combined with the narrow beamwidth of the large antenna, meant that it was unable to follow satellites in near earth orbit as they approached the zenith. The major modification required was to minimise this 'cone of silence'.

There were three possibilities:

(i) To install an auxiliary antenna to take over while the main antenna was slewed around zenith.

(ii) To provide additional tracking movement by controlled tilting of the Cassegrain sub-reflector.

(iii) To increase the azimuth slew rate.

With the aid of consultants, these alternatives were examined in some detail; (i) was too expensive, (ii) gave insufficient movement of the boresight, and (iii) presented some quite horrifying mechanical problems — mainly because the slew rate theoretically needed was in excess of $50°$ sec^{-1} and because the antenna weighs over 120 tonnes. In fact, by making use of the fact that the path-length attenuation of the signals from the satellite is approximately 12 dB less at the zenith than at the horizon, it was possible to track on the edge of the antenna beam rather than on the boresight. This could be controlled by a computer programme which would take over from the monopulse auto-follow system when the boresight elevation reached, say, $80°$. By this means the required slew rate could be reduced to $10°$ sec^{-1} or so — still fast for such a large structure, but possible.

Table 2.1
Leading Particulars of TCS Antenna

Antenna is a 18.3 m (60 feet) diameter praboloid reflector.

Gain at 2 GHz	: 48 dB
Beamwidth	: $0.6°$
Polarisation	: RH circular
Overall system equivalent noise temperature = $200°K$	
Figure of merit (G/T)	: 25 dB/$°K$

Preamplifier

Frequency range	: 2.195 to 2.305 GHz

Antenna tracking performance

Tracking modes

    (1) Autotrack

    (2) Autotrack with zenith pass control

    (3) Program track

    (4) Spiral scan

    (5) Manual positioning

Elevation rate	$2°$ sec^{-1} max
E levation acceleration	$2°$ sec^{-2} max
Slew rate	$10°$ sec^{-1}
Slew acceleration	$4.5°$ sec^{-2}

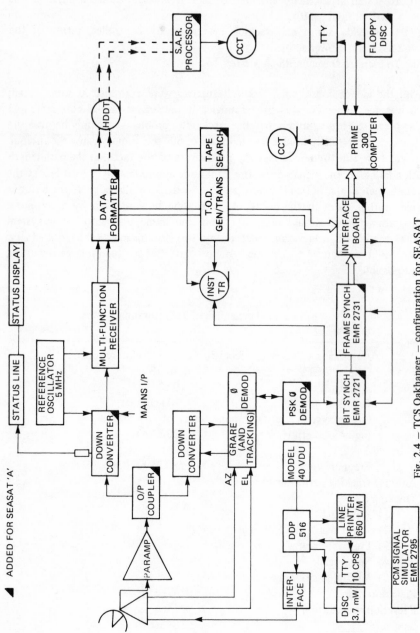

Fig. 2.4 – TCS Oakhanger – configuration for SEASAT

The other major modifications to the station are shown in Fig. 2.4. They consisted of:

(a) providing dual channel reception chains;
(b) providing a more versatile PCM telemetry demultiplexing system and a Prime 300 computer for initial telemetry processing;
(c) acquiring a specific SAR data-handling chain consisting of a coherent sampling and digitising equipment under development at the Applied Physics Laboratory at Johns Hopkins University, a serial PCM recorder which would operate at 10^8 bits sec^{-1}, and special timing equipment.

The leading particulars of the station in SEASAT configuration are given in Table 2.1.

2.3.2 Data processing

The meeting of the SEASAT Users Research Group of Europe (SURGE), held at the Institute of Oceanographic Sciences, Wormley on 5–6 July 1977, established that, except for SAR images, the vast majority of users required annotated raw data from SEASAT, so that they could perform their own evalation and validation experiments. This represented an achievable goal. The recommendation of the Interim Programme Board for Remote Sensing, which was made on 19 July of that year, that the UK offer in repect of SEASAT should be accepted, enabled RAE to enter into liaison with the relevant US laboratories, in particular JPL. It was a requirement of EARTHNET that the RAE archieve of SEASAT data should emulate the format of the JPL Master Sensor Data File (MSDF) so that JPL software could be used for further processing. It soon became clear that the JPL software was in a very early stage of development and that RAE would have, to a large extent, to proceed independently. The following suites were designed and written (in order of priority):

(i) Real-time software for data handling and recording at Oakhanger and for antenna control.
(ii) Software for compilation of the MSDF.
(iii) Software to produce Individual Sensor Data Records (ISDR), consisting of raw data formatted with satellite attitude, orbit and time information.
(iv) An Experimental SAR Processing Facility (ESPF).

In explanation of (iv) it should be said that it was assumed that EARTHNET would provide a bulk SAR processing facility, probably consisting of special-purpose hardware, which was likely to be inflexible in operation. RAE wished to explore the techniques of SAR processing in some detail and decided to develop the experimental processor, which was based on computer software, for its own purposes.

The remote-sensing data processing facilities at RAE were augmented in several stages to accommodate these and other commitments and the present configuration is shown in Fig. 2.5.

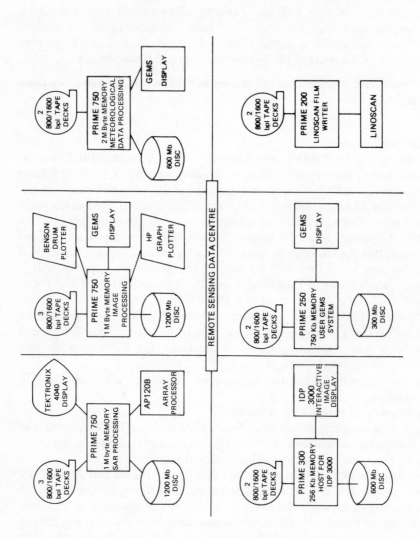

Fig. 2.5 – The remote-sensing data centre at RAE Farnborough – configuration

2.4 OPERATIONS

SEASAT was launched on 27 June 1978, and the present writer was able to report to the ESA Programme Board on the same morning that 'housekeeping' telemetry from the satellite had been successfully received and processed at Oakhanger during the first orbit.

The sensors were switched on and routine data acquistion began on 11 July. Between that date and the failure of the satellite on 10 October, data were acquired from 767 passes over Oakhanger. Acquisition of SAR data began on 4 August, and 51 passes, totalling about 250 minutes of usable data from the SAR, were recorded at high density. There was occasional deterioration of the signal/noise ratio of the SAR signal during zenith passes, but otherwise the cone of silence was successfully eliminated.

2.5 DATA PROCESSING

This topic is to be dealt with in subsequent chapters, but it is worth recording that the SEASAT archive of European data is still held by RAE. Request for data from sensors other than SAR have dwindled, but the interest in SAR is unabated. Because of the early end to the mission, the concept of a dedicated hardware SAR processor was not pursued. Instead, EARTHNET arranged with the Envirnomental Research Institute of Michigan (ERIM) for all data to be optically processed. Some hundreds of copies of these images have been supplied to users by RAE.

The digital data is transferred to computer-compatible tape at Oakhanger by using a two-port 300 m Byte disc as a buffer store (Fig. 2.6). The output from this operation is processed by the ESPF at RAE Farnborough and by the Macdonald Dettwiler Associates software processor which is operated by DFVLR at Oberpfaffenhofen. The ESPF (Fig. 2.7) was developed in collaboration with System

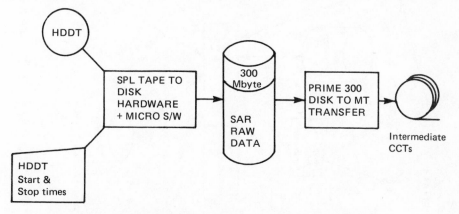

Fig. 2.6 – SEASAT SAR–data transfer to computer-compatible tape

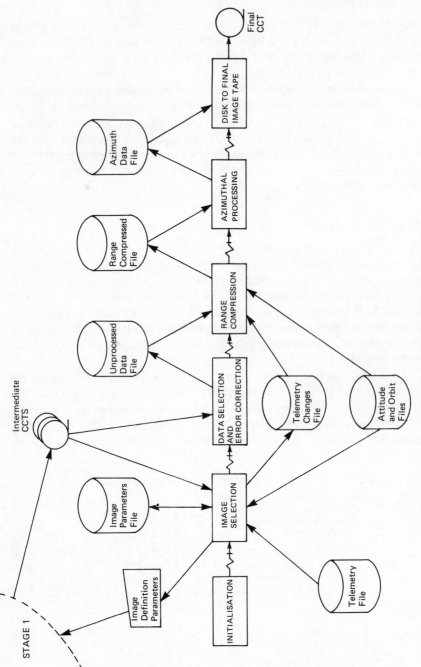

Fig. 2.7 – The RAE experimental SAR processing facility (ESPF)

Designers Ltd and was the first digital processor for spaceborne SAR to be developed outside North America. It has enabled fundamental work to be done on the properties of the SAR signals and the resulting images — examples of which are given at Figs 2.8 and 2.9.

Fig. 2.8 — Example of ESPF output — the wave field at the island of Foula

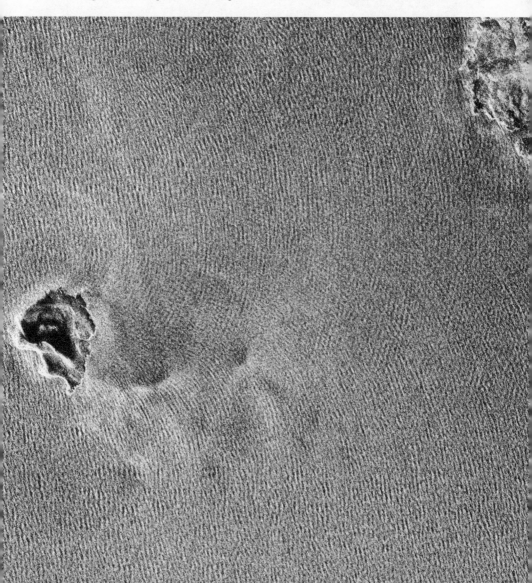

Fig. 2.9 – Example of ESPF output – the Suffolk coast at Ipswich

2.6 CONCLUSIONS

The obvious conclusion to be drawn is that the decisions to proceed without waiting for committee deliberations and to give absolute priority to data capture and recording were correct. If they had not been taken, no real-time date would have been acquired over Europe. There are other lessons to be learnt. The mission has been impaired because the collection of data from surface observations was, with a few worthy exceptions, not organised in time to take advantage of the few weeks of satellite operation.

In the case of RAE, because of the pressure of work on data capture and processing, we were unable to deploy the resources needed to develop the equipment for SAR calibration until it was, in the event, too late to use it. In future missions it will be essential to provide sufficient resources to enable the sensor data to be calibrated and verified, even if the mission is curtailed.

There were some advantages in the rushed development of the European ground segment. All decisions had to be made after brief consultation between the principals, thus obviating a great deal of paperwork and many meetings. This efficient procedure did not lead to any significant errors of judgement; the entire project was completed within estimated cost, and all major objectives were met within the required timescale. The danger of developing equipment too early – with the consequent risk of component obsolescence – was avoided. There was also a mixed blessing: it was not possible to adhere to the strict rules of European procurement policy; although this meant that most of the mission-specific equipment was of US origin, the project was relieved of the task of monitoring local development of these items.

The greatest advantage of the programme was the challenge it offered to the RAE team and its contractors. It was accepted with an enthusiasm and dedication which is gratefully acknowledgeed by the present writer.

Chapter 3

SEASAT: A key element of the EARTHNET programme

LIVIO MARELLI
EARTHNET Program Manager, European Space Agency, ESRIN, Frascati, Italy

3.1 INTRODUCTION

EARTHNET is the programme of the European Space Agency (ESA) responsible for providing an acquisition, pre-processing, archival, and distribution service of remote sensing data from space to the European user community. The programme, created by the ESA Council meeting at ministerial level in early 1977, has become part of the mandatory activities of the Agency as from 1980 and has the object of:

- fostering the development of a strong remote sensing user community in Europe.
- developing the nucleus of the ground segment of the forthcoming European remote sensing missions.

EARTHNET activities have been concentrated up to now on NASA space missions which can be classified into two broad categories, viz:

- Pre-operation application missions (LANDSAT).
- Experimental missions such as HCMM, NIMBUS 7, and SEASAT.

The decision of member countries to give priority to SEASAT acquisition within the EARTHNET programme stems from several facts:

- Pressure from a strong and enthusiastic team of European scientists federated in SURGE who had contacted NASA and proposed a set of experiments in conjunction with SEASAT.
- The great interest of Europe in microwave sensing from space (of which SEASAT is the first civilian example) particularly because of their all-weather and high-resolution capability.
- The fact that experiments with SEASAT SAR over Europe would only be possible if a direct reception and recording capability would be implemented, and, in parallel, the keen interest of several bodies in Europe to undertake this task on behalf of the Agency.
- The presence within NASA, and, particularly in the SEASAT team at JPL, a spirit of cooperation and genuine desire to share the potential results of the SEASAT mission with the international scientific community in the broadest way possible.

Against this background, the EARTHNET team, with the support of several groups and organisations in Europe, Canada, and the USA, organised its SEASAT element to collect, process, and distribute SEASAT data which has formed the basis of several of the experiment to be described in this book. The task proved more expensive than initially estimated in spite of the short life of the mission: in fact we are still working on the data set and we expect to continue for some time to come.

3.2 THE EARTHNET SEASAT ELEMENT

In 1977 with the support of the ESA Delegations and the encouragement of the scientific community, the Agency started the steps required to implement a SEASAT reception and processing capability in Europe, viz:
- Negotiation with NASA of a memorandum of understanding (MOU).
- Selection of the optimal facility to handle the mission in Europe.
- Specification and procurement of the necessary equipment for access to the satellite and pre-processing of its data.

The directive from member countries had been to select national facilities already in existence or planned, in order to reduce investments and to benefit from existing expertise: this philosphy was justified by the experimental character of SEASAT and the absence of an approved follow-on programme. The response of national organisations to the Agency's Announcement of Opportunity was major both in quality and quantity; after a complex techhical and financial evaluation the recommendation which received Delegation's support was for Oakhanger, UK to be the acquisition archive site, and the Royal Aircraft Establishment, Farnborough, to be the processing and distribution site.

A memorandum of understanding between the United Kingdom Department of Industry and ESA for the use of the facilities mentioned above was signed in February 1978.

A rush procurement action was undertaken in order to make Oakhanger suitable for acquisition and recording of all SEASAT downlinks, and it is worth stressing the valuable support received during this hectic phase from NASA, APL, and the UK Embassy in Washington.

The station was ready for the acquisition of the SEASAT low bit rate data stream by the time of launch, and the SEASAT SAR acquisition and recording chain came into operation in early August 1978.

Looking back we can appreciate that the effort to be ready in time paid off: because of the short lifetime of the satellite any slight delay in completing the work would have rendered the whole project useless.

A second aspect we had to solve was the handling of data acquired. It was decided to concentrate on the processing of SAR data, owing to the many unknowns which existed at the time in that domain; this point will be dealt with in the next section.

Concerning the non-SAR sensors of SEASAT, we explored with the RAE team the possibility of implementing in Farnborough the packages required to generate standard products with formats and quality equivalent to those of the JPL project.

Unfortunately, the approach selected by JPL to implement the Algorithms Development Facility for testing and validating those SEASAT products was not entirely suitable for a straight transfer to Farnborough. In fact, documentation was not available in the quality and quantity required, and most algorithms had not been finalised. It was decided initially to limit the effort in this area to the provision of products at the level of instrument data records (ISDR) and delay till the end of validation any further action.

Hardly 100 days after launch SEASAT failed: a plea to NASA to plan for a second launch of SEASAT came from many sources, but very soon it appeared that NASA would not embark on such an undertaking immediately, preferring first to evaluate the results of SEASAT 1. EARTHNET evaluated the impact of the SEASAT failure on the SEASAT element and received the approval of ESA's member states to continue the archival, processing, and distribution tasks at a lower budgetary level.

Since that time the activities related to SEASAT have lost some of the pioneering spirit of the early days. Nevertheless the effort of the scientific community to evaluate the data collected by SEASAT has continued at full pace in the USA, Canada and Europe. The achievement of these concerted efforts has continually increased, and we need only refer to the importance that SEASAT-related reports have taken in the main international remote sensing symposia in the last three years. The astonishing success of the altimeter and wind scatterometer have been used to provide the basis of the NOSS and ERS-1 missions.

The results of the SAR experiment are equally spectacular, if not more so, and the ERS-1 (European), MOS (Japanese) and RADARSAT (Canadian) project can be seen as direct derivatives of the SEASAT experience.

3.3 THE SAR PROCESSING ELEMENT

From the beginning of EARTHNET, SAR processing had been identified as technically the most complex and challenging task to be carried out within the programme, and it gave rise to a set of diversified proposals on the best way to handle it.

Controversial reports were produced on:
- Best algorithms (time domain, frequency domain etc.).
- Recommended technology (optical vs digital, special hardware vs traditional configurations, etc.).
- Expected costs and development time required.
- Volume and quality of products to be produced etc.

As an example, the estimated cost of a SAR processor for SEASAT grew by a factor of 15 within a year, in official reports. In a few months we had to catch up on the efforts that JPL and CCRS had put in over several years of SEASAT preparation. On the other hand we could rely on the expertise developed in Europe in this domain at industrial level, particularly in the domain of airborne SLAR and SAR.

Another factor which played a role in these phases was the idea that sooner or later SAR processing should move from ground to space, and thus the logical desire to favour, as much as possible, those technical solutions which could develop into spaceborne implementation.

Conscious that the problem was technically complex and comparatively expensive, we visited centres of excellence in the USA, Canada, and Europe whose contribution in clarifying the issue has to be acknowledged here. We had reached the stage of specifying the SEASAT SAR processor and evaluating the industrial responses to our request for proposals when SEASAT failed and it became apparent that we would not see a second SAR flying for a few years. ESA member countries decided not to go ahead with the procurement of a fully fledged SAR processor, and asked for alternative solutions. A second round of negotiations led us to propose a set of possible options out of which the following were retained:

- The bulk of the data would be optically processed at ERIM initially in a so-called survey mode, and in a precision mode thereafter for the subset considered most interesting.
- The software SAR processing package developed by MDA for CCRS/ SURSAT in Canada would be purchased and operated for EARTHNET by DFLVR at Oberpfaffenhofen.
- DOI/RAE would develop on national funds an HDDT-CCT conversion and a digital SAR processor. They also agreed to make it available to the Agency for experimental and production purposes.

All these facilities have been implemented and extensively used in the last years: in fact we are expecting to continue using them for several years to come. The limited experience available in this domain meant that the planning for the implementation of the tasks listed proved somewhat optimistic and, as a consequence, we experienced delays in data distribution and received expressions of concern from the users.

Annex 1 gives the statistics of SEASAT data acquired, processed, and distributed, and it shows that interest in these data is still high and the task is far from being completed. We can only imagine what size of job it would have been if SEASAT had continued flooding us with data for the whole of its nominal lifetime.

3.4 THE SEASAT EVALUATION ACTIVITIES

SEASAT offered an opportunity for a cerain amount of evaluation activities aimed at:

- Defining and developing specialised products (OFT, gridding etc.).
- Assessing the technical performance of the sensors.
- Identifying the usefulness of some products (particularly in the area of SAR) for specific applications.

Within ESA these activities focussed mainly on SAR, for a variety of reasons, viz:

- The low bit rate sensors were available at JPL where they could be analysed with the Algorithm Development Facility (ADF) and in any case were of global nature in character.
- The main problem related with SAR was, at that stage, the SAR processor and its impact on the products; while for non-SAR sensors the main issue was data calibration and correction which required a deeper knowledge of the sensor performance, as well as ground/sea-truth data availability.
- Financial and manpower resources were limited.

The evaluation tasks involved a set of studies carried out by several scientific and industrial groups, many of which were based in the UK, because of the close link required with the Oakhanger/Farnborough complex. Futhermore, a series of workshops were organised at ESA's establishment in Frascati in order to offer a forum to scientists and technologists active in SEASAT activites to exchange views and experiences.

The evaluation activities covered such applications as:

- Geology.
- Land use.
- Agriculture and forestry.
- Sea state (wave spectra).
- Bathymetry (in some special cases).
- Ice.

It is difficult today to evaluate the full value of all of those studies, since many were carried out while SEASAT data were just becoming available: often they produced more questions than had been posed in the first place.

3.5 SEASAT AS THE FORERUNNER OF A NEW GENERATION OF REMOTE SENSING MISSIONS

The outstanding achievements of SEASAT are now being acknowledged in the world-wide remote sensing community.

This was shown at the Venice COSPAR symposium ('Oceanography from Space'), at the IGARSS '81 symposium in Washington, and at an impressive series of meetings, workshops, and colloquia held on this mission in recent years. But perhaps the most obvious measure of SEASAT's success is the impact SEASAT had on plans world-wide in the area of remote sensing from space:

- The European Space Agency cemented its plans for the ERS-1 ocean/ice mission with a payload derived from SEASAT's.
- Japan plans for an Earth Resources Satellite carrying radar very similar to the one of SEASAT.
- Canada is planning RADARSAT.
- In the USA the NOSS programme was intended as an operational follow-on to SEASAT before it became the victim of NASA budgetary cuts.

In Europe the design of the ERS-1 ground segment has been influenced by the SEASAT experience, and we expect to use SEASAT data as a primary tool for simulating ERS-1 products. A typical example to show the help we derived from SEASAT for sizing ERS-1 ground components, is the bench-mark exercise we are carrying out on a variety of computer configurations using a subset of SEASAT SAR processor data in order to evaluate optimal architecture for this task as well as for estimating the performance of the configuration for the ERS-1 SAR. Clearly not everything can be achieved in this way, and we expect to complement our SEASAT experience with that of the Convair 580 Campaign carried out in the summer of 1981.

3.6 CONCLUSIONS

SEASAT has been an exciting and challenging experience for the Agency in general and EARTHNET in particular as well as, I am sure, for our colleagues of RAE and DFLVR.

SEASAT has offered us an opportunity for cooperating with NASA/JPL and with CCRS in an open and frank way. I wish to acknowledge the contributions we recieved from the SEASAT project at JPL, by NASA International Affairs Department, as well as well as from CCRS and the SURSAT office. I believe also that the European contribution to the success of SEASAT has been valuable both through SURGE and through the very many evaluation activities carried out on the SEASAT data set in Europe: we are particularly thankful to SURGE for having made this project possible.

SEASAT has been an example of how European cooperation at Agency as well as national level can work from the technical, organisational, and operational standpoint. The contribution of the SEASAT team at RAE should be acknowledged with special emphasis here. Of course, not everything went as smoothly and successfully as we would have liked. Perhaps data delivery has not been as prompt as planned, no pre-operational demonstration project proved possible owing to the short lifetime of the mission, and possibly, not enough effort was expended on non-SAR sensors. We feel, however, that the achievements outnumber the difficulties: it is clear the exercise has been very worthwhile, as is made evident by the continued interest that exists for the mission — which justifies the continuation of the programme of data analysis for some years to come.

Annex 1
EARTHNET SEASAT DATA ACQUISITION, PROCESSING AND DISTRIBUTION

1. Acquired at RAE Oakhanger   : 53 SAR passes
                                                2500 non-SAR passes
2. SAR optically processed at ERIM   : All SAR passes survey
                                                  processed
                                                  500 subsections to full
                                                  resolution
3. SAR digitally processed at RAE and DVFLR   : Some 500 scenes of 40 × 50
                                                  km representing
                                                  some 10% of available raw
                                                  data
4. SEASAT data distributed

	79	80	81	82 (JAN-MARCH)	TOTAL
ISDR CCTs	154	107	–	–	261
SAR CCTs	–	127	209	67	403
SAR OPT FILM	407	677	64	156	1304
SAR DIG FILM	–	267	406	101	774

PART 2

Scatterometer

Chapter 4

Highlights of the SEASAT-SASS program: A review

WILLARD J. PIERSON Jr., CUNY Institute of Marine and Atmospheric Sciences, The City College of the City University of New York, New York

4.1 INTRODUCTION

The prediction of things to come in remote sensing from Earth-orbiting space-craft is a difficult endeavour. One has the tendency to be either too conservative or too optimistic. As an example, the predictions of what might be done by means of an altimeter on a spacecraft in geodesy (Greenwood *et al.* (1969a) and in oceanography (Greenwood *et al.* (1969b)) have all come to pass, with the exception of the detection of tsunamis because none occurred during SKYLAB, GEOS-3, and SEASAT. These predictions were, however, far too conservative compared to what has been achieved with the SEASAT altimeter data on the geoid, on ocean currents, and on the tides. They were not extravagant; nor were they premature, because they were available as part of the documentation for the first altimeter on SKYLAB.

Examples of results with the SEASAT altimeter are Marsh *et al.* (1982) who claim a 6 cm precision for the geoid, which shows the effect of the Mendocino Fracture Zone and the Pacific Plate Boundary from the Gulf of Alaska south to Hawaii; the detection of 20 to 30 cm deviations in the level of the sea caused by transient cold rings south of the Gulf Stream by Cheney & Marsh (1981); and measurements of the tides by Cartwright & Alcock (1981) who write of detecting tidal amplitudes of 10 cm with future altimeters.

The GEOS-3 altimeter was a refinement of the altimeter (S193) on SKYLAB. The SEASAT altimeter was a refinement of its predecessor on GEOS-3. Similarly, the scatterometer on SKYLAB was a refinement of the radars on aircraft used in earlier proof-of-concept missions. The scatterometer on SEASAT was an improvement on the SKYLAB system with a design based on data from SKYLAB and newly developed aircraft instruments. Predictions made about the quality of the SEASAT SASS data were met with considerable scepticism (Pierson 1981). Spacecraft scatterometry is one instrument generation behind spacecraft altimetry. One wonders whether or not as dramatic an advance in instrumentation and precision as from GEOS-3 to SEASAT for altimetry will be possible from SEASAT to the next spacecraft to carry a scatterometer to measure the winds. The most dramatic future advance in scatterometry will be the ability to obtain a unique vector wind with no ambiguities (aliases) most of the time.

4.2 ACCOMPLISHMENTS OF THE SASS

The accomplishments of SEASAT and, in particular, of the SASS on SEASAT, have been documented in a special issue of *Science* (**204**, No. 4400), in a special issue of the *IEEE Journal of Oceanic Engineering* (**OE-5** No. 2), and in the book *Oceanography from Space* (Ed. J. F. R. Gower). A special issue of the *Journal of Geophysical Research* (**87**, No. C5) contains papers on all of the SEASAT instruments.

Publications on SEASAT SASS results, including four workshops, technical reports, and reprints from books and journals, form a stack many metres high. Many different wind fields based on various conventional reports such as transient ships and data buoys were compared with a number of different SEASAT SASS products from the earliest analyses on.

The final JASIN-SASS products were based on the extremely high quality JASIN data for winds averaged over a long time. The results of the final Storms Workshop are yet to appear. One needs a scorecard and a set of rules to aid in the interpretation of these various results and of the statistics on wind speed and wind direction. To compound the problem of interpretation, some results were corrected for the effects of the attenuation of large cloud drops and light rain; others were not. Some data on SASS winds based on the earlier model functions and algorithms were released for study fairly early in the program. Depending on the date of analysis of the SEASAT winds, one will expect to find that the earlier the data, the larger the r.m.s differences and the larger the biases between the meteorological data and the SASS.

Some of the SASS vector winds have been compared with wind fields generated from the analysis of data obtained solely by transient ships. Other data were compared with fields containing a few of the US National Data Buoy winds plus special observations from Weather Station *Papa*, the scientific research ship the *Oceanographer,* plus transient ships. Still other data were compared with fields and spot values obtained by the JASIN program.

The ever-increasing quality of the comparison data sets (they are not considered to be surface *truth*) and the ever-improving model functions, have led to progressive improvements in the ever-smaller values for the r.m.s differences between the SASS wind speeds and the meteorological wind speeds and in the r.m.s differences between the SASS wind directions and the meteorological wind directions.

The present minima are those from the final model function used for the production of the SEASAT final wind data, which was based on one more iteration with the JASIN data. These minima are r.m.s differences of 1.2 to 1.3 m/s, with negligable bias, for speed, and 17°, with negligable bias, for direction based on data for the full range of wind speed, aspect angles, and incidence angles.

The tables that compare the JASIN winds to the SASS winds reveal many interesting features upon closer inspection. The r.m.s wind direction differences are a strong function of wind speed and incidence angle. For incidence angles between $30°$ and $45°$ and for winds above 10 m/s, the r.m.s direction differences are frequently around $10°$. For light winds they tend to be over $25°$.

4.3 RADAR SCATTEROMETER DESIGN

A radar scatterometer is a unique kind of radar designed with the overriding objective of measuring the normalised radar backscattering cross-section as accurately as possible along rows of cells on the sea surface. One hundred watts of radiated power are transmitted in a fan beam with a long-duration pulse (milliseconds, compared to nanoseconds for an altimeter). The leading edge of the pulse is about two-thirds of the way back to the radar when the rear edge is turned off. The narrow beam-width in the vertical plane determines a long illuminated strip on the sea surface. This strip is chopped into pieces by means of Doppler filters since the Doppler frequency shift of the return signal is a function of the spacecraft velocity and the incidence angle of the various portions of the fan beam with the sea surface.

Each cell is sampled by sixty four pulses, and the return signals are heterodyned, amplified, detected, and integrated to obtain a measurement of the power in the return signal plus the noise in the system. The noise power is measured separately, and the received power is then calculated by subtracting the noise measurement from the measurement of the signal plus the noise. In this way the received power can be found even if it is only ten (and even five) per cent of the noise.

Mathematically,

$$\hat{P}_S = (P_R \hat{+} N) \tag{4.1}$$

$$\hat{P}_R = \hat{P}_S - \hat{N} = (P_R \hat{+} N) - \hat{N} \tag{4.2}$$

Both $\hat{P}_s$ and $\hat{N}$ are noise-like signals in that they fluctuate rapidly during the time that they are measured. Noise-like signals can only be stabilised by averaging them for a long time. This is accomplished in the SASS by means of both the very long duration transmitted pulse and the averaging of the returns from a succession of pulses.

The spacecraft carrying the SASS had to be stabilised. Its orientation in roll, pitch, and yaw, as well as its position, had to be known accurately so as to compute the location of the cells (or areas) on the sea surface from which the backscattered signals came. Small errors in measuring these quantities introduce small errors in the location of the cells and in the terms of the radar equation, and, consequently, in the calculation of the normalised radar backscattering cross-section (NRCS).

These considerations lead to the ultimate result that the received power in watts is a normally distributed random variable with an expected value given by the true value and a variance that is a function of the expected value of the noise and of the true value of the received power as in equations (4.3), (4.4) and (4.5) where $\hat{P}_R$ is a random variable and α and β are determined from the radar design and appropriate measurements.

$$f(\hat{P}_R) = (2\pi \, \text{VAR} \, (\hat{P}_R))^{-1/2} \exp - ((\hat{P}_R - P_R)^2 / 2 \, \text{VAR} \, (\hat{P}_R)) \tag{4.3}$$

$$E(\hat{P}_R) = P_R \tag{4.4}$$

$$\text{VAR} \, \hat{P}_R = \alpha \, (P_R + N)^2 + \beta N^2 \, . \tag{4.5}$$

The NRCS, usually designated by σ°, can be calculated from

$$\hat{\sigma}^\circ = \hat{P}_R R / P_T = \hat{P}_R \, R^* \tag{4.6}$$

where everything except $\hat{P}_R$ is calculated from the radar equation and the measured transmitted power. It follows that $\hat{\sigma}^\circ$ is also a normally distributed random variable with an expected value equal to the true value and a variance given by an equation similar to equation (4.5) as in equations (4.7), (4.8) and (4.9).

$$f(\hat{\sigma}^\circ) = (2\pi \, \text{VAR} \, (\hat{\sigma}^\circ))^{-1/2} \exp - ((\hat{\sigma}^\circ - \sigma^\circ)^2 / 2 \, \text{VAR}(\hat{\sigma}^\circ)) \tag{4.7}$$

$$E \, (\hat{\sigma}^\circ) = \sigma^\circ \tag{4.8}$$

$$\text{VAR} \, (\hat{\sigma}^\circ) = \alpha \, (\sigma^\circ + N R^*)^2 + \beta N^2 \, R^{*2} \, . \tag{4.9}$$

In Monte Carlo studies of the effects of sampling variability on the recovery of wind speeds and directions, given the model function, the theoretical value of σ° for a given wind speed and direction would be perturbed by means of equation (4.10) where t is a random number from a normal population with a zero mean and a unit variance.

$$\hat{\sigma}^\circ = \sigma^\circ + t \, (\text{VAR} \, (\hat{\sigma}^\circ))^{1/2}$$

$$= \sigma^\circ \, (1 + \frac{t \, (\text{VAR} \, \sigma^\circ)^{1/2}}{\sigma^\circ})$$

$$= \sigma^\circ \, (1 + t K_p) \tag{4.10}$$

The K_p value, often mentioned in the literature, is thus defined by (5.11).

$$K_p = (\alpha \, (1 + \frac{N}{P_R})^2 + \beta \, (\frac{N}{P_R})^2)^{1/2} \, . \tag{4.11}$$

It is a function of the inverse signal-to-noise ratio. The stronger the return signal the greater the variability of $\hat{\sigma}^\circ$. On the other hand, however, the weaker the return signal, the greater the percentage variability of $\hat{\sigma}^\circ$ about the true value. For $\sigma°$ truly zero, K_P becomes infinite, whereas the estimate of $\hat{\sigma}°$ (i.e. the one quantity measured for the cell) is normally distributed about a mean of zero.

Perfectly satisfactory estimates of $\hat{\sigma}°$ are obtained even for signal-to-noise ratios of 0.10 (or −10 dB) and of 0.05 (about −13 dB) in terms of a positive estimate of $\hat{P}_R$. Nevertheless, it is possible to obtain a negative number for equation (4.2) for areas of light winds. This made it difficult, after applying (4.6) to get $\hat{\sigma}°$, to convert the backscatter to the usual engineering quantity expressed in decibels. The Monte Carlo study by Pierson & Salfi (1982) used both negative and positive values of $\hat{\sigma}°$ in the simulation of a new scatterometer design by always working in antilog space.

4.4 THE MODEL FUNCTION

The model function is an empirical relationship between the backscatter value and the wind speed (V), wind direction (χ), and incidence angle (θ) of the radar beam with the sea surface. The wind speed is that wind that would have been measured at 19.5 metres in a neutrally-stratified atmosphere. The wind direction is relative to the pointing direction of the radar beam. It is assumed that

$$\sigma° = \sigma°(V,\chi,\theta) \qquad\qquad (4.12)$$

and that no other physical effects, such as the details of the gravity wave vector wave number spectrum, are important.

Graphs of $\sigma°$ in dB as a function of incidence angle (θ) for wind speeds of 23.6 and 4.6 m/s for both polarisation and up-wind ($\chi = 0$), down-wind ($\chi = 180$), and cross-wind ($\chi = 90°$ or $\chi = 270°$) are shown in Fig. 4.1 based on the SASS–1 final model function which used the JASIN data. The curves cover a range of nearly 50dB, and the SASS and future scatterometers thus require this dynamic range.

With such a large dynamic range, it was impracticable to work in antilog space, and a further assumption was used such that

$$\sigma°dB = 10\,(G(\chi,\theta) + H(\chi,\theta)\log_{10} V) \qquad\qquad (4.13)$$

with G and H even functions about $\chi = 0$. The functions, $G(\chi,\theta)$ and $H(\chi,\theta)$, can be represented either quasi-analytically or in the form of G-H tables with θ varying in $2°$ steps and χ varying over $0°$ to $180°$ in $10°$ steps. Single local maxima (not equal) were required at $0°$ and $180°$, and a local minimum was required at $90°$ (and $270°$).

The objective of determining the backscatter versus wind speed relationship is to predict the wind speed, given the backscatter measurement. If the meteorological wind speed V_M, and the meteorological direction χ_M, are known then the difference between V_M and V_R, the radar wind, ought to be minimised over the data set. The mathematics becomes difficult when (4.13) is used, since minimising $\log_{10} V_M - \log_{10} V_R$ weights small errors for light winds too heavily. If 4.13) is solved for V_R, the quantity Q can be minimised over a small range of values of χ and over the available data to determine the model function.

The weight by V_M^2 makes errors of constant ΔV nearly of equal importance over all wind speeds.

$$Q = \sum_n V_{Mn}^2 \left(\log_{10} V_{Mn} - (H)^{-1} \left((\sigma_n^o/10) - G\right)\right)^2 \qquad (4.14)$$

An extension of this concept was used in obtaining the various model functions developed by Pierson & Salfi (1978).

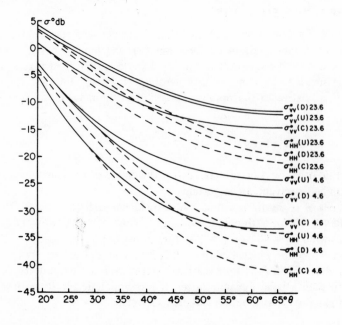

Fig. 4.1 – Backscatter as a function of incidence angle for the SASS–1 model function for winds of 23.6 and 4.6 m/s, upwind, downwind and cross-wind and both polarisations.

4.5 THE RECOVERY OF THE VECTOR WIND FROM PAIRS OF BACKSCATTER VALUES

As SEASAT passed a given area of the ocean, a measurement of $\hat\sigma^o$ for a given incidence angle θ_1 and radar beam pointing direction, say, χ_o, was made for one of the various modes. Later, by a few minutes, a second measurement of $\hat\sigma^o$ was made for nearly the same incidence angle and a radar beam pointing direction of $\chi_0 + \pi/2$ at nearly the same location. The result was two triplets of numbers as in

$$\hat{\sigma}^\circ_1, \theta_1, \chi_0$$

$$\hat{\sigma}_2, \theta_2, \chi_0 + \pi/2.$$

With wind directions measured clockwise from $\chi_0 = 0$, (5.13) can be inverted to yield

$$\log_{10} V = (H(\chi,\theta))^{-1} (\sigma^\circ \text{ dB}/10 - G(\chi,\theta)) . \qquad (4.15)$$

From the first set of measurements, σ°_1 and θ_1 are known but the wind direction is not. However, V(or $\log_{10} V$) can be graphed as a function of χ from 0 to 360°. For the second set of measurements the graph has to start with a $-90°$ phase shift.

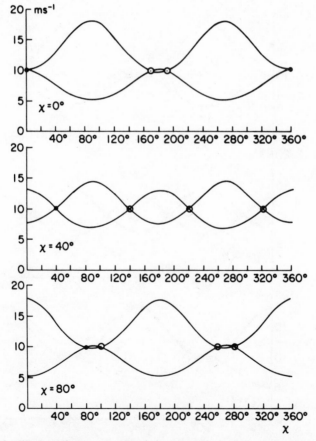

Fig. 4.2 – Wind speed versus aspect angle for two vertical polarisation measurements 90° apart for input conditions shown by the black dot.

Three examples are shown in Fig. 4.2 for conditions where the wind directions relative to beam 1 were 0°, 40° and 80° for noise-free measurements. Two different speeds for the same direction at the same place obviously cannot exist, so the only possible speeds and directions are those for which the curves cross. The input $\sigma°$ values were computed for a 10 m/s wind. The black dot represents the input conditions, and, of course, that value of speed and direction is recovered along with two or three false solutions (aliases or ambiguities). A similar analysis applies for horizontal polarisation pairs or for mixed vertical and horizontal polarisation pairs for other scanning modes.

Difficulties arise for χ near 0°, 90°, 180°, and 270°. Sampling variability can shift the two curves up and down relative to their true value, thus producing large changes in wind direction. The upper curve in the top part of Fig. 4.2, could shift up and the lower curve down by an amount such that the two curves

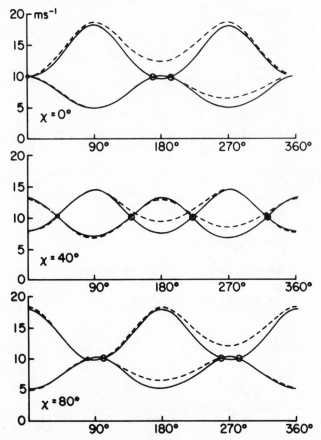

Fig. 4.3 – Wind speed versus aspect angle for two vertical and two horizontal polarisation measurements 90° apart.

would not intersect at all. A maximum likelihood estimate of the values of V (χ is always either $0°$ and $180°$ or $90°$ and $270°$) must then be found.

For another mode of the SASS, two vertical polarisation and two horizontal polarisation measurements were made for the same area. By the same analysis as before, the four curves that result for the noise-free case are shown in Fig. 4.3. One could hope that combined vertical and horizontal polarisation measurement might make it possible to eliminate some of the aliases†. However, to date, efforts to do so have not been successful, perhaps because the attempts were made only with the earlier model functions.

4.6 NEW SCATTEROMETER DESIGNS

New scatterometer designs have been considered. For one design, measurements are made for the same area of the ocean with a new pair of vertical and horizontal polarisation measurements $20°$ clockwise from beam 1. An example for the noise-free case is shown in Fig. 4.4. Two of the three aliases are clearly eliminated for some conditions. This design has been studied by Pierson & Salfi (1982) in

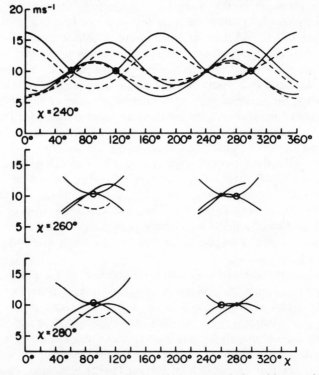

Fig. 4.4 − Wind speed versus aspect angle for a new design with two additional measurements $20°$ behind the first pair.

---

†Solely by means of the SASS data.

terms of a maximum likelihood estimator for the vector wind and in terms of other quicker ways to recover the vector wind. Other designs are possible, many of which promise the elimination of nearly all aliases.

4.7 MESOSCALE TURBULANCE

Monte Carlo studies of how the SASS on SEASAT would perform, and of what the effects of communication noise and attitude errors would be, were made, based on model functions derived from aircraft data. The simulated errors were a function of incidence angle and wind speed. The simulated r.m.s. errors were very small, being under 0.7 m/s and 7° for winds from 10 to 20 m/s.

For the final JASIN results the r.m.s. errors were about 1.3 m/s and 17°. For speed, $(1.3)^2 - (0.7)^2 = 1.2$ for the unexplained variance, and for direction, $(17)^2 - (7)^2 = 240$ for the unexplained variance.

The difference between the Monte Carlo simulations and the JASIN results requires an adequate explanation.

The explanation of the variance lies in carefully defining the winds that were actually measured versus the wind that ought to have been measured. The primary use of the vector winds from future remote sensing systems will be to make improved computer-based numerical weather predictions. Neither the conventional meteorological measurements of the wind by anemometers on transient ships and data buoys nor the SASS measurements on a pair-by-pair basis measure the synoptic scale wind required for numerical predictions. The winds measured by anemometers on transient ships are averaged for only two minutes. Those measured by data buoys are averaged for 8.5 minutes. For special cases during GOASEX, winds were averaged for 20 and 30 minutes. Only for JASIN were all of the anemometer winds averaged for quite long times.

A scatterometer trades area for time. The two-dimensional horizontal turbulence pattern over the areas sampled by the SASS is not well understood. Taylor's hypothesis can be used as an expedient to transfer from space to time. The result is that the SASS measurements have mean wind-speed-dependent effective anemometer averaging times that are not much better than those of conventional anemometers.

To obtain the synoptic scale wind, the solution of the problem is rather straightforward — once the problem is recognised as such (Pierson 1982). Conventionally measured winds ought to be averaged for as much as one hour, depending on synoptic scale conditions. Individual SASS winds ought to be pooled over areas of roughly 2° in latitude and longitude and averaged over speed and direction (or vectorially) to produce a super observation.

One of the two examples from Pierson (1982) of a super observation is described below in abbreviated form. The application of the theory requires the consideration of the mesoscale variability of the wind in terms of the frequency and wavenumber spectrum of the horizontal components of the fluctuations about the synoptic scale wind. The spectrum of U' where $U(t) = \overline{U} + U'(t)$ is normalised by $u_*^2 \, F(z/L)$ (with L the Monin-Obukov length) and varies as n^{-1} where n is the usual notation for frequency in the literature (i.e. Kaimal *et al.* 1972) for n corresponding to periods of several hours to a minute or so. The various numbers used in this example are all derived theoretically in Pierson (1982).

For this example, there were 16 SASS measurements grouped around a conventional measurement by the *Oceanographer*. The area covered was two degrees in latitude and longitude at about 49° N in the North Pacific. The *Oceanographer* obtained a wind of 15.5 m/s from 260° for a 30-minute average.

The winds from SEASAT, tabulated in increasing order by wind speed, are given in Table 4.1.

Table 4.1

Winds obtained by the SASS near the *Oceanographer* (m/s, degrees, meteorological convention)

SPEED	DIRECTION
14.1	262°
14.3	257°
14.5	265°
14.7	262°
14.8	261°
15.1	267°
15.3	271°
15.3	272°
15.4	259°
15.5	260°
15.6	259°
15.6	260°
15.8	254°
15.8	260°
16.0	267°
16.4	266°

From the tabulated data, the average value of the wind speed is 15.23 m/s. The unbiased estimate of the variance yields a standard deviation of 0.64 m/s. The 90% confidence interval on the estimate of the synoptic scale wind, with normality assumed, is

$$\overline{U}_s = 15.23 \pm 0.28 \text{ m/s} . \tag{4.16}$$

Similarly the tabulated SASS wind directions yield an average direction of 262.69°. The standard deviation is 4.96°. The 90% confidence interval on the synoptic scale wind direction is given by

$$\overline{\chi_8} = 262.69° \pm 2.17° . \tag{4.17}$$

SEASAT SASS winds clustered and averaged in this way (or perhaps vectorially) have been called super observations. Values can be produced every two or three degrees of latitude and longitude. The above implied accuracies for entire fields generated in this way would produce results far superior to the fields at present produced by conventional data.

The scatter in the SASS wind speeds and directions illustrated by Table 5.1 is the combined result of two separate effects. One is the sampling variability of the measurement of the noise-like backscattered power and of the spacecraft attitude so as to compute the NRCS. This effect is the only true 'error' of the SASS. What was measured was not measured incorrectly, but the measurement did, in fact, fluctuate about its correct value for statistical and physical reasons that cannot be eliminated. The other effect is the result of the relatively small area sampled by a SASS cell. The average of the wind over a particular SASS cell area was, in fact, different from the 30-minute average by the ship. That part of the SASS measurement was not an error. It was just a difference when compared to the ship measurement.

Although the 30-minute average by the ship was a substantial improvement over the conventional 2-minute average, an hour might have been better. The ship measurement thus also fluctuated in a random way about, say, a multiple ship, longer time average to represent the same area as in JASIN.

The individual SASS winds, the SASS super observation, and the *Oceanographer* wind all consequently differ to a greater or lesser extent from an idealised 'synoptic' scale wind measurement made for the area and time interval of interest.

The 'errors' of the SASS for this particular example can be represented theoretically by

$$U_R = 15.23 + (0.38)\, t_1 + (0.60) t_2 = 15.23 + (0.71) t_3 \tag{4.18}$$

and

$$\chi_R = 262.69° + (1.2°) t_4 + (6°) t_5 = 262.69° + (6.1°) t_6 , \tag{4.19}$$

where the t's are random numbers from a zero mean unit normal distribution. The first number represents mesoscale effects, and the second represents communication noise and attitude error effects. The two are indistinguishable in the actual data. The corresponding variances when compared to those estimated from the sample of sixteen values are well within the range of what could happen from small sample theory.

Similarly the standard deviations for the speed and direction of a 30-minute anemometer average are 0.42 m/s and 2.42° for neutral stability.

One is tempted to interpret the tables that compare SASS winds with conventional winds as if the differences were all errors assigned to the SASS. The above analysis showed that this is far from the true situation. For the ship

$$U_N = 15.5 + (0.42)t_7 \qquad (4.20)$$

and

$$\chi_N = 260° + (2.42°)t_8 \qquad (4.21)$$

Thus, $U_R - U_N \sim 15.23 - 15.5 + (0.38)t_1 + (0.60)t_2 - (0.47)t_7 \qquad (4.22)$

and, $\chi_R - \chi_N \sim 262.29° - 260° + (1.2°)t_4 + (6°)t_5 - (2.42°)t_8 \qquad (4.23)$

The r.m.s. difference over a larger number of similar observations would be

$$\text{r.m.s.}(V) = ((0.27)^2 + (0.38)^2 + (0.60)^2 + (0.47)^2)^{1/2} = 0.89 \text{ m/s} \qquad (4.24)$$

$$\text{r.m.s.}(\chi) = ((2.29)^2 + (1.2)^2 + (6°)^2 + (2.42)^2)^{1/2} = 7° . \qquad (4.25)$$

The example illustrates how difficult it was to get below 1.5 m/s in comparing SASS with conventional winds. Winds need to be routinely reported to greater accuracy. The super observations suppress these sources of inherent variability and yield a good estimate (in the statistical sense) of the synoptic scale wind.

4.8 THE WIND STRESS ON THE SEA SURFACE

The SEASAT SASS winds have been referred to an effective neutral stability wind at 19.5 m above the sea surface. The arguments for and against 19.5 m as opposed to 10 m continue unabated. To find the effective neutral wind from conventional reports required the various assumptions of the Monin-Obukov theory and some assumed drag coefficient. The work of Smith (1980) and Large & Pond (1981) had not yet appeared when this decision was made. The available drag coefficients produced wind stress values for the same 10 (or 19.5) m wind and neutral stability that differed by a factor of two. However, no matter which (of many) proposed drag coefficients and values of Von Karman's constant (0.35 to 0.41) were used, the resulting effective neutral winds at 19.5 m would agree within a few centimetres per second. This important point has not been fully appreciated in the recent literature.

In this way, the decision of which of many empirical relationships was correct could be avoided. If any one of the many proposed relationships had proved to be absolutely correct the SASS winds would have yielded the correct stress! The problem remains in remote sensing of how to find the actual wind at a given height, since the air-sea temperature difference is not at present a remotely sensed quantity.

The recent research of Smith (1980) and Large & Pond (1981) is truly impressive in this problem area. When agreement can be demonstrated between the Kolmogorov inertial range implied stress, and Reynolds flux stress, and the bulk aerodynamic stress to the levels recently demonstrated by high-quality, properly calibrated, essentially deep-ocean measurements, there is hope that this long-standing vexing problem area has at last yielded to the scientific method.

Given, for example, the drag coefficient proposed by Large & Pond (1981), the vector wind stress can be calculated from every single SASS de-aliased vector wind. However, error propagation produces strange effects when wind speeds are squared and cubed. For example, it turns out to be far more accurate for most major oceanographic applications to compute the stress from a super observation. The alternative would be to compute the vector stress from each SASS wind and then average the vector stresses. An error analysis by both methods, using the data in Table 4.1, is most instructive.

At present, and during most of the SEASAT programme, there has been a spirited discussion concerning whether or not backscatter can be more directly related to the vector wind stress than it can to the vector wind.

The latest contribution to this subject is by Liu & Large (1981). Among their conclusions was the statement that "This case study shows no significant difference between the correlation of σ° versus U_N and σ° versus u_*." There are numerous reasons for this result, one of them being the fact that the sources of scatter in the backscatter measurements to be compared to a u_* value all have parallels to the sources of scatter in the wind measurements described above. Another is that the authors had to use the SASS-1 wind model function to transform the actual backscatter values into a set of derived backscatter values for various ranges of the other parameters so as to obtain a large enough sample.

Eventually, it may be possible with data from future spacecraft to conduct what might be a definitive experiment. The procedure would be to obtain sets of measurements, averaged over a sufficiently long time, of the following quantities,

$$U_a, U_N, u_*, \chi_w, \chi_s, \theta, \sigma^\circ, T_A - T_S, q, \rho, a$$

where U_a is the average wind at the known anemometer height, U_N is the effective neutral wind at some fixed height, u_* is the friction velocity as measured from $\langle u' \, w' \rangle$ (and perhaps also the Kolmogorov range), χ_w is the wind direction relative to the pointing direction of the radar beam, χ_s is the direction of the wind stress relative to the pointing direction of the radar beam (not always the same as χ_w), θ is the incidence angle of the radar beam, σ° is the backscatter value, and $T_A - T_S$, q, ρ, and a, are respectively the air sea temperature difference, water vapor content, air density, and anemometer height. The various atmospheric parameters would have to vary over an extensive range of U_a from calm to 25 to 30 m/s, $T_A - T_S$ from $-5°C$ to $+5°C$, and χ_w over $180°$.

Given this data set, one could then try to predict U_a by means of an equation of the form of equation (4.26) with, say, the results of Large & Pond (1981) to provide the required interrelationships.

$$U_{aR} = U_a \, (\sigma°, \theta, \chi_W ; T_A - T_S, q, \rho, a). \tag{4.26}$$

For the SASS on SEASAT, the assumption was made that

$$U_{NM} = U_N \, (u_*, T_A - T_S, q) \tag{4.27}$$

and an attempt to predict the form of (4.28) followed to obtain the model function:

$$U_{NR} = U_N \, (\sigma°, \chi_W, \theta) \, . \tag{4.28}$$

Since, for this hypothesised experiment, u_{*M} is known independently of the bulk aerodynamic equations, one could predict (4.29).

$$u_{*R} = u_* \, (\sigma°, \chi_S, \theta) \tag{4.29}$$

from the same data.

This might not even be quite right. Perhaps one really needs (4.30):

$$\tau_R = \tau(\sigma°, \chi_S, \theta, \rho) \, . \tag{4.30}$$

For each model, there would be a meteorologically observed quantity to match with the radar-derived quantity, and various functional forms for the right-hand sides of (4.26), (4.28), and (4.30) could be tried so as to minimise differences of the form

$$\sum_n (U_{NMn} - U_{NRn})^2 = \text{MIN} \tag{4.31}$$

for equation (4.28), for example, with corresponding equations for the other quantities.

After this, any radar-derived quantity can be used to compute any one of the other quantities. The results can then be compared to the appropriate meteorological measurement. The results of such an experiment may or may not be conclusive. The sources of scatter identified above propagate through such procedures in extremely intricate ways that require detailed analysis and careful interpretation.

It is not at all clear to the writer that either wind stress, or u_*, is to be preferred to the actual wind in the interpreatation of backscatter. One should note, however, that the measured values of the wind stress scatter by a non-negligible amount about the value of the stress computed from the bulk aerodynamic equations. It would be most interesting if the results of (4.29) or (4.30) agreed with the actual stress measurements better than the results of (4.28) when used to compute the stress.

Backscatter from the sea surface varies with the wind, or an effect of the wind, because the short waves on the ocean increase with height as the wind increases. The slopes of the waves from 10 m to 1 m long are important because these waves tilt the roughened capillary waves toward and away from the radar. The increase in height of the short capillary waves is also important because the Bragg scattering mechanism, corrected for slope, is the dominant mechanism.

The wind stress is the downward flux of momentum to the ocean surface. The transfer of momentum to the ocean from the air cannot generate waves. The downward flux of momentum acts on the air above the water for fully rough flow to generate pressure fluctuations on the water that in turn generate the waves. The energy from the wind must first go into the waves and then be transformed by rectification into ocean currents. The concept of a turbulent shearing stress directly generating currents without waves as an intermediate step requires the production of vorticity and either transitional or smooth flow. The residual effect, if any, of shearing stress during fully rough flow on the generation of currents is not well understood.

In any case, only some small part of the total momentum flux goes directly into maintaining in an equilibrium state those wave spectral components dominantly responsible for radar backscatter.

Observational evidence and theory both suggest that the wind input to any part of the spectrum is related to the mean wind speed at some height, say 5 m, and the phase speed of the waves as in U_5/C. Thus it is not established that u_*, or u_*^2 (which represents all of $<u' w'>$), rather than the wind alone, should be better correlated with backscatter.

4.9 ULTIMATE RESOLUTION

For most remote sensing applications, higher resolution and greater accuracy are synonymous with better data. Two examples are the progressive improvements in altimetry and the most recent geostationary cloud imagery compared to the earliest imagery. For scatterometry, this axiom may not necessarily be true.

Higher resolution of a wind field is needed for the study of hurricanes and tropical storms. Higher resolution would also help near coastlines, around and between islands, near fronts, and in areas of patchy rainfall (given comparable passive microwave resolution).

Most of the time over most of the ocean, higher resolution only obtains incorrect values for the mesoscale wind variability which is masked by the increased sampling variability in the measurement of backscatter. This mesoscale variability must be filtered out to define the synoptic scale. Its details are neither resolvable nor predictable with the present (or potential) numerical models for meteorological predictions.

4.10 CONCLUDING REMARKS

In this brief review of the highlights of the SEASAT-SASS Program, an attempt has been made to describe some of the important concepts of the SASS and some of the decisions that were made during the program as to methods for relating the wind and the backscatter. Ideas that were set forth crudely in the 1960s have progressed to the point of practical global application. As in all science, the work is unfinished. Scientists have been trying to measure the winds, the waves, and other effects of the winds for centuries. It is not to be expected that the definitive answers to the full understanding of radar scatterometry have been found. There is much that still needs to be learned.

However, there is no reason to wait until all is understood perfectly. Enough has been proved to make it undeniably clear that radar scatterometry can measure the winds over all the oceans of the Earth twice a day, and that these winds are already superior to the conventionally measured winds when processed correctly and used for numerical weather prediction and ocean models. Further improvements over the next few years and the next few decades will follow, based on a deeper understanding of the physical processes that are involved and on better data processing and analysis procedures.

ACKNOWLEDGEMENTS

The contributions to the SASS program at City College were the result of the close collaboration of Mr Robert Salfi and me as evidenced, in part, by the many reports we have co-authored. To design a model or a procedure is one thing. To get it to work on a computer on masses of data is still another.

Discussions of many theoretical aspects of these problems with Dr Mark Donelan and with members of the S-Cube group (O'Brian, *et al.* (1982)) have contributed to the understanding needed for this paper.

This paper has been prepared under the sponsorship of the National Aeronautics and Space Administration under Contract NAGW-266. It is based on research done, both past and present, for many sponsors including the NASA Jet Propulsion Laboratory, the NASA Langley Research Center, and at NASA Headquarters.

Travel to Britain to present this material and participate in the discussion meeting was co-sponsored by the SEASAT Users' Research Group of Europe (SURGE), and by the Science Division of the City College of New York.

REFERENCES

Cartwright, D. E. & Alcock, G. A. (1981) On the precision of sea surface elevations and slopes from SEASAT altimetry of the Northeast Atlantic Ocean *in* Gower J. F. R. Editor *Oceanography from space* Plenum press. 885–896.

Cheney, R. E. & Marsh, J. E. (1981): Oceanographic evaluation of geoid surfaces in the Western North Atlantic *in* Gower J. F. R. Editor *Oceanography from space* Plenum press. 855–864.

Greenwood, J. A., Nathan, A., Neumann, G., Pierson, W. J., Jackson, F. C. & Pease, T. E. (1969a): Radar altimetry from a spacecraft and its potential applications to geodesy. *Remote Sensing* (of environment) 1, 1, 59–70.

Greenwood, J. A., Nathan, A., Neumann, G., Pierson, W. J., Jackson, F. C. & Pease, T. E. (1969b): Oceanographic applications of radar altimetry from a spacecraft. *Remote Sensing* (of environment) 1, 1, 71–80.

Kaimal, J. C., Wyngaard, J. C., Izumi, Y. & Cote, O. R. (1972): Spectral characteristics of surface-layer turbulence. *Quart. J. Roy. Meteorol. Soc.* 98, 563–589.

Large, W. G. & Pond S. (1981): Open ocean momentum flux measurements in moderate to strong winds. *J. Phys. Oceanogr.* 11, 324–336.

Liu, W. T. & Large, W. G. (1981): Determination of surface stress by SEASAT-SASS: a case study with JASIN data. *J. Phys. Oceanogr.* 11, 1603–1611.

Marsh, J. G., Cheney, R. E., Martin, T. V. & McCarthy, J. J. (1982): Computation of a precise mean sea surface in the Eastern North Pacific using SEASAT altimetry. *EOS* March 2, 1982. Vol. 63, No. 9, 178–179.

O'Brien, J. J., *et al.* (1982): Scientific opportunities using satellite wind stress measurements over the ocean (Report of the Satellite Surface Stress Working Group) Nova University/N.Y.I.T. Press, Ft. Lauderdale, Fla.

Pierson, W. J. (1981): Winds over the ocean as measured by the scatterometer on SEASAT *in* Gower J. F. R. Editor. *Oceanography from space* Plenum Press. 563–571.

Pierson, W. J. (1982): The measurement of the synoptic scale wind over the ocean. NASA Contractor Report 166041, Langley Research Center, Hampton, Va.

Pierson, W. J. & Salfi, R. E. (1978): The theory and data base for the CUNY SASS wind vector algorithm. Part 1 of a final report to JPL and NEPRF, Contracts 954411 and N00014-77-G-0206 (Including Appendices A to H) CUNY Institute of Marine and Atmospheric Sciences.

Pierson, W. J. & Salfi, R. A. (1982): Monte Carlo studies of ocean wind vector measurements by SCATT: Objective criteria and maximum likelihood estimates for removal of aliases, and effects of cell size on accuracy of vector winds. NASA Contractor Report 165837-1 Langley Research Center, Hampton Virginia 23665.

Smith, S. D. (1980): Wind stress and heat flux over the ocean in gale force winds. *J. Phys, Oceanogr.* 10, 709–726.

Chapter 5

Validation and applications of SASS over JASIN

T. H. GUYMER

Institute of Oceanographic Sciences, Wormley, Godalming, Surrey

5.1 INTRODUCTION

Routine measurements of near-surface wind at sea often lack the spatial and temporal coverage necessary to resolve adequately synoptic-scale features of the atmosphere (particularly in the tropics) and to provide surface forcing for oceanic models. The number of fixed stations is extremely small and, although many merchant ships transmit weather observations for dissemination by meteorological centres, these are inevitably concentrated along major shipping routes. Additional problems are those of calibration and exposure errors which vary from platform to platform. Satellites, on the other hand, offer great potential for worldwide coverage by a minimum of sensors.

Winds have been derived from sequences of visible and infra-red images using the motion of identifiable cloud elements, but errors in extrapolating such winds to the surface and the lack of suitable coverage, especially in mid-latitudes, render this an unsatisfactory technique for many purposes. Microwave scattering techniques using satellite-borne sensors, however, are capable of providing estimates of near-surface wind velocity in all weather conditions.

SEASAT carried a fan-beam scatterometer operating at a frequency of 14.6 GHz (Grantham *et al.* 1977, Johnson *et al.* 1980, Jones *et al.* 1979). Basically, the instrument relies on the fact that the strength of the radar backscatter, which at the range of incidence angles used (25-55°) is due to Bragg scattering, is related to very short gravity waves. These in turn are in equilibrium with the near-surface wind speed. Since the backscatter is also anisotropic, scatterometer measurements at different azimuths (for SEASAT at ± 45° and ± 135° to the subsatellite track) can be combined to yield wind direction but with some ambiguities. The design specification for the SEASAT-A Satellite Scatterometer (SASS) required an accuracy of ± 2 ms^{-1} in speed and ± 20° in direction over the 4-26 ms^{-1} wind speed range. Inferred winds must therefore be validated against more conventional data, of high quality, and under as wide a variety of conditions as possible. It is also important that circumstances under which the scatterometer fails to give realistic winds are identified and possible causes investigated. Owing to the fortuitous coincidence of the SEASAT operational phase with the field experiment of the Joint Air-Sea Interaction (JASIN) project this has proved possible.

The paper describes the results of the SEASAT–JASIN comparisons, discusses the probable causes of an anomaly, and presents some examples of the usage of such data in the Northeast Atlantic.

5.2 COMPARISON WITH JASIN WINDS

The JASIN experiment took place from July to September 1978 in the NE Atlantic about 300 km northwest of Scotland in deep water (Pollard *et al.* 1983). Fourteen research ships, many with wind-measuring capabilities, participated, and thirty-five mooring systems were deployed. Three aircraft operated from their base at RAF Macrihanish and made measurements through the atmospheric boundary layer, including the turbulent fluxes. Fig. 5.1 shows the locations of wind-measuring platforms used in this study, the positions of sensors on each being given in Royal Society (1979).

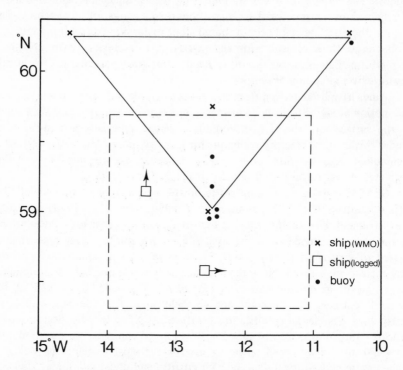

Fig. 5.1 – Location of surface wind-measuring platforms used in the study. The broken lines mark the Hydrographic Survey Area, and the 200 km meteorological triangle is also shown.

Four of the ships were designated meteorological ships and occupied stations at the corners and, during Phase 2, the centre of a 200 km triangle. These vessels rendezvoused several times during the experiment to intercompare their meteorological sensors. Winds varied from 3-17 ms^{-1}, which was the same range as during the actual observational phases. During the intercomparisons the ships adopted the same position relative to the wind as when on station (head to wind except for *Gardline Endurer* at the NW corner, which was beam on). This was in an attempt to minimise errors due to changes in the air flow over the ships. (Experiments were, however, conducted to quantify this behaviour and showed that, with relative wind directions of greater than 30° from the bow, wind speeds could be 20% low (e.g. Parkes 1978). In a comparison of hourly observations made according to World Meteorological Organisation specifications some systematic differences were apparent (Macklin & Guymer 1980). For example, Hecla's anemometer tended to overspeed, and there was a change in the bias of *John Murray's* wind direction which was traced to a movement in the wind-vane mounting. Unexplained differences are shown in Fig. 5.2 and indicate that uncertainties in the four-ship system were ± 0.7 ms^{-1} (both extremes due to *Gardline Endurer*) and ± 5°. The ships also carried automatic logging systems, but problems in retrieving reliable wind directions from these systems caused them to be removed from the current analysis. Logged values from two of the oceanographic ships, *Tydeman* and *Discovery*, were well-behaved and were used to increase the number of comparisons.

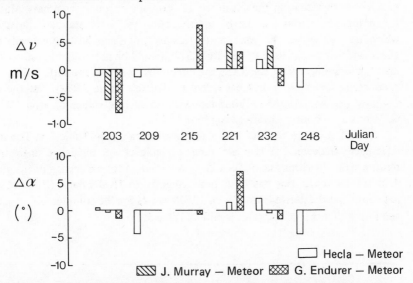

Fig. 5.2 – Difference between WMO wind speed (Δv) and wind direction ($\Delta \alpha$) measurements after systematic biases (e.g. due to sensor height, calibration error) have been removed. *Meteor* winds have been used as a standard. The day numbers refer to the occasions on which the intercomparisons were made.

Most of the meteorological buoys were situated a few kilometres apart at the southern corner of the 200 km triangle. It was assumed that over periods longer than a day real horizontal differences on this scale would be small. Three-day mean winds were examined for significant differences between platforms, and some surprisingly large differences were found e.g. a 40° directional bias on B4, but it was possible to construct correction schemes to reduce the errors to ± 0.3 ms^{-1} and ± 5°. One of the buoys, W2, was adopted as an arbitrary standard, and the ship observations were related to this using *Meteor* as a transfer standard in the case of the WMO observations since her station position was close to W2. The absolute accuracy of the measurements has proved difficult to ascertain, and even subsequent calibrations and intercomparisons on land have failed to resolve the question (Weller, personal communication).

The development of algorithms to obtain wind velocity from the backscatter measurements had involved some tuning to surface data, principally those obtained during the later Gulf of Alaska SEASAT experiment (JPL 1979). JASIN data were withheld from algorithm developers and thereby provided the first independent assessment of the scatterometer's wind-measuring capabilities. A workshop to assess the comparisons for 23 selected passes and two algorithms was held in March, 1980 (JPL 1980), and results were summarised by Jones *et al.* (1981).

An example of the results is given in Fig. 5.3 in which 60-minute means of autologged data, corrected for stability to 19.5 m, are plotted against SASS winds, each point representing the mean for all comparisons on a given pass. Most of the points lie within the design specifications and have standard deviations which are well within the same limits. However, one pass has anomalously high winds which are discussed in section 6.3. It should be noted that because of the directional ambiguity of the SASS the directions closest to the surface data have been used. However, it has been shown (Wurtele *et al.* 1982) that the true direction can be selected without reference to surface values on over 70% of occasions by synoptic pattern recognition techniques.

Overall statistics for the two algorithms are given in Table 5.1. The main differences between the two are their dependence on incidence angle. As a result a new algorithm, named SASS-1, was devised for processing of the global data set. Although this has been tuned slightly to JASIN data, Offiler (1981) has compared the derived winds with JASIN winds for the complete set of passes and found that similar statistics resulted (Table 5.1).

Facing page: Fig. 5.3 – Comparison of SEASAT scatterometer winds derived with the Wentz-7 algorithm with JASIN surface winds. For the latter 60-minute means of automatically-logged ship and buoy data were used. Each point represents the mean of all comparisons on a given pass; 1 refers to the anomaly on Rev. 557.

SASS

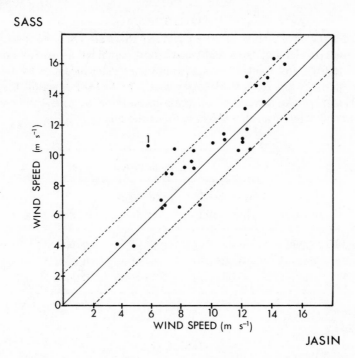

JASIN

SASS

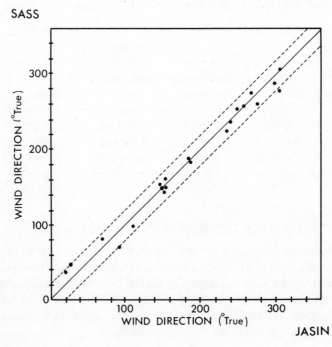

JASIN

Table 5.1a

Mean and standard deviation of differences (SASS–JASIN) in wind speed (m/s) for different model functions and polarisation combinations of fore and aft beams. The CWK and Wentz-7 model functions are independent of JASIN data, and the statistics are for 23 selected passes. The SASS-1 function has been tuned slightly to take account of comparisons on these passes and the statistics have been computed from all SASS passes over the area.

	Polarisation		
Model function	Both beams horizontal	Both beams vertical	Mixed
CWK	−0.44±1.58	0.50±1.26	−0.04±1.41
Wentz–7	0.40±1.61	0.84±1.60	0.62±1.55
SASS–1	0.03±1.72	−0.11±1.61	0.04±1.73

Table 5.1b

As Table 6.1a but for wind direction (°) differences.

	Polarisation		
Model function	Both beams horizontal	Both beams vertical	Mixed
CWK	1.2±17.9	3.4±15.6	0.9±17.1
Wentz–7	−0.4±18.5	1.5±16.6	−0.6±17.1
SASS–1	0.7±18.3	1.6±17.2	1.3±17.2

5.3 ANOMALOUS WIND ESTIMATES ON 4 AUGUST 1978

The scatterometer wind field for the region containing the JASIN area is shown in Fig. 5.4 (Guymer *et al.* 1981). Directional ambiguities have been removed by reference to surface pressure analyses. A general E to NE flow is indicated, except in the west, and speeds varied from < 3 ms^{-1} in the north to 10 ms^{-1} in the south. However, in a small region near the southern corner of the triangle, winds were over 12 ms^{-1} greater than surface measurements.

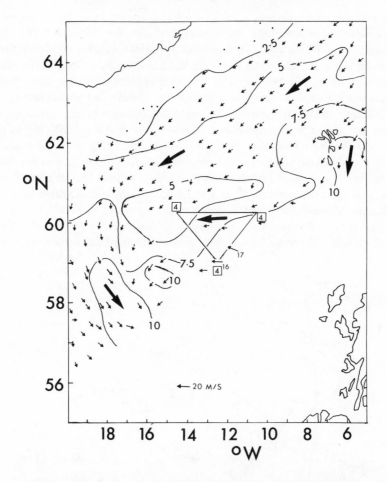

Fig. 5.4 – SASS wind field for Rev. 557 at 2316 GMT, 4 August. Wind speed has been contoured in ms^{-1}, and the arrows indicate the direction of the flow after removal of directional ambiguities. The numbers in boxes indicate WMO wind speeds recorded by the corner ships. (taken from Guymer *et al.* (1981) by permission of Macmillan Journals Ltd.)

A synoptic chart constructed for 1800 GMT (Fig. 5.5) five hours before the anomaly, indicated a shallow low to the north of Scotland and an occluded front advancing southwestwards with a pronounced veer of wind across it. Subsequent analyses revealed that this front reached the southern corner just before midnight. The most significant feature on NOAA-5 images was a cloud mass to the southeast of the JASIN area which had a cumuloform appearance. Its development and movement during the evening have been estimated from successive IR images including those from the Visible and Infra-Red Radiometer on SEASAT (Fig.

5.6). Its propagation velocity of ~ 10 ms⁻¹ towards 280° is similar to mid-level winds measured by the JASIN radiosondes (Taylor *et al.* 1983b) and implies that the system would have been over *Meteor* at 2300 GMT. Taylor *et al.* (1983a) have discussed the associated liquid water distribution given by the SEASAT passive microwave radiometer. Between 2100 and midnight three ships near the southern corner reported thunder and one also observed a heavy rain shower lasting for 15 minutes, though no quantitative measurements were taken. A radiosonde ascent not long after the anomalous SEASAT pass but about 200 km to the northwest has been plotted in the form of a tephigram (Fig. 5.7) and indicates the thermodynamic structure of the environment of the storm. Convection from the surface is capable of reaching only 930 mb (= 800 m), further ascent being prevented by an isothermal layer 1 km thick which was associated with the occlusion. However, and this is a crucial point, parcels of air at 800 mb, where the air is nearly saturated, if displaced upwards would quickly reach condensation and then follow the wet adiabatic (as shown) maintaining an excess temperature

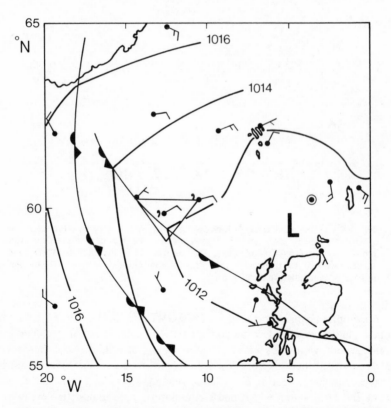

Fig. 5.5 – Surface synoptic chart at 18 GMT, 4 August, adapted from the Daily Weather Report of the UK Meteorological Office.

over their environment up to 500 mb and reaching a maximum height of 400–450 mb. The initial impetus for such motion can be plausibly supplied by convergence at the front. Strong downdraughts would not be expected with this type of storm, nor were any observed, so it seems unlikely that the SASS winds reflect an actual wind event at the surface.

Thus it appears that the anomaly resulted from increased backscatter due to precipitation in the thunderstorm. Guymer *et al.* (1981) conclude that rain rates of 10 mm hr^{-1} over ~ 100 km^2 would be required, and this is consistent with the 15-minute shower and the 10 ms^{-1} propagation speed that were observed. It is not clear, however, whether rain (or possibly hail) in the atmospheric column or striking the sea surface is the dominant mechanism, but it seems likely that the effect of the latter is to reduce the small-scale surface roughness. The possible effect of convective rainfall on wind estimates from scatterometers has to be taken into account when interpreting the data.

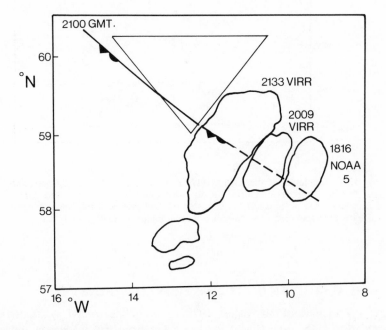

Fig. 5.6 – Approximate outlines of the convective cloud system associated with the thunderstorm over the southern corner of the JASIN triangle constructed from successive IR satellite images.

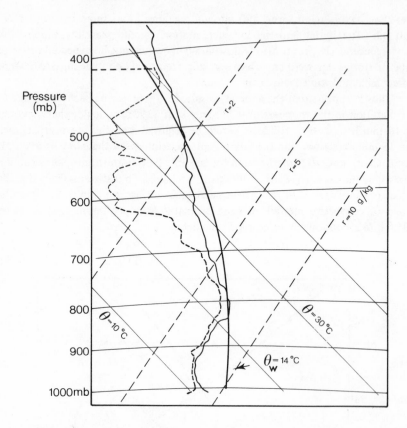

Fig. 5.7 – Tephigram of the 2351 GMT/4 August radiosonde ascent from *Gardline Endurer.* Sloping solid lines are dry adiabatics, i.e. constant θ, and sloping dashed lines are of constant mixing ratio (r). The curved solid line is a moist adiabatic corresponding to a wet-bulb potential temperature (θ_W) of 14°C.

5.4 30/31 AUGUST: AN OCCLUDING FRONTAL SYSTEM

During the evening of 30 August 1978 an occluding frontal system passed through the JASIN array (Fig. 5.8). The warm front had made slow progress east but the cold front moved steadily at 8 ms⁻¹ from the northwest. The fronts were well-defined on a NOAA-5 IR image at 2030 GMT and the warm sector was relatively cloud-free.

A SASS pass at 2311 GMT (Fig. 5.9) gives good coverage of the system and shows a southerly flow ahead with winds from the WSW in the warm sector and veering to north of west behind the cold front. Winds are light in the ridge

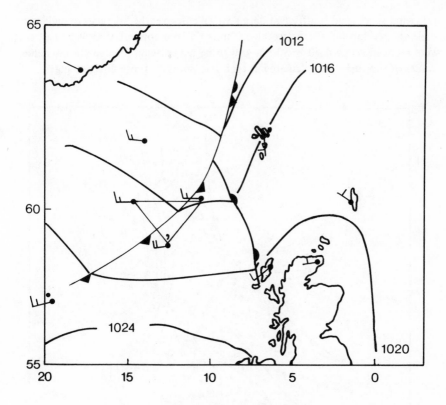

Fig. 5.8 – Surface synoptic chart at 00 GMT/31 August.

several hundred kilometres ahead of the occlusion, and strongest ahead of the surface fronts. The SASS winds enable the fronts to be placed more accurately than with conventional data. Such a density of wind measurements suggests that it would be meaningful to examine the horizontal variation of the wind stress curl, given by

$$\text{curl } \tau = \frac{\partial \tau_y}{\partial x} - \frac{\partial \tau_x}{\partial y}$$

The SASS data as supplied provide estimates of the friction velocity U_* ($\equiv \tau/\rho$) calculated from the 19.5 m neutral stability wind using a constant flux layer model, from which curl τ can be calculated. (Liu & Large (1981) discuss the possiblity of deriving stress directly from backscatter measurements using some high-quality dissipation data to determine the surface stress. However, Pierson (1983) argues that stress may not be as well-correlated with backscatter as is wind speed, and insufficient data are available over the full range of incidence angle to develop an alternative algorithm.) For the computation, values of U_* were reduced on to a 50 × 50 km grid using a Laplacian interpolation scheme. Values of the curl were calculated at each grid point by finite differencing.

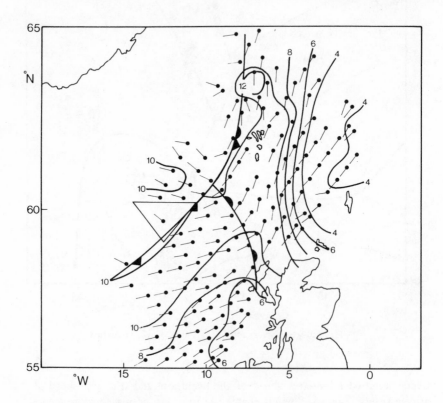

Fig. 5.9 – SASS wind field at 2311 GMT/30 August. Wind speed has been contoured in ms^{-1}, directions from which the wind is blowing are also indicated (from Guymer *et al.* 1983)

The wind stress curl can be related to the mean vertical velocity in the friction layers of the atmosphere and ocean. In steady, horizontally homogeneous flow the momentum equations can be solved to give an expression relating the divergence to the vertical gradient of the wind stress curl. By integrating to a level of zero stress, i.e. the top of the friction layer in the air, and invoking continuity

$$W_f = \frac{1}{\rho f} \, \text{curl} \, \tau$$

where W_f is the vertical velocity due to frictional convergence (sometimes called Ekman pumping), ρ is the air density, and f is the Coriolis parameter. Values of W_f at the base of the oceanic mixed layer are smaller by the ratio of the density of air to that of water. Other non-frictional effects may, of course, also contribute to the vertical velocity. Comparisons with the divergence method using radiosonde winds (Guymer *et al.* 1983) show that in JASIN Phase 2 Ekman pumping accounts for a significant proportion of the total vertical motion at the top of the boundary layer.

The vertical velocity distribution at 2311 GMT, calculated as above, is plotted in Fig. 5.10. Maximum values of 0.8 cms^{-1} are found with the greatest descent some 200–300 km ahead of the occluded part of the front, i.e. to the west of the ridge axis. Upwards motion occurs along or just ahead of the fronts, especially near the occlusion point. Much of the warm sector air is weakly descending. In general, the distribution is consistent with the synoptic situation, and values agree well with a time series of W_f using corner-ship winds if Taylor's hypothesis is used.

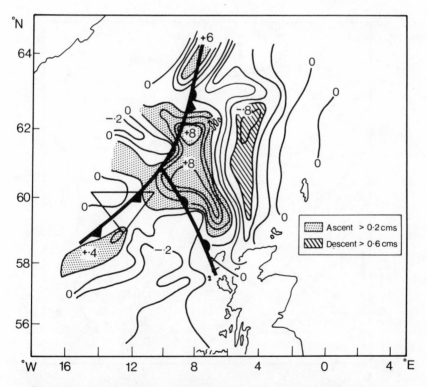

Fig. 5.10 – Spatial variation of the vertical velocity at the top of the atmospheric friction layer at 2311 GMT/30 August. Values were calculated from the curl of the wind stress using the wind field shown in Fig. 5.9.

The map of W_f may also be compared with the distribution of integrated water vapour (Q_v), liquid water (Q_L) and rain rate derived for the same occasion by Taylor *et al.* (1983) from SMMR data. Low values of Q_v were found in the ridge, and the highest occurred in the warm air just to the south of the 200 km triangle. The Q_L maximum, however, was further NE near the occlusion. Rain rate also reached peak values at this location with values of 0.8 mm hr^{-1}. Taylor *et al.* showed these results to be consistent with WMO present weather observations and, at the NE corner, with rain-gauge measurements.

Such results imply that warm, moist air ahead of the cold front is being lifted and some of the water vapour is being converted to liquid water as it moves towards the occlusion point. There it experiences additional uplift, and rain is produced.

5.5   5 SEPTEMBER: AHEAD OF AN OCCLUDED FRONT

At the time of the SASS wind field to be presented an occluded front was approaching from the south-west (Fig. 5.11). Winds ahead of the front attained

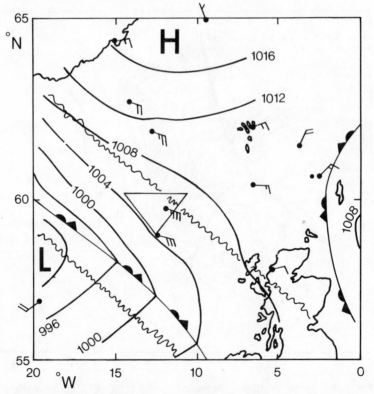

Fig. 5.11 – Surface synoptic chart at 06 GMT/5 September. The band of cloud associated with the occluded front to the SW is marked.

gale force and were more or less parallel to the front so that it became slow-moving during the day. A well-defined band of cloud, about 400 km wide according to NOAA-5 images, accompanied the front and coincided with a maximum in the Q_v distribution (Taylor *et al.* 1981). Early on the 5th the feature was still quite active and rain was detected by the SMMR (Taylor 1983). Just ahead of the band of thick cloud, skies were much clearer particularly near the Hebrides (58° N, 6° W) where even low cloud was present in only small amounts.

The swath covered by the SASS (Fig. 5.12) unfortunately did not extend as far southwest as the surface front. Nevertheless, the increase of winds as the front is approached can be clearly seen with values up to 16 ms^{-1}. Northeasterly winds further to the east are also indicated. Vertical velocities (Fig. 5.13) deduced as in section 5.4 cannot be so clearly related to the synoptic scale as on 30/31 August. Indeed, the 'C' shaped pattern of upward motion in the north is somewhat surprising since there was an absence of middle- and high-level cloud.

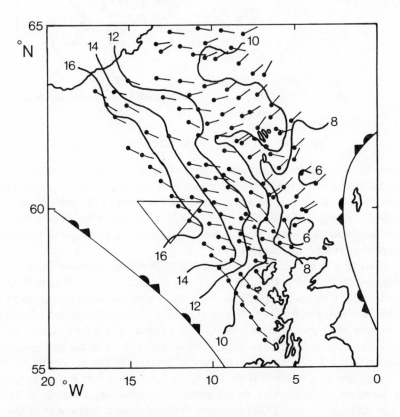

Fig. 5.12 – SASS wind field at 0813 GMT/5 September. See caption to Fig. 5.9.

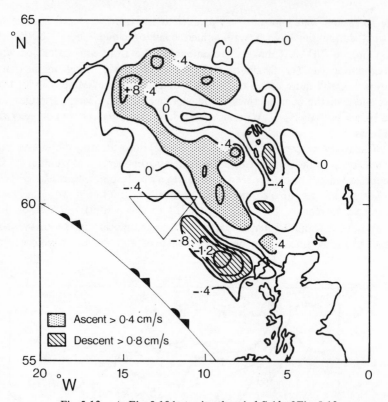

Fig. 5.13 − As Fig. 5.10 but using the wind field of Fig. 5.12.

Generally, there was much stratiform cloud at low levels in this region. Substantial descent occurred to the west of the Hebrides, consistent with the low cloud amounts, and in a similar position relative to the front as the descending air on the 30/31 occasion. Taylor & Guymer (1982) noted descending motion a similar distance ahead of a warm front a few days earlier on the basis of divergence measurements from radiosondes.

This case is also of interest because of a Synthetic Aperture Radar (SAR) image which lies within the scatterometer swath about 100 km from its SW edge (Lodge 1982). Although significant swell was present, no waves are visible on the image. Instead, north of 59° N and extending for several hundred kilometres, i.e. where wind speeds exceed 14 ms^{-1}, are streaks which lie within 5° of the wind direction given by the scatterometer and have a horizontal spacing of 2-3 km. Allan & Guymer (1983) have studied this and another similar occasion and postulate that the streaks are due to a modulation of the surface roughness by longitudinal circulations in the atmospheric boundary layer. In an earlier study Allan & Guymer (1981) showed that the general appearance of the sea surface on SAR images could be broadly related to the SASS wind speed.

5.5 CONCLUDING REMARKS

The accuracy of the SEASAT scatterometer winds, once the directional ambiguity has been removed, is at least as great as that obtained with conventional surface-based instrumentation (± 1.6 ms^{-1} and $\pm 18°$), but with much superior spatial sampling. One particular anomaly has shown that care is needed in interpreting SASS data when close to heavy rain cells, but it should also be appreciated that it was the discovery of the anomaly that supplied the impetus for studying the thunderstorm. Two other case studies have shown that wind and wind stress curl fields are consistent with synoptic scale events. The value of the SASS in interpreting SAR images is also apparent.

These results were obtained during a field experiment where unusually detailed meteorological and oceanographic measurements were available. Such data sets are vital for the calibration and interpretation of the new techniques of microwave remote sensing. The analyses give confidence in using scatterometer measurements in data-sparse regions and so improving our understanding of the spatial variability of the wind field in relation to hurricanes, fronts, cloud clusters, etc., and of improving the results of forecasting models. On longer time-scales the wind stress distribution from SASS can be used to study aspects of ocean circulation with particular relevance to climate.

REFERENCES

Allan, T. D. & Guymer, T. H. (1981) A preliminary evaluation of SEASAT's performance over the area of JASIN and its relevance to ERS-1 *Proc. EARSel-ESA, Symp.*, Voss. ESA-Sp-167, 119-127.

Allan, T. D. & Guymer, T. H. (1983) SEASAT measurements of wind and waves on selected passes over JASIN, *Int. J. Rem. Sensing* (in press).

Grantham, W. L., Bracalente, E. M., Jones, W. L., & Johnson J. W., (1977) The SEASAT-A Satellite Scatterometer, *IEEE J. of Oceanic Eng.* **OE-2** 200-206.

Guymer, T. H., Businger, J. A., Jones, W. L. & Stewart, R. H., (1981) Anomalous wind estimates from the SEASAT scatterometer, *Nature*, **294** 735-737.

Guymer, T. H., Businger, J. A., Katsaros, K. B., Shaw, W. J., Taylor, P. K., Large, W. G., & Payne, R. E. (1983) Transfer processes at the air-sea interface. Results of the Royal Society Air-Sea Interaction Project (JASIN 1978). *Phil. Trans. Roy. Soc. Lond.* **A380**, 253-273.

Johnson, J. W., Williams, L. A. Jr., Bracalente, E. M., Beck, F. B., & Grantham, W. L. (1980) SEASAT-A Satellite Scatterometer Instrument Evaluation, IEEE J. of Oceanic Eng. **OE-5** 138-144.

Jones, W. L., Boggs, D. H., Bracalente, E. M., Brown, R. A., Guymer, T. H., Chelton, D., & Schroeder, L. C. (1981) Evaluation of the SEASAT Wind Scatterometer, *Nature* **294** 704-707.

JPL (1979) SEASAT Gulf of Alaska Workshop Report. Vol. I, Panel Reports. JPL Internal Document 622-101, Jet Propulsion Laboratory, Pasadena.

JPL (1980) SEASAT–JASIN Workshop Report, Vol. I: Findings and Conclusions, JPL Publication 80–62, Jet Propulsion Laboratory, Pasadena.

Liu, W. T., & Large, W. G. (1981) Determination of surface stress by SEASAT-SASS: A case study with JASIN data. *J. Phys. Oceanogr.* **11** 1603-1611.

Lodge, D. (1983) Sea-surface features revealed by SAR (this volume).

Macklin, S. A. & Guymer, T. H. (1980) Inter-platform comparisons of JASIN WMO observations, *JASIN News* **15** (unpublished manuscript), Institute of Oceanographic Sciences, England.

Offiler, D. (1981) Surface wind measurements from satellites – a comparison of SEASAT scatterometer data with JASIN winds. Met. 019 Branch Memorandum No. 64 (unpublished), Meteorological Office, Bracknell, England.

Parkes, G. S. (1978) Errors in the measurement of pressure – on land and at sea, MSc dissertation (unpublished) Univ. of London.

Pierson, W. J. (1983) Highlights of the SEASAT-SASS Programme: A review (this volume).

Pollard, R. T., Guymer, T. H. & Taylor, P. K. (1983) Summary of the JASIN 1978 Field Experiment. Results of the Royal Society Air-Sea Interaction Project (JASIN 1978). *Phil. Trans. Roy. Soc.* Lond. A308, 221-230.

Royal Society (1979) Air-Sea Interaction Project: *Summary of the 1978 Field Experiment,* Royal Society, London.

Taylor, P. K. (1983) The Scanning Multichannel Microwave Radiometer – an assessment (this volume).

Taylor, P. K., Katsaros, K. B., & Lipes (1981) Determinations by SEASAT of atmospheric water and synoptic fronts. *Nature* **294** 737-739.

Taylor, P. K., Guymer, T. H., Katsaros, K. B., & Lipes, R. G. (1983a) Atmospheric water distributions determined by the SEASAT Multichannel Radiometer, in *Proc. Symp. Variations in the Global Water Budget,* 10–15 August 1981, Oxford, D. Reidel, Dordrecht, 93-106.

Taylor P. K. & Guymer, T. H. (1983) The structure of an atmospheric warm front and its interaction with the boundary layer. Results of the Royal Society Air-Sea Interaction Project. (JASIN 1978). *Phil. Trans. Roy. Soc.* Lond. A308, 341-358.

Taylor, P. K., Grant, A. L. M., Gunther, H., & Olbruck, G. (1983b) Mass momentum, sensible heat and latent heat budgets for the lower atmosphere. Results of the Royal Society Air-Sea Interaction Project (JASIN 1978). *Phil. Trans. Roy. Soc.* Lond. A308, 275-290.

Wurtele, M. G., Woiceshyn, P. M., Peteherych, S., Borowski, M., & Appleby, W. S. (1982) Wind direction alias removal studies of SEASAT Scatterometer-derived wind fields. *J. Geophys. Res.* **87, C5**, 3365-3377.

PART 3

Synthetic Aperture Radar

Chapter 6

Imaging ocean surface waves by synthetic aperture radar – a review

W. ALPERS
Institut für Geophysik, Universität Hamburg, and
Max-Planck-Institut für Meteorologie, Hamburg, FRG

6.1 INTRODUCTION

The synthetic aperture radar (SAR) is an active remote sensing instrument, i.e. it illuminates the scene by it own source. The wavelengths of the electromagnetic waves usually employed by SAR lie in the millimetre to decimetre range ('microwaves'). The backscattered power from small scene elements (pixels) is detected and the values from each scene element are put into a 2-dimensional array to form a radar image.

SAR operates on a moving platform and transmits a series of short coherent pulses to the ground in a direction perpendicular to the flight pass. The image resolution in range (across-track) direction is determined by the effective pulse width. The fine image resolution in azimuth (along-track or flight) direction is achieved by utilising the motion of the platform to synthesise a long antenna.

In contrast to the real aperture radar (RAR), which measures the 'instantaneous' amplitude of the radar return from each target, the synthetic aperture radar (SAR) measures the amplitude and phase of this return over a finite time interval T (the SAR integration time). In order to convert the data received by a synthetic antenna into 'equivalent' real antenna data, one has to eliminate the time factor in the measurements. This can be achieved if the time variation of the phase of the return signal is *a-priori* known. If the scene contains only stationary targets, then such *a-priori* knowledge exists, since the distance radar-target can be predicted as a function of time, provided the path and the velocity of the platforms are known. However, if the scene contains targets which exhibit unknown motions, then this *a-priori* knowledge of the time variation of the phase does not exist. Such a situation is encountered when SAR is applied to ocean surface wave imaging. Strictly speaking, the SAR principle cannot be applied in this case. But fortunately, these target motions are often relatively small such that their effect on SAR imaging can be treated as small perturbations to stationary target imaging.

Most of the discussion in the literature concerning SAR imaging of ocean surface waves has revolved around a description of the wave motions (Elachi & Brown 1977; Jain 1978; Shemdin *et al.* 1978; Shuchman *et al.* 1978; Teleki *et al.* 1978; Alpers & Rufenach 1979; Swift & Wilson 1979; Raney 1980; Valen-auela 1980; Harger 1980; Alpers & Rufenach 1980; Rufenach & Alpers 1981; Alpers *et al.* 1981; Raney 1981; Tucker 1983; Hasselmann & Alpers 1983; Alpers 1983). Since this aspect of the SAR imaging mechanism is most controversial, we shall mainly discuss in this paper the question of how the motion of the ocean surface affects the SAR performance. Section 6.5 contains a general discussion of the effect of target motion on SAR imaging, and in section 6.6 these results are applied to ocean surface wave imaging (see Hasselmann & Alpers 1983).

In section 6.2 we review basic concepts of the SAR ocean wave imaging mechanism as based on the two-scale scattering and hydrodynamic model. Section 6.3 and 6.4 contain a brief description of the cross-section modulation due to hydrodynamic and electromagnetic interaction. Finally, in section 6.7 the main results of this paper concerning the influence of wave motion on SAR performance are summarised.

6.2 BASIC CONCEPTS OF THE SAR OCEAN WAVE IMAGING MECHANISM

Satellite and air-borne synthetic aperture radars normally operate at angles of incidence between 20° and 70° for which the microwave return from the sea surface is predominantly due to Bragg backscattering from short surface ripples. Longer waves are seen by the radar because they modulate the backscattering due to the short surface ripples. The modulation can be described in terms of a two-scale model in which the sea surface is represented as a superposition of short Bragg scattering ripples superimposed on longer gravity waves. Bragg scattering is applied locally in a reference system lying in the tangent plane ('facet') of the long waves and moving with the local long wave *orbital velocity* (Bass *et al.* 1968; Wright 1968; Keller & Wright 1975; Alpers & Hasselmann 1978).

Three processes contribute to long-wave imaging:
 (i) modulation of the energy of the short Bragg scattering ripples through interactions between the ripple waves and the long gravity waves (hydrodynamic interactions);
(ii) changes in the effective angle of incidence relative to the local facet normal; this modifies the radar return by changing both the Bragg backscattering coefficient and the resonant Bragg wavenumber (electromagnetic interactions and tilt modulation.
(iii) the temporal variations of the facet parameters (facet position, direction of facet normal) and the Bragg backscattering coefficients of the facets during the finite integration time in which the SAR sees a facet (motion effects).
The first two processes are important for both real aperture and synthetic aperture radars, while the third process affects only synthetic aperture radars.

The cross-section modulation due to hydrodynamic and the electromagnetic interactions is usually assumed to be linearly dependent on the long-wave field. The modulation is described mathematically by a modulation transfer function (MTF) $R(k)$, defined by (Alpers & Hasselmann 1978):

$$\sigma = \sigma_0 + \delta\sigma = \sigma_0 \ [1 + \int (R\,(K)\,z\,(K)\,e^{\,i(KX\,-\,\omega t)} + \text{c.c.})\,\mathrm{d}K] \quad (6.1)$$

σ is the normalised radar cross section and $z(K)$ the Fourier transform of the surface elevation ζ associated with the long waves:

$$\zeta = \int \ (z(K)\,e^{\,i(KX\,-\,\omega t)} + \text{c.c.})\,\mathrm{d}K \ . \tag{6.2}$$

K and ω are the wave vectors and radian frequencies of the large-scale wave field, and c.c. stands for complex conjugate. Since we assume that the cross-section modulation depends linearly on the long-wave field ζ, this implies that $R(K)$ does not depend on $z(K)$. This is a rather rash assumption, but it has proved to be a useful one, at least for modelling microwave backscattering at the sea surface for low to moderate sea states. If only one ocean wave were present, e.g. a monochromatic swell, then (6.1) could also be written as

$$\sigma = \sigma_0 \ [1 + |R|\,\zeta_0 \cos\,(KX - \omega t + \delta)] \tag{6.3}$$

where ζ_0 is the wave amplitude and

$$\delta = \arctan\,\frac{\text{Im}\,(R)}{\text{Re}\,(R)} \tag{6.4}$$

the phase of the ocean wave-radar MTF.

In our model the complex MTF $R(K)$ is a sum of two terms, one describing the modulation due to hydrodynamic interaction (R^{hydr}) and the other the modulation due to electromagnetic interaction, which is also called tilt modulation (R^{tilt}),

$$R = R^{\text{hydr}} + R^{\text{tilt}} \tag{6.5}$$

Often it is convenient to introduce non-dimensional MTF's M by

$$R = K \cdot M \ . \tag{6.6}$$

In the literature dealing with measurements of the ocean wave − radar MTF the values of M are usually given.

The effect of wave motion on SAR ocean wave imagery can be evaluated analytically by noting that the SAR integration time (typically 0.1–3 s) is generally short compared with both the period of the long waves (8–16 s), and the intrinsic hydrodynamic interaction time of the backscattering ripples in the reference frame of the moving facets (typically several seconds). Thus the temporal variations of the facet parameters and the complex backscattering can be expanded in a Taylor series with respect to the integration time. The dominant terms in the expansion are found to arise from the radial (slant range) components of the facet (orbital) velocity and acceleration (Alpers & Rufenach 1979; Swift & Wilson 1979; Valenzuela 1980). We shall neglect here the decorrelation effects arising from the intrinsic hydrodynamic interaction time (Raney 1981). These may be expressed in terms of an equivalent Bragg resonance line broadening. Wind-wave tank measurements (Keller & Wright 1975) indicate that the intrinsic Bragg line broadening is generally small compared with the Dopppler broadening associated with the facet motion effects.

After correction for motion effects, the SAR image obtained for a moving sea surface can be related to the image of an equivalent time-independent ('frozen') surface whose complex backscattering coefficient $r(X)$ can be represented as the product

$$r(X) = w(X) \, m^{\frac{1}{2}} (X) \tag{6.7}$$

of a random, statistically homogeneous, complex white noise process $w(X)$, which represents the backscattered return due to uniformly distributed, small-scale (Bragg scattering) ripples, and a slowly varying function $m^{\frac{1}{2}}(X)$, which describes the modulation of this return by the long gravity waves. The representation of the small scale backscattering coefficient $w(X)$ as a zero correlation scale random process is standard. It is valid provided the correlation scale of the backscattering elements is small compared with the SAR resolution, which is normally satisfied (Rufenach & Alpers 1981).

6.3 THE HYDRODYNAMIC MODULATION

The hydrodynamic contribution of the cross-section modulation is characterised by a non-uniform distribution of the short waves with respect to the long ocean wave field, which is attributed to interactions between short and long waves. At present, no really satisfactory theory exists on hydrodynamic modulation. However, simplified theoretical models have been developed by Longuet-Higgins & Stewart (1964); Bretherton & Garrett (1969); Keller & Wright (1975); Garrett & Smith (1976); Alpers & Hasselmann (1978); Valenzuela (1979); and Phillips (1981). These theories are based on a weak interaction theory (WKB-type interaction theory) and can be valid only for a relatively smooth sea, where nonlinear effects, which lead to a steepening of waves and eventually to wave breaking, are unimportant. Note also that non-uniform surface wind drifts that might be induced by the wave field (Phillips & Banner 1974) are not included in this hydrodynamic interaction theory.

Basically, in these theories the non-uniform distribution of short waves on the long waves is attributed to the fact that the short waves are positively or negatively 'strained', depending on whether they propagate in convergent or divergent surface flow sections. The orbital motion associated with the long ocean waves produces this variable current. In the first case, the spectral energy density of the Bragg scattering waves is increased, while in the second case it is decreased.

Values of the modulation transfer function based on this theory have been calculated by Keller & Wright (1975) and Alpers & Hasselmann (1978). Assuming that the short-wave spectrum exhibits a k^{-4} wavenumber dependence, the dimensionless MTF M^{hydr} is always equal to or smaller than 4.5. Measured values of M^{hydr} in the ocean often lie above this theoretical value (Wright *et al.* 1980; Plant *et al.* 1982, Alpers *et al.* 1983), but nevertheless, the order of magnitude seems to be correct. Modulation measurements also show that M^{hydr} depends on wind speed and ocean wave frequency. For a discussion on the cross-section modulation for high sea states the reader is referred to the paper by Alpers *et al.* (1981).

6.4 THE TILT MODULATION

The tilt modulation can be calculated from Bragg scattering theory by using the two-scale wave model (Alpers *et al.* 1981). We divide the total modulation transfer function in parallel and perpendicular components, $M_{\parallel}$ and $M_{\perp}$, respectively:

$$M_{\parallel} = -i\,|k|^{-2}\,k_{\parallel}\,R$$

$$M_{\perp} = -i\,|k|^{-2}\,k_{\perp}\,R \tag{6.8}$$

$M_{\parallel}$ describes the modulation for waves travelling in look direction of the antenna, and $M_{\perp}$ the modulation for waves travelling perpendicular to the look direction. Fig. 6.1 shows a plot of $iM_{\parallel}^{tilt}$ and $iM_{\perp}^{tilt}$ for both polarisations (HH and VV) as function of incidence angle. These plots apply for X-band (10 GHz), but they are not significantly different for other microwave bands.

Note that the tilt modulation is larger for HH polarisation than for VV polarisation, although the cross-section itself is smaller for HH polarisation than for VV polarisation (see Wright 1968; Valenzuela 1978). Furthermore, the plots show that the tilt modulation for waves propagating perpendicular to the plane of incidence (i.e. in cross-range direction) is roughly one order of magnitude smaller than for waves travelling parallel to the plane of incidence (i.e. in range direction). Note also that the tilt modulation transfer function is a purely imaginary quantity, which means that tilt modulation and wave amplitude are out of phase by 90°. The sign is such that for an up-wave ('up-wind') looking antenna maximum tilt modulation occurs at the forward face of the wave.

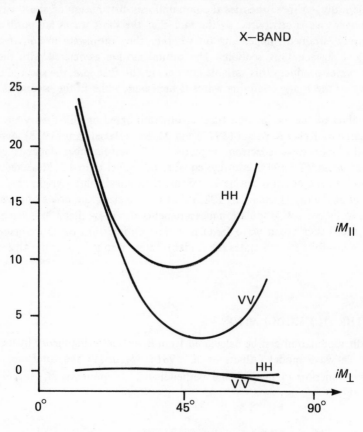

Fig. 6.1 – Dimensionless tilt modulation transfer functions (MFTs) as function of incidence angle. $M_{\parallel}^{tilt}$ is the MTF in up-wave, and $M_{\perp}^{tilt}$ the MTF in cross-wave direction (for X-band). The indices HH and VV denote horizontal and vertical polarisations.

6.5 THE INFLUENCE OF TARGET MOTION ON SAR PERFORMANCE

The phase history of a stationary facet (target) passing through the antenna beam of a side-looking radar is given by

$$\phi(t) = \frac{b}{2}\, t^2 \tag{6.9}$$

where

$$b = \frac{2k_0 V^2}{R} \qquad (6.10)$$

is a system constant, the 'chirp rate', k_0 the radar wavenumber, V the platform velocity, and R the slant range. The phase is recorded over a time interval from $-T/2$ to $+T/2$, where T is the coherent integration time. The SAR processor expects this quadratic phase history, and by matched filtering positions the facet at zero Doppler or zero boresight.

6.5.1 Azimuthal image shift
If the facet has a constant radial velocity component, u_r, then the quadratic phase history is not affected. However, the time at which zero Doppler is encountered is changed relative to the stationary facet case. This time difference Δt is given by

$$\frac{d}{dt} \phi(t) \big|_{t = \Delta t} = 2k_0 u_r. \qquad (6.11)$$

Inserting equations (6.9) and (6.10) into equation (6.11) yields

$$\Delta t = \frac{2k_0}{b} u_r \qquad (6.12)$$

A SAR processor expecting stationary facets therefore erroneously allocates the azimuthal position x of the moving facet to the image position

$$x' = x + \Delta x \qquad (6.13)$$

where

$$\Delta x = V \cdot \Delta t = \frac{R}{V} u_r \qquad (6.14)$$

is the azimuthal image shift. Note that Δx is independent of the radar wavenumber and integration time.

Ample evidence of this effect can be found in SAR imagery. A ship travelling in cross-track direction, for example, appears as displaced in the platform flight direction relative to its wake in the SAR image.

6.5.2 Azimuthal image smear
If the facet is subject to an acceleration in slant-range direction, then an additional quadratic term enters into the phase history:

$$\phi_a(t) = 2k_0 \frac{a_r}{2} t^2 \qquad (6.15)$$

where a_r is the radial acceleration and $|t| \leqslant \dfrac{T}{2}$.

Owing to the additional linear chirp generated by the acceleration, a processor, which is tuned for stationary facets, is mismatched or out-of-focus for this particular facet. The result is a smearing of the azimuthal target position. The magnitude of the azimuthal smearing is given by the variation δx in azimuthal image shift experienced by the facet due to the variation $\delta u_r = a_r T$ of the radial facet velocity during the integration time T:

$$|\delta x| = \left| \frac{R}{V} a_r \, T \right| \tag{6.16}$$

The exact formula for the net degraded azimuthal (one-look) resolution ρ_a' can be derived from SAR system theory and is given by

$$\rho_a' = [\rho_a^2 + \rho_{acc}^2]^{\frac{1}{2}} \tag{6.17}$$

where

$$\rho_a = \frac{\pi}{k_0} \frac{R}{V \cdot T} \tag{6.18}$$

is the full-bandwidth (one-look) azimuthal resolution for stationary targets and

$$\rho_{acc} = \frac{\pi}{2} |\delta x| = \frac{\pi}{2} \left| \frac{R}{V} a_r \cdot T \right| \tag{6.19}$$

represents the smearing induced by the facet acceleration.

We note that the acceleration induced azimuthal image smear depends linearly on the coherent single look integration time T.

Eq. (6.17) is obtained formally by computing the width ρ_a' of the convolution of two Gaussian functions of widths ρ_a and ρ_{acc}, respectively. (The factor $\pi/2$ enters in eq. (6.19) because ρ_a is given here by the width of the azimuthal image intensity response function, which is normally defined as

$$\exp\left(-\frac{\pi^2}{\rho_a^2} x^2\right) \quad \text{rather than} \quad \exp\left(-\frac{1}{4\rho_a^2} x^2\right) \quad).$$

Incoherent multi-look processing has no effect on the total acceleration smearing. It merely divides it into N azimuthally displaced subsections, where N is the number of looks. The formal expression for the degraded net azimuthal resolution for N looks is given by

$$\rho'_{aN} = [\,(N\rho_a)^2 + \rho^2_{acc}\,]^{\frac{1}{2}}.$$ (6.20)

6.6 THE EFFECT OF ORBITAL MOTIONS ON SAR OCEAN WAVE IMAGERY

6.6.1 Velocity bunching

The azimuthal displacement Δx due to radial facet velocities contributes to the imaging of the long waves through the alternating concentration and spreading of the apparent position of the backscatter elements within the long wave pattern, the so-called 'velocity bunching' effect. Velocity bunching is determined by the variation of the number of facet images per unit length in azimuth direction, which is proportional to

$$c = \frac{R}{V}\,\frac{\mathrm{d}u_r}{\mathrm{d}x_0}$$ (6.21)

where x_0 is the azimuthal coordinate in the ocean plane (see Fig. 6.2). For small $|c|$, the effect is linear and can be characterised in the same way as the hydro-dynamic and electromagnetic interactions by a linear transfer function. However, for large $|c|$ (typically $|c| \geqslant \pi/2$) velocity bunching is a non-linear process and leads to image distortions. Such situations are often encountered in SAR ocean wave imagery. The effect is largest for azimuth travelling waves and vanishes for range travelling waves. The non-linearity gives rise not only to a distortion of the SAR image spectrum, but also to a shift of the spectral peaks towards lower azimuthal wavenumbers (Alpers 1983).

6.6.2 Azimuthal image smear

The azimuthal image smear due to radial facet accelerations generally degrades the SAR image by reducing the image contrast and filtering out small scale features in the image. However, because of the spatially variable nature of the acceleration smearing, (see Fig. 6.2), it is possible to reduce the smearing for some facets while enhancing it for others by suitable refocusing. Thus in some situations the net image quality can be enhanced by suitable focus adjustments.

In contrast to velocity bunching the acceleration effects are a basically non-linear phenomenon, and cannot be expressed by a linear transfer function even for small variations of ρ'_{aN}. The acceleration effects also differ from the orbital velocity effects through their linear dependence on the integration time (see eq. (6.19)).

Velocity bunching and azimuthal image smear together constitute, to first order, the motion part of the SAR imaging mechanism. For small integration times the azimuthal image smear due to acceleration can become negligible compared with velocity bunching (e.g. for ERS-1). However, for the SEASAT SAR it is found generally to be a non-negligible factor in the imaging process.

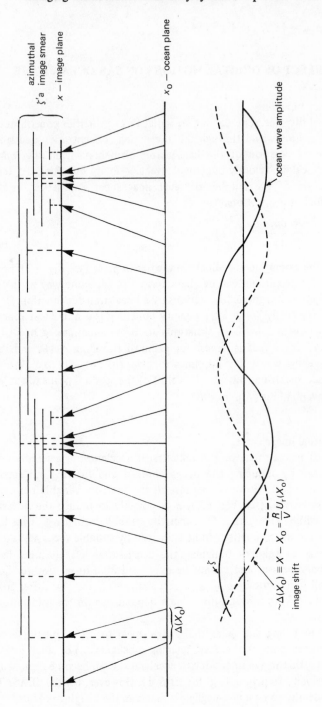

Fig. 6.2 — Azimuthal image shift and image smear associated with a monochromatic ocean wave. The equally spaced facets on the ocean appear non-uniformly distributed on the SAR image. In addition, they are smeared non-uniformly in azimuth direction.

6.7 DISCUSSION

Although the response of synthetic aperture radar (SAR) to moving point targets has been studied for more than ten years (see, Kelly & Wishner 1969, Raney 1971), the application of these results to ocean wave imaging has created considerable discussion. The difficulties encountered revolve around the proper kinematical description of the sea surface. This in turn depends on the proper multiscale dynamical description of the wave field, extending from long gravity waves down to short backscattering ripples. Some basic facts regarding the SAR imaging of ocean surface waves can be summarized as follows:

(1) The SAR response to a moving ocean wave field can be described to first order by a superposition of SAR responses to independent moving scatter elements on the ocean surface. Thus the SAR theory for moving point targets (Raney 1971) is applicable to SAR ocean wave imaging. The dominant influence of the motion of the backscattering surface on SAR imaging is the azimuthal displacement of the apparent positions of individual backscattering elements induced by the component of the facet velocity directed towards the radar.

(2) The scene coherence time relevant for SAR ocean wave imaging is determined by the intrinsic hydrodynamic interaction time of the Bragg scattering waves in the reference frame of the moving facet and by the orbital velocity spread of the facets within a resolution cell (Tucker, 1983). This may be estimated to be of the order of a few seconds. It follows that the finite scene coherence time of the ocean suface usually contributes less to the total azimuthal image smear than the orbital facet acceleration.

(3) For SARs with long integration times and large R/V-ratios (R = slant range, V = platform velocity) the orbital acceleration is an important factor in SAR ocean wave imaging. In particular, the C-band SAR proposed for the European Remote Sensing Satellite (ERS-1) may be expected to perform better for ocean wave imaging than the SEASAT L-band SAR because the acceleration effects are reduced by a factor 0.14 corresponding to the ratio of the integration times.

REFERENCES

Alpers, W., & Rufenach, C. L. (1979) The effect of orbital motions on synthetic aperture radar imagery of ocean waves, *IEEE Trans. Antennas Propagat.*, **AP-27** (5), 685–690.

Alpers, W., Ross, D. B. & Rufenach, C. L. (1981) On the detectibility of ocean surface waves by real and synthetic aperture radar, *J. Geophys. Res.* **86** 6481–6498.

Alpers, W. R. & Rufenach, C. L. (1980) Image contrast enhancement by applying focus adjustment in synthetic aperture radar imagery of moving ocean waves, in *SEASAT-SAR Processor, ESA SP 154,* pp. 25–30, ESA Sci. and Tech. Publ. Br., ESTEC, Nordwijk, Netherlands.

Alpers, W. (1983) Monte Carlo simulations for studying the relationship between ocean wave and synthetic aperture radar image spectra, *J. Geophys. Res.*, (in press).

Alpers, W., & Hasselmann, K. (1978) The two-frequency microwave technique for measuring ocean wave spectra from an airplane or satellite. *Boundary Layer Meteorol.* **13**, 215-230.

Alpers, W., & Jones, W. L. (1978) The modulation of the radar backscattering cross section by long ocean waves, in *Proc. 12th International Symposium on Remote Sensing of Environment*, Manila, Philippines, pp 1597-1607, Environmental Res. Inst. of Michigan, Ann Arbor.

Alpers, W., Schröter, J., & Keller, W. C. (1983) Measurements of the ocean wave-radar modulation transfer function at 4.3 GHz, (in preparation).

Bass, F. G., Fuks, I. M., Kalinykov, A. I., Ostrowsky, I. E., & Rosenberg, A. D. (1968) Very high frequency radio wave scattering by a disturbed sea surface, *IEEE Trans.* **AP-16** 554-568.

Bretherton, F. P., & Garret, C. J. (1969) Wavetrains in inhomogeneous moving media, *Proc. Roy. Soc. Ser. A* **302** 529-554.

Elachi, C. E., & Brown, W. E. (1977) Models of radar imaging of the ocean surface waves, *IEEE Trans. Antennas Propagat.* **AP-25** 84-95.

Garret, C., & Smith, J. (1976) On the interaction between long and short surface waves, *J. Phys. Oceanogr.* **6** 925-930.

Harger, R. O. (1980) The synthetic aperture radar image of time variant scenes, *Radio Science* **15**, 749-756.

Hasselmann, K., & Alpers, W. (1983) The response of synthetic aperture radar to ocean surface waves, *Proc. of the Symposium on Wave Dynamics and Radio Probing of the Ocean Surface*, Miami, Florida, May 1981, (in press).

Jain, A. (1978) Focusing effects in synthetic aperture radar imaging of ocean waves, *Appl. Phys.* **15** 323-333.

Keller, W. C., & Wright, J. W. (1975) Microwave scattering and straining of wind generated waves, *Radio Sci* **10** 139-147.

Kelly, E. J., & Wishner, R. P. (1969) Matched-filter theory for high velocity accelerating targets, *IEEE Trans. Military Electronics Systems* **AES-5** 98-105.

Longuet-Higgins, M. S. & Stewart, R. W. (1964), Radiation stress in water waves, a physical discussion with applications, Deep Sea Res. **11**, 529-562.

Phillips, O. M., & Banner, M. L. (1974) Wave breaking in the presence of wind drift and swell, *J. Fluid Mech.* **66** 625-640.

Plant, W. J., Keller, W. C., & Cross, A. Parametric dependence of ocean wave-radar modulation transfer functions, *J. Geophys. Res.* (in press).

Raney, R. K. (1971) Synthetic aperture imaging radar and moving targets, *IEEE Trans. Aerospace and Electronic Systems* **AES-7** (3), 499-505.

Raney, R. K. (1980) SAR response to partially coherent phenomena, *IEEE Trans. Antennas Propagat.* **AP-28** No. 6, 777-787.

Raney, R. K. (1981) Wave orbital velocity, fade and SAR response to azimuth waves, *J. Oceanic Eng.* **OE-6**, 140–146.

Rufenach, C. L., & Alpers, W. (1981) Imaging ocean waves by synthetic aperture radars with long integration times, *IEEE Trans. Antennas Propagat.*, **4P-29**, 422–428.

Shemdin, O. H., Brown, W. E. Jr., Staudhamer, F. G., Shuchman, R., Rawson, R., Zelenka, J., Ross, D. B., McLeish, W., & Berles, R. A. (1978) Comparison of *in-situ* and remotely sensed waves off Marineland, Florida, *Boundary-layer Met.* **13** 225–234.

Shuchman, R. A., Kasischke, E. S., & Klooster, A. (1978) *Synthetic aperture radar ocean wave studies,* Final report number 131700-3-F, September 1978, Environmental Research Institute of Michigan, Box 8618, Ann Arbor, MI 48107.

Swift, C. T. & Wilson, O. R. (1979) Synthetic aperture radar imagery of ocean waves, *IEEE Trans. Antennas Propagat.* **AP-27**,, 725–729.

Teleki, P. G., Shuchman, R. A., Brown, W. E., McLeish, W., Ross, D. B., & Mattie, M., (1978) Ocean wave detection and direction measurements with microwave radar, *Proc. OCEANS 78,* Washington, D. C., September 6–8, 1978, 639–648.

Tucker, M. J. (1983) The imaging of waves by satellite-borne synthetic aperture radar: the effects of the sea surface motion, (submitted to *J. Geophys. Res.*)

Valenzuela, G. R. (1978) Theories for the interaction of electromagnetic and ocean waves — A review, *Boundary Layer Meteorol.* **13** 61–85.

Valenzuela, G. R. (1980) An asymptotic formulation for SAR images of the dynamical ocean surface, *Radio Sci.* **15** 105–114.

Wright, J. W. (1968) A new model for sea clutter, *IEEE Trans. Antennas Propagat.* **AP-16** 217–223.

Wright, J. W., Plant, W. J., Keller, W. C., & Jones, W. L. (1980) Ocean wave-radar modulation transfer functions from the West Coast Experiment, *J. Geophys. Res.* **85** 4957–4966.

Chapter 7

Can optical measurements help in the interpretation of radar backscatter?

M. S. LONGUET-HIGGINS, FRS
Department of Applied Mathematics and Theoretical Physics, University of Cambridge, UK,
and Institute of Oceanographic Sciences, Wormley, Surrey, UK

7.1 INTRODUCTION

The backscatter of electromagnetic radiation from the sea surface in the X– and L– bands presumably depends on the spectral density of sea-surface waves at wavelengths of a few centimetres, and on how the intensity of these waves is related locally to the longer surface waves that are present. It depends, in other words, on the 'long-wave short-wave interaction'. In this chapter we indicate how it may be possible to derive information concerning this interaction from remote observations at optical wavelengths.

In some classical observations of the sunlight from a wind-roughened sea surface Cox & Munk (1956) studied the 'glitter pattern' of the sun, as seen from a high-flying aircraft at a fairly high angle of incidence. From such observations one can infer the probability density $p(\zeta_x, \zeta_y)$ of the two components of the surface slope ζ_x and ζ_y. One of Cox & Munk's findings was that the variance of the slope component ζ_x in the upwind-downwind direction was proportional to the surface windspeed (measured at a height of 41 ft).

A second but less well-known result was that the distribution of ζ_x was significantly skewed so that the area of maximum intensity in the glitter pattern was not where it would be if the surface were smooth, but was displaced from it by an angle Δ, say, sometimes of order $10°$. In a recent paper (Longuet-Higgins 1982†) the author showed that a skewness of this magnitude could not be due to viscous damping of the waves. It could, however, be explained on the assumption that the shorter waves, which contribute most to the slope variance, were modulated in phase and amplitude by the longer waves in the spectrum. The observed magnitude of Δ therefore gives some information regarding this modulation.

In the following sections we summarise the theory of this effect (section 7.2) and how it is confirmed by observation (section 7.3). A brief discussion is given in section 7.4.

---

†To be referred to as Paper I

7.2 A TWO–SCALE MODEL

Consider the simplest situation when sun, wind, and wave direction are in the same vertical plane, as in Fig. 7.1. It can be seen that the angle Δ (after correction for background radiation and variable surface reflectance) is related to the value $(\zeta_x)_m$ of the slope at the mode, or maximum, of the distribution, by

$$\Delta = 2(\zeta_x)_m \tag{7.1}$$

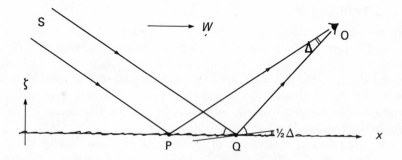

Fig. 7.1 – The reflection of rays from the sun S towards an observer O, when wind and waves are in the same direction.

Moreover it can be shown (see I) that if $p(\zeta_x)$ is nearly Gaussian, then $(\zeta_x)_m$ is related to the nth moments μ_n of the distribution by

$$(\zeta_x)_m = -\mu_3/2\mu_2 . \tag{7.2}$$

(It is not necessary that the moments be calculated explicitly; only that $p(\zeta_x)$ be given approximately by a Gram-Charlier series over the central part of the distribution.) From the two last equations it follows immediately that in general

$$\Delta = -\mu_3/\mu_2 . \tag{7.3}$$

Now suppose that short (capillary or capillary-gravity) waves ride on a random sea of much longer waves, as in Fig. 7.3. Suppose also that the steepness of the shorter waves is modulated (for reasons to be discussed later) in relation to the

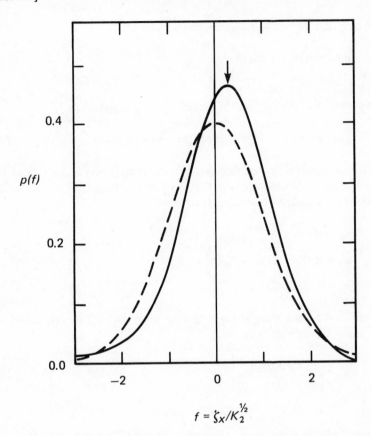

$$f = \zeta_x/K_2^{1/2}$$

Fig. 7.2 – Schematic diagram of the distribution of up-wind slope. Compare Cox & Munk (1956), Fig. 15.

longer waves. To fix the ideas, let the short waves be steeper, on average, when riding on the forward faces of the longer waves than they are when on the rear faces, and write

$$\zeta = a \cos \theta + a' \cos \theta' \tag{7.4}$$

where a and θ denote the amplitude and phase function of the longer waves, with wavenumber

$$k = \theta_x \tag{7.5}$$

so that a and k vary slowly with x and t. Primed symbols a' and k' denote corresponding quantities for the short waves, and we assume

$$a' k' = \alpha + \beta \, ak \cos (\theta - \gamma) \tag{7.6}$$

where α, β, and γ are constants.

By hypothesis $k' \gg k$, so that the surface slope ζ_x is given by

$$\zeta_x = ak \sin \theta - \{\alpha + \beta \, ak \cos (\theta - \gamma)\} \sin \theta'. \tag{7.7}$$

From this, the moments μ_n of the distribution of $p(\zeta_x)$ may be calculated by averaging ζ_x^n first with respect to the fast phase θ' and then with respect to the slow phase. Thus we obtain (see Paper I, section 6)

$$\mu_1 = 0$$
$$\mu_2 = \tfrac{1}{2}s^2 + \tfrac{1}{2}(\alpha^2 + \tfrac{1}{2}\beta^2 s^2) \tag{7.8}$$
$$\mu_3 = -3/2 \, \alpha \beta \, s^2 \sin \gamma$$

where s denote the r.m.s. value of the long-wave steepness ak. From equation (7.3) the angle Δ is then given by

$$\Delta = \frac{3 \, \alpha\beta \sin \gamma}{1 + \alpha^2/s^2 + \tfrac{1}{2}\beta^2}. \tag{7.9}$$

It will be noticed that when $\alpha\beta \sin \gamma$ is positive, as in Fig. 8.3, then Δ is also positive. Physically this is because the central part of the slope distribution, where $|\zeta_x|$ is relatively small, is affected more by the *rear* slopes of the longer waves than by the forward slopes, where $|\zeta_x|$ is on the whole larger. But on the rear slopes of the longer waves, there is a positive contribution to ζ_x from the longer waves. This tends, on balance, to make the mode $(\zeta_x)_m$ become positive, hence the angle Δ positive also.

In view of equation (7.7) it is useful to define a 'depth of modulation' of the shorter waves as

$$\delta = \beta s/\alpha, \tag{7.10}$$

s being the r.m.s. value of ak. Then, if most of the slope variance is contributed by the short waves, that is $\alpha^2 \gg s^2$ equation (7.9) becomes

$$\Delta = \frac{3\delta}{1 + \tfrac{1}{2}\delta^2} \, s \sin \gamma. \tag{7.11}$$

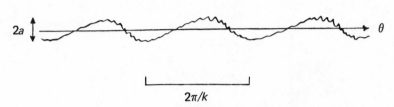

Fig. 7.3 – A modulated train of short waves riding on the surface of longer waves.

In other words, the apparent angular displacement Δ is independent of the mean-square ripple steepness α^2, and depends only on the r.m.s. slope s of the longer waves, together with the phase angle γ and relative depth δ of the short-wave modulation.

Assuming $0 < \delta < 1$, the factor $3\delta/(1 + \frac{1}{2}\delta^2)$ in equation (8.11) is always less than 2. Hence the inequality

$$\Delta \leqslant 2s. \tag{7.12}$$

7.3 COMPARISON WITH OBSERVATION

In the observations by Cox & Munk (1956) the wind direction differed from the direction of the 'significant waves' by an angle ψ which usually was small, but in two cases exceeded $90°$. Assuming the wind direction to be the same as that of the short waves, the simplest generalisation of equation (7.12) to arbitrary angles ψ would be

$$\Delta = \frac{3\delta}{1 + \frac{1}{2}\delta^2} \sin \gamma \, (s \cos \psi). \tag{7.13}$$

Fig. 7.4 shows a plot of the observed angles Δ against the variable $s \cos \psi$. The broken line corresponds to the bound

$$|\Delta| \leqslant 2s|\cos \psi|. \tag{7.14}$$

According to (7.14) we would expect the observation to lie somewhere between the broken line and the horizontal axis. This in fact they do. Some, indeed, are actually quite near to the broken line, suggesting that the model is essentially correct.

Even more striking is the fact that when $s \cos \psi$ is negative, that is the wind is opposite to the significant waves, then Δ also is negative. This implies that the short waves are steeper on the forward slopes of the significant waves, even when the latter are opposite to the wind.

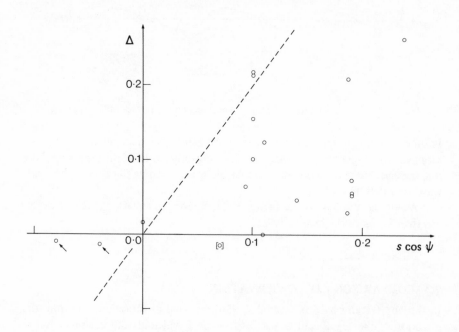

Fig. 7.4 – A plot of Δ against $s \cos \psi$, from the data of Cox & Munk (1956). (For details see Longuet-Higgins 1982, Table 1.)

In Paper I a dynamical model for the short waves is proposed tentatively in which the short-wave energy E' satisfies an equation of the form

$$\frac{\partial E'}{\partial t} + \frac{\partial}{\partial x} [E'(c'_g + U)] + S\frac{\partial U}{\partial x} + D = G .\qquad (7.15)$$

Here U denotes the orbital velocity of the longer waves, c'_g and S the group-velocity and radiation stress for the short waves, and D and G are quantities representing viscous dissipation of the short waves and direct generation of short-wave energy by the wind. On the assumption that $G = KE'$, with K a constant, and that the wave-field is in a quasi-steady state, it is shown that the phase angle γ satisfies

$$\tan \gamma \doteqdot \frac{4}{5} \frac{K}{\sigma} \tag{7.16}$$

where σ is the radian frequency of the longer waves, If long and short waves are in the same direction, then $\sigma > 0$ and so $0° < \gamma < 90°$; the short waves are steepest on the forward face of the long waves, as observed. If, on the other hand, the long waves are opposed to the short waves, then $-90° < \gamma < 0°$ also in agreement with observation.

7.4 DISCUSSION

The two-scale model suggested here, in spite of its roughness, does appear to give results in quantitative agreement with observation. The model has two unknown parameters γ and δ, corresponding to the phase angle and the depth of short-wave modulation. In Fig. 7.3 every plotted point yields an estimate of the ratio

$$\frac{'\Delta}{s \cos \psi} \doteqdot \frac{3\delta}{1 + \frac{1}{2}\delta^2} \sin \gamma \tag{7.17}$$

and hence a relationship between γ and δ. In particular cases when the plotted point lies on the broken line we should have necessarily $\delta = 1$ (full modulation) and $\gamma = 90°$ (short waves steepest at mean level of longer waves). Generally, neither of these is satisfied. But by combining the information from other sources with that given by (7.7), we may be better able to understand the behaviour of the short waves and hence the processes of wave generation; also the reflection of incident radiation from the sea surface.

REFERENCES

Cox, C. & Munk, W. (1956) Slopes of the sea surface deduced from photographs of sun glitter. *Bull. Scripps Inst. Oceanogr.* 6 401–488.
Longuet-Higgins, M. S. (1983) On the skewness of sea-surface slopes. *J. Phys. Oceanogr.* (in press).

Chapter 8

Some properties of SAR speckle

B. C. BARBER
Space department, Royal Aircraft Establishment Farnborough, Hampshire GU146TD

8.1 INTRODUCTION

Synthetic aperture radar image speckle is the grainy, granular noisy phenomenon which is observed in SAR images. It is characteristic of any coherent imaging system and has been extensively studied in optical imaging systems – see Goodman (1975). Fig. 8.1 shows clearly what is meant by SAR speckle. It is a SEASAT image from pass 882 and shows an area of desert in the Goldstone, California area.

In the case of optical systems the speckle is caused by the roughness of the object. This is shown in Fig. 8.2 (which has been taken from Goodman's paper). It will be seen that the rough surface has been replaced by a number of point scatterers. Each point produces a point spread function in the image plane. The variance of the heights of the points is much greater than the wavelength of the light. Hence each point spread function is completely dephased, and at each point on the image we have an isotropic random walk as shown in Fig. 8.3. The speckle then has circular normal (first order) statistics. Thus in optical speckle it is the roughness of the object which produces the image speckle. The analysis of optical speckle has progressed far and many further details will be found in Dainty (1975).

It is the object of this chapter to investigate the corresponding situation in coherent radar image speckle. It will be shown that there are some differences between the way in which optical speckle is formed and the way in which radar image speckle is formed.

Only stationary objects are considered here. Randomly moving surfaces produce speckle on account of the random motion, and an analysis of the speckle produced by random movers will be found in Raney (1980) or Ouchi (1981).

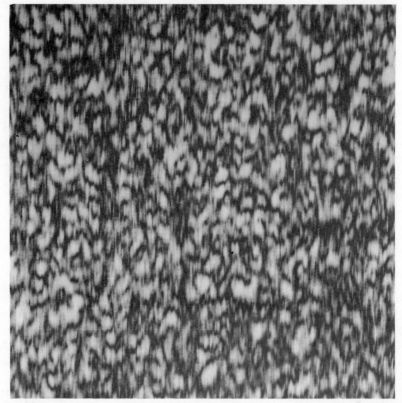

↑Azimuth/Range →

Fig. 8.1 − Approximately uniform speckle (area of rocks, boulders, and sand)

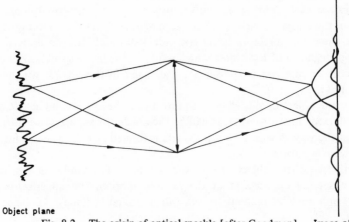

Object plane

Fig. 8.2 − The origin of optical speckle [after Goodman]   Image plane

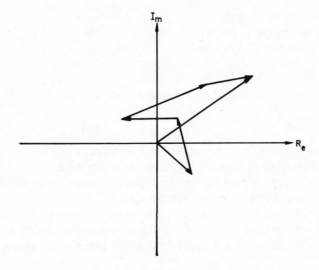

Fig. 8.3 – Two-dimensional random walk

8.2 THE POINT SPREAD FUNCTION OF A SAR

It will be evident from the remarks made in the Introduction that the point spread function is of primary importance, and that function will now be derived for a SAR system. Usually it is shown that the SAR imaging process is a two-dimensional convolution for a rectangular aperture giving a point spread function which, in the case of a uniformly weighted range and azimuth spectrum (aperture), is a two-dimensional 'sin x/x' or sinc x function. This, however, is not the entire story. There are, in addition, phase terms. Usually one is concerned with the SAR image after demodulation in which all phase terms are removed. A typical SAR processor produces an image in which each pixel is complex valued and has both an amplitude and a phase. The final operation which the processor performs is to take the modulus of each pixel, and this is demodulation. In the case of an optical processor this is implicit in the production of the image on a screen, but in the case of a digital processor the pixels produced have complex values, and an additional demodulation operation is performed.

A careful derivation of the point spread function in a complex valued image will now be made. All the first order phase terms will be included, that is to say those phase terms with a period comparable with, or less than, the resolution, i.e. the spatial length scale of the point spread function. Long period phase terms in the point spread function are neglected.

8.2.1 The range direction

Consider just one pulse, the nth pulse say, and an echo of unit amplitude from a point on an absorbing background.

The received echo is

$$E(t) = \exp i2\pi[f_0 t_n - f_0 2R(t)/c + \alpha(t_n - t_D)^2] \qquad (8.1)$$

see Corr (1979)

f_0	is the centre frequency of the transmitted range chirp,
t_n =	time measured from the beginning of the nth pulse,
t_D =	the round trip time from radar to point to radar for this particular pulse
$R(t)$	is the range of the point from the radar.
$R(t)$	is assumed constant for one pulse but varies from pulse to pulse. Thus for this pulse $2R(t)/c = t_D$, t being the overall time coordinate.

The transmitted chirp has been implicitly put into time symmetric form

$$A(t) = \exp i2\pi t(f_0 + \alpha t) \qquad -\frac{T}{2} \leqslant t \leqslant \frac{T}{2}, \qquad (8.2)$$

where T is the pulse duration.
and $\alpha = \text{bandwidth}/2T$.

A number of approximations have been made in (8.1). Briefly these are:

(1) When a signal is received as a result of scattering from a target and there is a relative velocity between the transmitter/receiver and the target, the frequency of the received signal will be changed by a scaling factor β where $\beta = 1 + 2v/c$ and so the signal spectrum is transformed:

$$\Psi(f) \rightarrow \Psi(f/\beta)\sqrt{|\beta|} . \qquad (8.3)$$

It is assumed that $f/\beta = f(1 - 2v/c)$ and in (8.3) $\sqrt{|\beta|} = 1$, and that the corresponding scaling in the time domain is $t \rightarrow \beta t$. This is all justified by the fact that the relative velocity of the radar, v, is much smaller than the velocity of light, c.

(2) Although the signals received at the spacecraft were transmitted a few milliseconds earlier when the spacecraft was at another position it is assumed that the round trip range and Doppler frequency shift are those corresponding to the time of reception. An examination of the approximation shows that the two main effects are an error of order 1 metre in the range direction resulting in very slight defocusing in the azimuth direction (totally negligible for a system such as SEASAT) and an image shift of order 10 metres in the azimuth direction with no corresponding defocus.

Returning to equation (8.1) we have $t_D = 2R(t)/c$ and so

$$E(t) = \exp i2\pi[f_0 t_n - f_0 t_D + \alpha(t_n - t_D)^2]. \tag{8.4}$$

The processor mixes the received signal coherently to translate the r.f. centre frequency f_0 down to the 'offset video frequency' f_1 (approximately 10 MHz for SEASAT), and then the received echo is:

$$E(t) = \exp i2\pi[f_1 t_n - f_0 t_D + \alpha(t_n - t_D)^2]. \tag{8.5}$$

It should be noted that the signal has been expressed as an analytic signal with a single-sided spectrum. In reality at this stage of the processing it is still only in real form and is converted into complex form after range compression. This, however, makes absolutely no difference to the analysis and makes it a good deal tidier.

Range compression is performed on each pulse by correlating the pulse against a replica of the transmitted chirp translated to the offset video band. Let $\tau_R = t_n - t_D$ and use the symmetrical form of the chirp replica (8.2). Then we correlate

$$\exp i2\pi[f_1(\tau_R + t_D) - f_0 t_D + \alpha \tau_R^2] = \exp i2\pi[f_1 t_D - f_0 t_D] \exp i2\pi[f_1 \tau_R + \alpha \tau_R^2] \tag{8.6}$$

against (8.2) for $-T/2 \leqslant t \leqslant T/2$. This gives:

$$g(\tau_R) = \exp i2\pi[f_1 t_D - f_0 t_D] \int_{-T/2}^{T/2} \exp i2\pi[f_1 t + \alpha t^2 - f_1(\tau_R + t) - \alpha(\tau_R + t)^2] \, dt \tag{8.7}$$

where $g(\tau_R)$ is the point spread function in the range direction. So

$$g(\tau_R) = \exp i2\pi f_1 t_n \exp(-i2\pi f_0 t_D) \frac{\sin \pi \Delta F_R \tau_R}{\pi \Delta F_R \tau_R}, \tag{8.8}$$

where ΔF_R = the video bandwidth = $2\alpha T$.

After the range compression process the offset video frequency f_1 is mixed down to zero. The output from range compression for this particular pulse is then:

$$g(\tau_R) = \exp(-i2\pi f_0 t_D) \frac{\sin \pi \Delta F_R \tau_R}{\pi \Delta F_R \tau_R} \tag{8.9}$$

In (8.9) $g(\tau_R)$ has been multiplied by a constant ($= 1/\sqrt{A}$, where A is the time bandwidth product) in order to normalise it.

8.2.2 The azimuth correlation

The effects of range walk are included in this analysis; range walk produces a phase term in the point spread function which can be significant for some purposes, and the objective is to give a rigorous derivation of the point spread function.

Azimuth correlation is essentially the same process as range correlation. In this case the frequency modulation results from the Doppler effect and the change in t_D from pulse to pulse. The change in round trip time t_D from pulse to pulse gives a rotating vector via the phase term in (8.9). The rate at which the vector rotates represents the Doppler frequency. In a sense, then, each pulse samples the phase of the point, and over a number of pulses the rate of change of phase is built up and one can then meaningfully refer to a frequency.

The phase term in (8.9) gives the 'phase history' of the point.

Now

$$t_D = \frac{2R(t)}{c} = \frac{2}{c} \ (a_0 + a_1 t + a_2 t^2 + \ldots) \tag{8.10}$$

where the range to the target point has been expanded in a Taylor series about $t = 0$ (i.e. the centre of the synthetic aperture). a_0 is the slant range from radar to point at $t = 0$, a_1 is the slant range velocity at $t = 0$, and a_2 is one half of the slant range acceleration at $t = 0$. In practice (at least in the RAE processor) all terms up to and including the cubic in t are taken, but for analytical convenience only the terms up to the square will be taken here. The cubic is very small and results in an almost negligible phase error over the synthetic aperture for SEASAT. It will be observed that the finite value of a_1 at the centre of the aperture is a result of the rotation of the Earth, i.e. range walk.

We now correlate in the azimuth direction and at a slant range time displacement from t_D corresponding to the time t_n (i.e. at $\tau_R = t_D - t_n$). Now for a small change in the slant range the first order changes to the range polynomial are Δa_0 in a_0, $\Delta a_0 (a_1/a_0)$ in a_1, and $\Delta a_0 (a_2/a_0)$ in a_2. For lengths comparable with the point spread function scale it can easily be shown that the changes to a_1 and a_2 are utterly negligible. So we consider just the change Δa_0 in a_0; thus in the tensor which arises from the Taylor expansion about $a = a_0$ and $t = 0$ we retain the terms:

$$R'(t) = a_0 + \Delta a_0 + a_1(t + \tau_A) + a_2(t + \tau_A)^2$$

where $\Delta a_0 = c\tau_R/2$,

and, in addition, for a displacement d in the azimuth direction there will be a shift in the time corresponding to $d/v = \tau_A$ where v is the radar velocity.

So, the actual phase history of the point is $\exp(-i2\pi f_0 t_D)$ with t_D given by (8.10), and the phase history against which it is correlated by the processor is $\exp(-i2\pi f_0 t'_D)$ with t'_D given by:

$$t_D' = \frac{2}{c} [a_0 + \Delta a_0 + a_1 (t + \tau_A) + a_2(t + \tau_A)^2].$$ (8.11)

It should be noted that these various approximations have been introduced here in order to make the analysis tractable and tidy, with negligible error. In practice digital processors compute the phase history (8.11) from the spacecraft orbit, the Earth rotation and geometry, even taking into account such things as the non-sphericity of the Earth. In the RAE processor the phase history (8.11) is computed for each image point (pixel).

Correlating (8.9) against $\exp(-i2\pi f_0 t_D')$, then, gives:

$$g(\tau_R,\tau_A) = \frac{\sin\pi \Delta F_R \tau_R}{\pi \Delta F_R \tau_R} \int_{-T_A/2}^{T_A/2} \exp(-i2\pi \frac{2f_0}{c} [a_0 + a_1 t + a_2 t^2])$$

$$\times \exp(i2\pi \frac{2f_0}{c} [a_0 + \Delta a_0 + a_1 (t - \tau_A) + a_2(t - \tau_A)^2])dt$$ (8.12)

where a symetric Doppler frequency band has been taken. An unsymmetric band simply introduces a constant phase term which is not relevant for our present discussion. So

$$g(\tau_R,\tau_A) = \exp(i\frac{4\pi f_0}{c} [\Delta a_0 + a_1 \tau_A + a_2 \tau_A^2])\frac{\sin \pi \Delta F_R \tau_R}{\pi \Delta F_R \tau_R} \frac{\sin \pi \Delta F_A \tau_A}{\pi \Delta F_A \tau_A},$$ (8.13)

ΔF_A is the Doppler bandwidth. The point spread function $g(\tau_R,\tau_A)$ can be expressed in lengths in azimuth (x) and slant range (y) by means of $\tau_A = x/v$ and $\tau_R = 2y/c$. The point spread function in (8.13) has been scaled by dividing by $\sqrt{A} \sqrt{B}$ where A and B are range and azimuth time-bandwidth products. In (8.12) T_A is the synthetic aperture length (in time). Usually the a_0 term in the range polynomial is ignored in the azimuth correlation process, and this results in an additional phase term of $\exp i4\pi f_0(a_0/c)$ being present. This does not, however, affect our present argument since it results in a carrier which is common to the whole image.

8.2.3 The phase terms

From (8.13) it will be seen that the two principal phase terms in the point spread function before demodulation are:

$$\exp i \frac{4\pi f_0}{c} \Delta a_0 ,$$ (8.14)

and

$$\exp\ i\ \frac{4\pi f}{c}\ a_1 \tau_\text{A}\ .\tag{8.15}$$

The first phase term varies in the slant range direction, and Δa_0 is the difference in slant range between the actual range to the scatterer and the range to a point on the image (both at $t = 0$). The other phase term varies in the azimuth direction, and $\tau_\text{A} = x/v$ where x is the distance in azimuth between the centre of the point spread function and the point at which it is measured, and v is the radar velocity. The period of the first term (8.14) is $c/2f_0 = \lambda_0/2$ which is 0.118 metres for SEASAT, and the period of the second term (8.15) is $cv/2f_0a$, which for the image in Fig. 8.1 is about 25.6 metres.

8.2.4 The spectrum

The point spread function has an associated spectrum: the Fourier transform in the range and azimuth directions. Fourier transforming (8.13) gives a rectangular spectrum centred on f_0 of width ΔF_R in the range direction, and centred on $2f_0a_1/c$ and of width ΔF_A in the azimuth direction.

Naturally, ΔF_R is the video bandwidth and ΔF_A is the Doppler bandwidth in the azimuth direction. In the range direction the spectrum is centred on the radar mean carrier frequency. In the azimuth direction the offset frequency $2f_0a_1/c$ is caused by the Earth's rotation. The Doppler frequency is not zero in the centre of the synthetic aperture, because the Earth rotates.

A wide video bandwidth implies fine slant range resolution and therefore a wide spatial spectrum in the slant range direction. Likewise a wide Doppler bandwidth implies a wide spatial spectrum in the azimuth direction. It is easy to link the temporal and spatial spectra. A carrier of f_0 implies a spatial frequency of $2f_0/c$ cycles per metre in the slant range direction, and a Doppler carrier of $2f_0a_1/c$ implies a spatial frequency of $2f_0a_1/cv$ cycles per metre in the azimuth direction. Likewise the corresponding spatial bandwidths are $2\Delta F_\text{R}/c$ and $\Delta F_\text{A}/v$ in the slant range and azimuth directions.

Therefore in the imaging process the synthetic aperture radar system is only sensitive to spatial frequencies in the object in the range $2f_0/c \pm \Delta F_\text{R}/c$ in the slant range direction and $2f_0a_1/cv \pm \Delta F_\text{A}/2v$ in the azimuth direction. All other frequencies in the object are ignored by the system. The same conclusion can be arrived at in a rather less intuitive manner by using the well-known theorem that the image is formed by a two-dimensional convolution of the object and the point spread function for a linear imaging system (see Papoulis (1968) p 20).

The above argument amounts to a statement of the Bragg scattering principle. The radar is sensitive only to those Bragg scatterers with a spatial frequency ($= 1/\lambda$ where λ is the Bragg scatter wavelength) of $2f_0/c \pm \Delta F_\text{R}/c \cong 8.5 \pm 0.06$ cycles per metre and $2f_0a_1/cv \pm 0.06$ cycles per metre in range and azimuth directions. We have assumed a square resolution cell, i.e. azimuth resolution = slant range resolution. The term $2f_0 a_1/cv$ varies depending on the latitude. It may range from 0 at the 'top' of the orbit (nearest to the poles) to 0.05 in the North Atlantic to 1 in the English Channel to 2 in the Mediterranean, in round numbers. A fuller account of these matters will be found in Barber (1982) where some implications for the imaging of the sea surface are examined. We how turn to the role of the spectrum in determining the speckle properties.

8.2.5 Speckle and the spectrum

Consider a point in the object plane. The point may be regarded as a delta function, and its associated spectrum contains all wave numbers

$$2\pi\delta(x - x_1) \quad \leftrightarrow \quad \exp ix_1 k \tag{8.16}$$

where $k = 2\pi n = 2\pi/\lambda$. (See Papoulis (1968) p. 99).

The SAR system has a pass band $F(k)$ and then the spectrum 'seen' by the radar is $F(k) \exp ix_1 k$ where in this case $F(k)$ is the rectangular function discussed in the previous section. The variable x_1 is either a slant range coordinate for the range spectrum or an azimuth coordinate for the azimuth spectrum. If another point is added at a coordinate x_2 and so on for n points then the spectrum of the sum of the points as seen by the SAR is:

$$F'(k) = \frac{1}{2\pi} F(k) \sum_{j=1}^{n} \exp ix_j k. \tag{8.17}$$

If the position of the points are random then the phases in (8.17) are random and have a uniform probability density over $(-\pi, +\pi)$ and as a result the powers of each vector add to give

$$\langle |F'(k)|^2 \rangle = |F(k)|^2 = S(k) \tag{8.18}$$

to within a multiplicative constant. The power spectrum of the speckle is then the same as the radar pass band. Also, since the random walk in (8.17) is isotropic then each frequency component in the spectrum will have circular normal statistics and then so will the resulting speckle. This is an alternative viewpoint to the formation of the speckle to that explained in the Introduction.

The autocorrelation of the speckle $R(x)$ is simply the Fourier transform of the power spectrum $S(k)$ (see Papoulis (1968) p. 270), and this is the Fourier transform of the radar power pass band. Thus by Fourier transforming the speckle power spectrum in the image in both range and azimuth directions we can recover the point spread function of the radar imaging system. The fundamental reason for this is that a very large number of randomly phased points looks to the radar like 'white noise' with an autocorrelation which is a delta function. A more extensive discussion of these arguments will be found in Uscinski, Ouchi, Thomas (1980).

8.3 SPECKLE AND THE SCATTERING POINTS

Having worked out a rigorous expression for the point spread function including those phase terms which vary significantly over the p.s.f. we now consider the result of adding many such p.s.f's together.

8.3.1 Scatterers on a flat surface
Consider the situation where we have an object consisting of a large number of scattering points on a flat absorbing surface — see Fig. 8.4. Each produces a point spread function on the image which may typically be 20 metres between the −3 dB points. These points are randomly distributed on the surface, and so the point spread functions are also randomly distributed on the image. The image is sampled at a particular position (the 'pixel value') and the phase of each point spread function is given by $\exp i4\pi f_0 \, (\Delta a_0 + a_1 \tau_A)/c$. Let us examine Δa_0. It is different for each scatterer, is random, and on average is several metres. This is much greater than the period of this phase term which is about 0.118 metres.

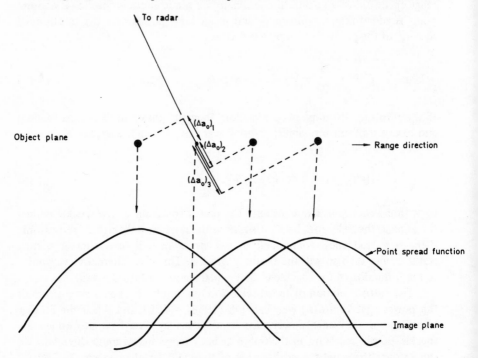

Fig. 8.4 − Scatterers on a plane absorbing surface

Thus the point spread functions are completely dephased and an isotropic random walk is formed. Speckle is then obtained with the usual circular normal statistics. That is, the real and imaginary parts of each complex valued pixel have independent Gaussian distributions with zero mean and the same variance. Note that a rough surface is not required. SAR speckle can be produced by scatterers imbedded in a flat surface.

8.3.2 Scatterers on a rough surface

A rough surface can be modelled in a simplistic (but effective) way by replacing it by a three-dimensional array of points. It is evident that in a SAR imaging process it is the distance of a point form the radar (the slant range) which matters. Two points which have the same range and are situated at the same azimuth coordinate (but have an angular separation) will be imaged at the same point on the image – see Fig. 8.5. Therefore our three-dimensional array of scatterers will be imaged as a collection of points on a notional flat object plane, each point being projected on to the plane along a line of constant range (Fig. 8.6).

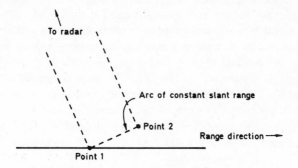

Fig. 8.5 – Two points superimposed by the radar

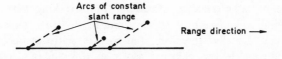

Fig. 8.6 – Points mapped on to a plane

Therefore variations in the height of the scattering points are translated into variations in their positions on the equivalent surface. Periodic variations in the height of the scatterers will result in periodic clumping on the image and periodic variations in the speckle mean Fig. 8.7. It is not difficult to show (although it will not be done here) that the mapping of height into position is not linear, and in consequence if the points are distributed on the surface of a sine wave, say, then the resultant periodic variation of density in the image is not sinusoidal. The image does, however, have the same period as the object. When the slope of the sine wave is small and the angle of incidence is large then the mapping is approximately linear.

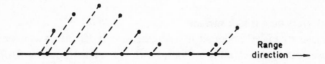

Fig. 8.7 – Periodic bunching of scatterers

In any event for a rough object surface an equivalent flat surface can be defined (in principle). In forming the image of the rough surface via the equivalent flat surface the vertical roughness scale is mixed with the horizontal roughness scale in the range direction. There is in general no one-to-one correspondence between a rough surface and its equivalent flat surface of scattering points. It therefore seems that one cannot measure the roughness of a surface by using the same techniques as are used in measuring the roughness of optical surfaces, by measuring the contrast of the resulting speckle (see Dainty 1975). The statistics of SAR speckle are always circular normal because the period of the slant range phase term $\lambda_0/2 = 0.118$ metres is very much less than the slant range resolution length. Even perfectly flat surfaces can, in principle, produce speckle; although such a surface cannot produce any backscatter (unless it has dielectric fluctuations, say) it is a useful idealisation and demonstrates an important principle in SAR imaging.

Finally, it should be pointed out that it has been implicity assumed in the arguments above that there are a large number of scattering centres in each resolution cell. This may not be true. Indeed in the case of the sea surface it is not even likely in some circumstances. For example, in the case of SEASAT, there could be large specular reflections from breaking waves in a rough sea and high wind and there may be only one or two such large reflections in a resolution cell. Locally, therefore, the speckle statistics may not be circular Gaussian and could be very elliptical. However, this kind of effect is very difficult to detect because such statistical behaviour is not stationary over the image.

8.4 SOME MEASUREMENTS OF SPECKLE STATISTICS

Some measurements have been made on the speckle shown in Fig. 8.1. It is in fact an area of the desert around Goldstone from SEASAT rev 882. The ground range and azimuth resolutions are 19 metres and 25 metres respectively. The Doppler bandwidth was 239.55 Hz and is symmetrical about the centre of the synthetic aperture; the corresponding angular width of the subaperture was 0.2120 degrees. No sidelobe reduction was used in the processing. The image was processed digitally with a pixel spacing of 3 metres and an area 512 pixels square of uniform speckle with no noticeable modulation was taken. There were, therefore, about 5000 speckles in the speckle field. Fig. 8.1 is in fact slightly smaller than the area analysed. Fig. 8.8 and Fig. 8.9 show the probability density functions of the real and imaginary parts of each pixel. It will be seen that they are reasonable zero mean Gaussian densities. The variance of the imaginary values is about 3.5% greater than the variance of the real values, and the ratio of the mean to the standard deviation is 2.8×10^{-4} for the real values and 4.2×10^{-5} for the imaginary values. Gaussian functions with the same variance are superimposed, and it will be seen that the histograms are slightly more pointed than the corresponding Gaussians.

So far only the first order statistics of the speckle have been examined and finally a brief examination of the second order statistics (the autocorrelation and power spectrum) of the speckle in Fig. 8.1 is made. In section 8.2.5 it was shown that if the spatial autocorrelation of the scatter is a delta function (corresponding to white spatial noise) then the autocorrelation of the speckle gives the point spread function of the system. This is quite general, for any imaging system, and is also valid for a SAR system when the white noise object is in any kind of motion, see Raney (1980) and Uscinski, Ouchi, & Thomas (1980).

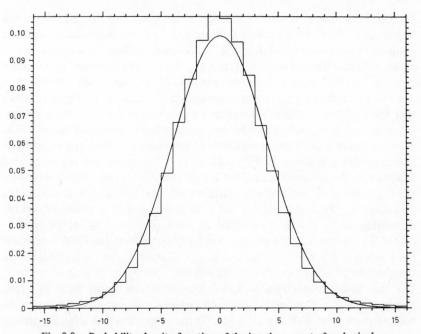

Fig. 8.8 – Probability density function of the imaginary part of each pixel

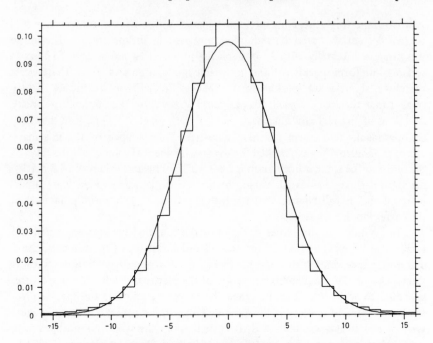

Fig. 8.9 – Probability density function of the real part of each pixel

Fig. 8.10 shows the auto-correlation of the speckle in Fig. 8.1, and it will
be seen that we do indeed have a point spread function. A few hundred metres
away from the area of desert in Fig. 8.1 a corner reflector was positioned. A
corner reflector behaves as a point, and Fig. 8.11 shows the actual point spread
function obtained from this corner reflector. The y axis is azimuth direction
and the x axis the range direction. Comparison of Fig. 8.11 and Fig. 8.10 shows
that the sidelobes in azimuth direction are greatly attenuated in the autocorrelation in comparison with the sidelobes in the actual point spread function. The
autocorrelation is the Fourier transform of the power spectrum and the Doppler
power spectrum is shown in Fig. 8.12. Here is the reason for the attenuated
sidelobes in the autocorrelation, for the spectrum is tapered. This is a genuine
effect produced by the scatterer. That no sidelobe reduction was used in the
processing will be seen from Fig. 8.11. In fact Fig. 8.11 was taken from the
same image as Fig. 8.1. In view of the narrow angular width of the sub-aperture
(about 0.21 degrees) it will be seen that backscatter from the Goldstone desert
has a rather peaky, noisy nature. This is typical, and if the sub-aperture is moved
then, usually, different results are obtained with most types of random scatterer.
The fact that the spectrum in 8.12 is tapered symmetrically is, in this case,
purely fortuitous. One should be very careful in interpreting the results of
autocorrelations on SAR speckle fields. On the other hand such a result evidently
contains information about the scatterer.

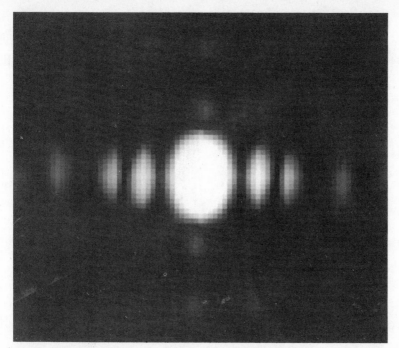

↑Azimuth/Range →

Fig. 8.10 – Autocorrelation of speckle in Fig. 8.1

↑Azimuth/Range →

Fig. 8.11 – Point spread function obtained from a corner reflector

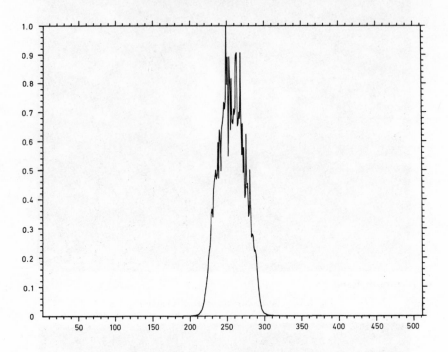

Fig. 8.12 – Doppler power spectrum

8.5 CONCLUSIONS

The point spread function of a conventional digital SAR processor has been derived including the phase terms. It has been shown that even for scatterers embedded in a flat surface the resulting image speckle has circular normal first order statistics. The conclusion is that the surface roughness can have little effect on the type of first order statistics of speckle. This is in contrast to the speckle obtained in optical imagery. Of course, the mean of the speckle can be affected by the backscatter but not the functional form of the first order statistics.

On the other hand an example is given where the second order statistics in the azimuth direction do seem to be affected by the scatterer, the Doppler spectrum being peaked around the centre of the synthetic aperture. This kind of effect should, however, be treated with caution owing to the narrowness of the angular width of the synthetic aperture and the irregularity of the angular spectrum of the backscattered waves from such scatterers.

REFERENCES

Barber, B. C. (1982) The role of Bragg scatters in image sea surface currents by SAR. RAE Technical Report (in preparation).

Corr, D. G. (1979) *Experimental SAR processor study and system implementation manual.* Report C0875 prepared under contract A926/502 for Space Department, RAE Farnborough.

Dainty, J. C. (1975) Laser speckle and related phenomena. *Topics in Applied Physics,* Vol 9. Springer-Verlag.

Goodman, J. W. (1975) Statistical properties of laser speckle patterns. In Dainty, see above.

Ouchi, K. (1981) Statistics of speckle in synthetic aperture radar imagery from targets in random motion. *Optical and Quantum Electronics* **13** 165–173.

Papoulis, A. (1968) Systems and transforms with applications in Optics. McGraw-Hill.

Raney, R. K. (1980) SAR processing of partially coherent phenomena. *Int. J. Remote Sensing* **1** No. 1 29–51.

Uscinski, B. J., Ouchi, K. & Thomas, J. O. *Speckle in synthetic aperture radar (SAR) imagery.* Oxford Computer Services Report prepared under Contract A57A/1504 for Space Department, RAE Farnborough.

Chapter 9

The effect of a moving sea surface on SAR imagery

M. J. TUCKER
Institute of Oceanographic Sciences, Crossway, Taunton, Somerset, UK

9.1 INTRODUCTION

Many SEASAT–SAR images show waves, some showing rather beautiful pictures of waves refacting round islands or other interesting features.

Several mechanisms are known by which waves can be imaged, and these are reviewed briefly by Alpers in Chapter 6 of this book. Most of the mechanisms produce 'real cross-section modulation' and are seen by all types of radar, but there is one imaging mechanism which operates only with a SAR. This is due to the interaction of the moving sea surface with the aperture synthesis process. This can produce either imaging or smearing of along-track (azimuthal) travelling waves, and is the subject of this chapter.

An extensive review of what is known about imaging is given in a recent paper by Alpers *et al.* (1981).

9.2 BASIC CONCEPTS

There are a few concepts which need to be understood before the effects of sea-surface motion on the imaging of waves can be considered.

9.2.1 Bragg resonance

If the sea surface is rippled by a light breeze with no long waves present, then microwave radar backscattering is due to that component of the wave spectrum which resonates with the radar wavelength (Fig. 9.1). This 'Bragg resonant wave' has its crest at right angles to the radar range direction and has a wavelength λ_B given by

$$\lambda_B = \lambda_R/(2 \sin \theta_i) \tag{9.1}$$

where λ_R is the radar wavelength
$\quad \theta_i$ is the angle of incidence
For SEASAT SAR, $\lambda_B \sim 34$ cm.
This phenomenon is well known, particularly in the case of h.f. radar.

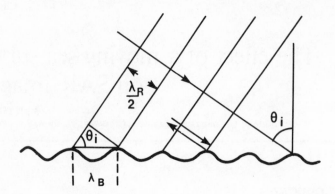

Fig. 9.1 – Bragg resonance. Coherent backscatter is given by waves whose crests are perpendicular to the radar incidence and whose wavelength λ_B is given by $\lambda_B = \lambda_R/(2 \sin \theta_i)$.

9.2.2 With long waves present

If long sea waves are also present, the simple concept of Bragg resonance no longer applies generally: for example, θ_i relative to the local wave slope varies (Fig. 9.2). However, it can be thought of as applying locally (see below), and then the waves modulate the amplitude of the backscatter by a number of mechanisms as mentioned in the Introduction. Thus, if we could take an instantaneous radar 'snapshot' of the sea surface, we would see the backscatter modulated by the longer waves and thus be able to get a picture of them. I shall call this the 'Primary scattering field': that is, before the interaction of motion effects with the aperature synthesis process has been considered.

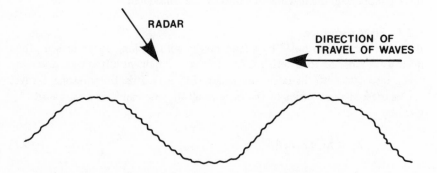

Fig. 9.2 – When longer waves with amplitudes large compared to the radar wavelength λ_R are present, Bragg resonance can no longer be considered to apply to a single component of the wave height spectrum. For example, from the formula in Fig. 9.1, λ_B varies with the local angle of incidence.

9.2.3 The primary scene element

It has been stated above that even with long waves present, Braggs scattering can be considered to apply locally. Further examination of the variation of roughness within a patch using the known characteristics of the wave spectrum shows that a patch about 5 Bragg wavelengths across can legitimately be considered in this way, and should produce backscattering equivalent to a single target at its centre, this target being carried about by the particle velocities in the longer waves present. It can be plausibly argued that this equivalent target can be considered as constant during a typical aperture synthesis time.

Thus, the sea can be considered as a collection of 'Primary scene elements', with many of them in each resolution cell of the radar (Fig. 9.3).

The average backscattering cross-section of the primary scene elements within a resolution cell is modulated by the long waves present as described briefly above.

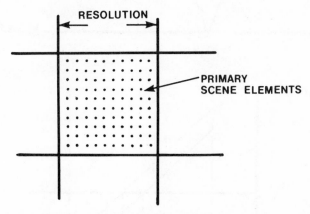

Fig. 9.3 — The backscatter pattern can be considered as coming from primary scene elements of which there are many per resolution cell.

9.2.4 Azimuthal offsets due to range velocities

If one considered the effect of target motion using a SAR, it turns out that for the sort of velocities of present interest (of order 1 m/s) the only component which matters is the range component v_r of the velocity. Such velocities have the odd effect of producing an azimuthal offset in the apparent position of the target, by the following mechanism.

As the real radar beam flies past a fixed target its range first shortens and then increases again, producing a Doppler shift as shown in (Fig. 9.4). The process of aperture synthesis effectively looks for such signals and, if detected, measures the time t at which the Doppler shift is zero: the target is then at right angles to the radar path. If, however, the target is moving towards the radar there is an extra Doppler shift which alters the time of zero Doppler, thus producing an apparent azimuthal offset.

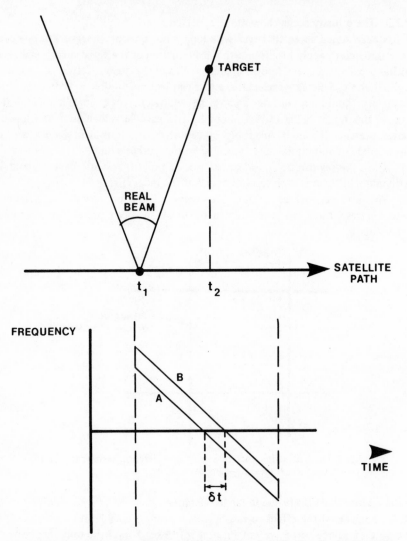

Fig. 9.4 — As the radar flies past a fixed target, the range first shortens and then lengthens, giving a Doppler frequency as shown by curve A. If the target is moving towards the radar an extra Doppler is added as in curve B, changing the apparent time of minimum range, and hence of azimuthal position.

The magnitude χ of this offset is given by the surprisingly simple formula

$$\chi = (R_s/V_s)\, v_r$$

where R_s is the range to the target

   V_s is the radar platform velocity

For SEASAT, $R_s/V_s \sim 115$ s.

9.2.5 The linearity of the aperture synthesis process

The aperture synthesis process (not including the final detector) is linear in the sense that the law of superposition applies: all targets pass through it without interacting.

Thus, we can take all our primary scattering elements through it independently and add those which fall in each resolution cell just before final detection.

Statistically, it is the expected backscattering cross-section σ_e of the primary scene elements which add to give the measured cross-section in each resolution cell.

9.3 THE EFFECT OF THE SEA-SURFACE VELOCITIES

9.3.1 General

The full effect of these in the general case is a difficult nonlinear problem, and no complete analytic solution is available. It can in principle be tackled by numerical simulation, and Hasselmann & Alpers (see Chapter 6 by Alpers in this book) have obtained some useful and interesting insights using one-dimensional modelling. Full two-dimensional modelling over a representative range of sea states is quite a formidable exercise, however. Here, we shall look at two extreme cases for which analytic solutions are available.

9.3.2 Long low swell travelling azimuthally

Consider a long low azimuthally-travelling swell under a surface lightly rippled by local winds.

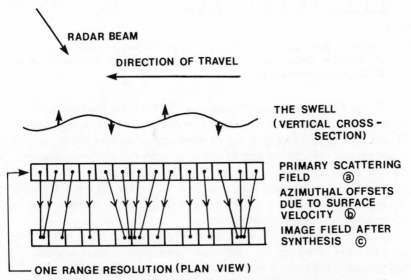

Fig. 9.5 – A low swell travelling azimuthally (along-track) produces offsets in the backscattering elements which cause them to bunch, thus imaging the swell. (Only one primary scattering element shown per resolution cell.)

Since satellite-borne radars have steep angles of incidence, the range component of sea-surface velocity for azimuthally-travelling waves is only a few per cent less than the vertical component.

Thus, the scattering will be moved sideways as shown in Fig. 9.5.

This is a potent imaging mechanism and has been analysed by Alpers *et al.* (1981). However, the linear analysis soon breaks down as the waves steepen, and analytical solutions become difficult.

9.3.3 A short wavelength sea with no swell

The situation is now as in Fig. 9.6. The range component of velocity varies within the resolution cell, and after the aperture synthesis processing, the back-scattering energy from one resolution cell in the primary field can be spread over several in the image field.

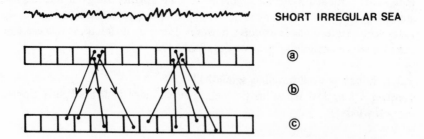

Fig. 9.6 — With only a short steep sea present, the primary scattering elements are offset in a random manner, producing a smearing of any patterns present in the primary scattering field.

If one assumes a standard formulation for the wave spectrum such as that due to Pierson & Moskowitz, it is possible to calculate the r.m.s. vertical component of velocity. For SEASAT, this will be within three per cent of the r.m.s. range component of velocity because of the steep angles of incidence, whatever the directional characteristics of the spectrum. Considering components in the spectrum with wavelengths less than than 60 m, this gives, for a 6.5 m/s wind for example, a mean-square velocity of 0.176 $(m/s)^2$ and a corresponding r.m.s. azimuthal offset of approximately 48 m.

It is possible to calculate the corresponding spatial filter. It is a Gaussian low-pass filter acting on the azimuthal components of wave number, with a half-intensity (3dB) cut-off corresponding to a wavelength of 364 m.

One SEASAT image analysed by Seichter (1981: his Fig. 5) corresponds to such a local windspeed with some swell present. It appears to show an azimuthal low-pass filter effect with a visually estimated cut-off at 330 m (Fig. 9.7).

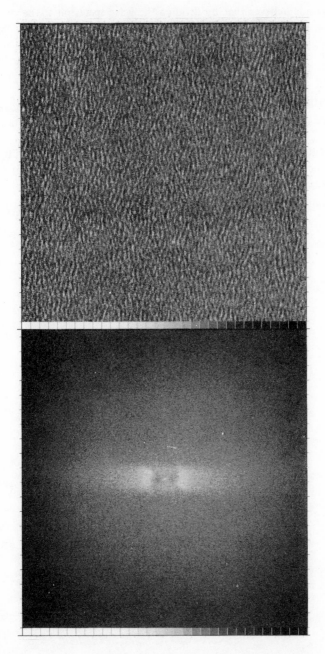

Fig. 9.7 — The image is 12.8 km square and the 2-D Fourier transform has scale of k ranging from −40 km⁻¹ (corresponding to 25 m wavelength) to +40 km⁻¹. The range direction is vertical.

9.3.4 A mixed sea

The present author believes that by dividing the wave spectrum at the correct frequency, it is possible to consider the effects of long and short wavelengths separately, but he has not yet been able to prove this conclusively.

REFERENCES

Seichter, H. (1981). Two-dimensional power spectra of SEASAT SAR imagery *SAR image quality:* selected papers from a Workshop at ESRIN, Frascati, Italy, 11–12 December 1980. European Space Agency, p. 55.

Alpers, N. R., Ross, D. B. & Rufenach, C. L. (1981). On the dectectability of ocean surface waves by real and synthetic aperture radar. *Journal of Geophysical Research* **86**, No. C7, 6481–6498.

Chapter 10

Theory of SAR ocean wave imaging

S. ROTHERAM

Marconi Research Centre, West Hanningfield Road, Great Baddow, Chelmsford Essex UK

10.1 INTRODUCTION

With the proposed launch of a number of spaceborne SARs such as the European Space Agencies ERS-1, the question of the relation between waves on the sea surface and features in SAR images is of crucial importance. There are two approaches to this question. The direct problem is to explain SAR images of a given sea surface. This is slowly yielding to analysis, and many results are given here. The inverse problem is to deduce the structure of the sea surface from SAR data. Some first steps in the solution of this problem are presented.

One aim of this work has been to adopt a more rigorous approach. The foundations of the theory are in sections 10.2, 10.3, and 10.4 which describe sea surface representations and electromagnetic scattering theory. The main results for the direct problem are in section 10.6 on linear imaging and section 10.7 on nonlinear imaging. The linear theory is fairly complete but the nonlinear theory has only been developed for a single long ocean wave. Linear inverse theory is discussed in section 10.8. This is only a tentative theory which ignores speckle, noise, and nonlinearity. SAR images can be constructed in a number of alternative ways, and these are pointed out in section 10.9. They include phase, autocorrelation function, and refocused images. A case study of a SEASAT SAR image is given in section 10.10. Its spectrum shows both harmonics and banding. The linear inverse theory is applied to the image in a reconstruction of the sea surface.

10.2 SEA SURFACE REPRESENTATIONS

An important aspect of sea imaging is to find representations of the sea surface including the interactions between long and short waves. The wave action method (Alpers 1978, Valenzuela 1979) has been much used for this purpose. It gives the modification of the short-wave wavevector, frequency and energy spectrum by the long waves. The amplitude modulation of the energy spectrum is known as straining. Source terms have been included in these approaches but the results are controversial. It is necessary to account for resonant interactions, wave breaking, viscous dissipation and the input from the wind.

In this work a slightly different approach is adopted. It proves convenient to formulate the theory of scattering from a rough surface in terms of the actual realisations of the surface rather than in terms of the spectrum which is a second moment of the surface. Thus explicit representations of the wave-field are required. These are obtained using perturbation expansions of the Navier-Stokes equations. Only the source-free case has been considered, but the results agree with the wave action method in the source free case.

Let $\mathbf{r} = (x,y,z) = (\rho,z)$ be the position vector with z vertical and $\rho = (x,y)$ a 2-dimensional horizontal vector. Except where given a special symbol such as ρ, 2-dimensional horizontal vectors will be distinguished by the suffix t. For example the wavevector $\mathbf{k} = (k_x,k_y,k_z) = (\mathbf{k}_t,k_z)$ with $\mathbf{k}_t = (k_x,k_y)$ the horizontal component. The rough sea surface can be described by a functional relationship of the form

$$z = \sigma\zeta(\rho,t) \tag{10.1}$$

where t is time and ζ is a random function. For convenience a factor σ, the root mean square roughness height, has been extracted to facilitate expansions, and consequently $\langle \zeta^2 \rangle = 1$ where $\langle \, . \, \rangle$ denotes assembly average. To first order the surface can be represented by a system of non-interacting linear waves $\zeta_1(\rho,t)$ obeying a dispersion relation. This linear wave field is the first term of a perturbation expansion in which higher order terms represent nonlinear interactions. The perturbation expansion has the form

$$z = \sigma\zeta_1(\rho,t) + \sigma^2 \zeta_2(\rho,t) + \ldots \tag{10.2}$$

The linear wave field has the Fourier expansion

$$\zeta_1(\rho,t) = \int \frac{d^2k_t d\omega}{(2\pi)^3} \tilde{\zeta}_1(\mathbf{k}_t,\omega) \exp(-i\mathbf{k}_t.\rho + i\omega t) \tag{10.3}$$

in which the Fourier component $\tilde{\zeta}_1(\mathbf{k}_t,\omega)$ for wavevector $\mathbf{k}_t = (k_x,k_y)$ and angular frequency ω obeys a dispersion relation. For deep water gravity waves this is $\omega^2 = gk_t$, where g is the acceleration of gravity, and so $\tilde{\zeta}_1(\mathbf{k}_t,\omega)$ can be written

$$\tilde{\zeta}_1(\mathbf{k}_t,\omega) = 2\pi\delta(\omega - g^{\frac{1}{2}}k_t^{\frac{1}{2}})\tilde{\zeta}_1(\mathbf{k}_t) + 2\pi\delta(\omega + g^{\frac{1}{2}}k_t^{\frac{1}{2}})\tilde{\zeta}_1^{*}(-\mathbf{k}_t) \tag{10.4}$$

in which $\delta(.)$ denotes the Dirac δ-function and an asterisk denotes complex conjugate. Eqn (10.3) becomes

$$\zeta_1(\rho,t) = \int \frac{d^2k_t}{(2\pi)^2} \tilde{\zeta}_1(\mathbf{k}_t) \exp(-i\mathbf{k}_t.\rho + ig^{\frac{1}{2}}k_t^{\frac{1}{2}}t) + cc \tag{10.5}$$

where cc denotes complex conjugate. Often it is simpler to use eqn (10.3) and leave the dispersion relation understood.

The second order field $\zeta_2(\rho,t)$ can be represented in terms of the first order field by

$$\zeta_2(\rho,t) = \int \frac{d^2\mathbf{k}_{1t}d\omega_1 d^2\mathbf{k}_{2t}d\omega_2}{(2\pi)^6} \; \tilde{\zeta}_1(1)\tilde{\zeta}_1(2) \exp\left[-(\mathbf{k}_{1t}+\mathbf{k}_{2t}).\rho + i(\omega_1+\omega_2)t\right]\tilde{T}(1,2)$$

$$(10.6)$$

$$\tilde{T}(1,2) = \frac{(\mathbf{k}_{1t}+\mathbf{k}_{2t})}{2(\omega_1+\omega_2)}\cdot\left[\frac{\omega_1}{k_{1t}}\mathbf{k}_{1t}+\frac{\omega_2}{k_{2t}}\mathbf{k}_{2t}\right] - \frac{\omega_1\omega_2}{k_{1t}k_{2t}}\frac{(\mathbf{k}_{1t}.\mathbf{k}_{2t}-k_{1t}k_{2t})\,|\mathbf{k}_{1t}+\mathbf{k}_{2t}|}{[g|\mathbf{k}_{1t}+\mathbf{k}_{2t}|-(\omega_1+\omega_2)^2]}$$

$$(10.7)$$

in which the shorthand $1 \equiv \mathbf{k}_{1t},\omega_1$ etc. has been used. This expansion agrees with that given by Barrick (1977) in the theory of h.f. sea sensing. Such a direct relation between the first and second order fields is only possible in the source free case.

It is useful to divide the linear wave field $\zeta_1(\rho,t)$ into long waves $\zeta_l(\rho,t)$ and short waves $\zeta_s(\rho,t)$ separated at some convenient wavenumber k_m. With σ_l and σ_s defined so that σ_l and σ_s have unit variance one has

$$\sigma_l\tilde{\zeta}_l(\mathbf{k}_t) = \sigma\tilde{\zeta}_1(\mathbf{k}_t)\,H(k_m - k_t)$$

$$(10.8)$$

$$\sigma_s\tilde{\zeta}_s(\mathbf{k}_t) = \sigma\tilde{\zeta}_1(\mathbf{k}_t)\,H(k_t - k_m)$$

$$(10.9)$$

where $H(.)$ is the Heaviside unit step function. One can write the two-scale expansion of the wave field

$$z = \sigma_s\zeta_{10} + \sigma_l\zeta_{01} + \sigma_s^2\zeta_{20} + \sigma_s\,\sigma_l\zeta_{11} + \sigma_l^2\zeta_{02} + \ldots$$

$$(10.10)$$

which is derived in an obvious way from eqn. (10.2). Although the use of two-scale expansions in microwave scattering theory is mathematically convenient, it is not essential and the theory can be framed independent of scales.

If one groups terms with the same powers of σ_s in infinite subseries there results

$$z = \sigma_l\zeta_{01} + \sigma_l^2\,\zeta_{02} + \ldots + \sigma_s(\zeta_{10} + \sigma_l\zeta_{11} + \ldots) + \sigma_s^2(\zeta_{20} + \sigma_l\zeta_{21} + \ldots)$$

$$(10.11)$$

$$= \zeta_0' + \sigma_s\zeta_1' + \sigma_s^2\zeta_2' + \ldots$$

$$(10.12)$$

One finds that the subseries $\zeta_1'\zeta_2'$ etc. converge poorly because of the Doppler shifting of the short waves by the long waves. This can be overcome by recasting $\zeta_1'\zeta_2'$ etc. as Rytov series. Only ζ_1' is the interest here, and this takes the form

$$\zeta_1'(\rho,t) = \int \frac{d^2 k_{1t} d\omega_1}{(2\pi)^3} \, \tilde{\zeta}_s(1) \exp\left[-ik_{1t}.\rho + i\omega_1 t + \sum_{m=1}^{\infty} \sigma_1{}^m \chi_{1m}(1,\rho,t)\right]$$

$$(10.13)$$

$$\chi_{11}(1,\rho,t) = \int \frac{d^2 k_{2t} d\omega_2}{(2\pi)^3} \, \tilde{\zeta}_1(2) \exp\left(-ik_{2t}.\rho + i\omega_2 t\right) 2\tilde{T}(1,2)$$

$$(10.14)$$

Thus eqn (10.12) is a two-scale perturbation − Rytov expansion of the wave field.

If the frequency ω_2 of the long wave is much smaller than the frequency ω_1 of the short wave, one can find an expansion of $\tilde{T}(1,2)$ in powers of ω_2/ω_1

$$\tilde{T}(1,2) = \frac{k_{1t}}{2}\left[\cos\theta_{12} - \frac{\omega_2}{2\omega_1}\sin^2\theta_{12} + \frac{\omega_2{}^2}{\omega_1{}^2}\{1 - \tfrac{1}{2}\cos^2(\theta_{12}/2)\sin^2\theta_{12}\} + \ldots\right]$$

$$(10.15)$$

where θ_{12} is the angle between k_{1t} and k_{2t}. Use of this in eqn (10.14) converts the Rytov expansion in eqn (10.13) into a WKB expansion, and eqn (10.12) into a two-scale perturbation-WKB expansion. This is the representation used in the SAR imaging theory. The first term of eqn (10.15) represents a Doppler shift. The second term originates in the short-wave dispersion relation in which both the length of the wavenumber and the gravitational acceleration are perturbed by the long waves. The third term contains hydrodynamic amplitude modulations. Eqn (10.15) is consistent with wave action results in the source-free case.

Some statistical properties of the short-wave field will be required. It will be assumed that $\zeta_s(\rho,t)$ is a stationary and homogeneous Gaussian random function with wavenumber-frequency spectrum $\tilde{E}_s(k_{1t},\omega_1)$ and wavenumber specturm $\tilde{E}_s(k_{1t})$ defined by

$$<\tilde{\zeta}_s(k_{1t},\omega_1)\tilde{\zeta}_s{}^*(k_{2t},\omega_2)> = (2\pi)^3 \, \delta\,(k_{1t}-k_{2t},\omega_1-\omega_2)\,\tilde{E}_s(k_{1t},\omega_1) \qquad (10.16)$$

$$<\tilde{\zeta}_s(k_{1t})\,\tilde{\zeta}_s{}^*(k_{2t})> = (2\pi)^2 \, \delta(k_{1t})\,\tilde{E}_s(k_{1t})$$

$$\tilde{E}_s(k_{1t},\omega_1) = 2\pi\,\tilde{E}_s(k_{1t})\,\delta\,(\omega_1-g^{\frac{1}{2}}k_{1t}{}^{\frac{1}{2}}) + 2\pi\tilde{E}_s(-k_{1t})\,\delta\,(\omega_1 + g^{\frac{1}{2}}k_{1t}{}^{\frac{1}{2}}).$$

$$(10.17)$$

Here $\tilde{\zeta}_s(k_{1t},\omega_1)$ is Hermitian, $\tilde{E}_s(k_{1t},\omega_1)$ is real and symmetric, whilst $\tilde{E}_s(k_{1t})$ is real. There is no necessary relation between $\tilde{E}_s(k_{1t})$ and $\tilde{E}_s(-k_{1t})$.

10.3 SCATTERING FROM ROUGH SURFACES

In this section scattering from the general rough surface in eqn (10.1) is considered for horizontal polarisation in a scalar wave approximation. The Green's function for a point impulse satisfies

$$\left[\nabla^2 - \frac{n^2(r,t)}{c^2}\frac{\partial^2}{\delta t^2}\right]G(\mathbf{r},t|\mathbf{r}_0,t_0) = \delta(\mathbf{r}-\mathbf{r}_0,t-t_0) \tag{10.18}$$

in which c is the velocity of light and the refractive index $n(\mathbf{r},t)$ is n_a for $z > \sigma\,\zeta(\rho,t)$ and n_b for $z < \sigma\zeta(\rho,t)$. Solutions are to be developed for G subject to the boundary conditions of continuity of G and its normal derivative at the surface.

By applying Green's theorem to the two regions above and below the surface, two mixed surface-volume integral equations can be derived. Using the boundary conditions these can be combined in many different ways. One form results from eliminating the surface integrals. Upon Fourier transformation it gives for the transform $\tilde{G}(\mathbf{k}_1,\omega_1|\mathbf{k}_0,\omega_0)$, or $\tilde{G}(1|0)$ for short,

$$\left[\frac{\omega_1(n_a{}^2+n_b{}^2)}{c^2}\right] - k_1{}^2 \quad \tilde{G}(1|0) = (2\pi)^4\delta(1-0)$$

$$+ \frac{i(n_b{}^2-n_a{}^2)}{c^2}\int\frac{d^3\mathbf{k}_2\,d\omega_2}{(2\pi)^2}\;\omega_2{}^2\,P\!\left(\frac{1}{k_{1z}-k_{2z}}\right)\tilde{S}(1-2)\,\tilde{G}(2|0) \tag{10.19}$$

in which the shorthand $1-0 = \mathbf{k}_1 - \mathbf{k}_0, \omega_1-\omega_0$ etc. has been used, $P(.)$ is the Cauchy principal value distribution and $\tilde{S}(\mathbf{k},\omega)$ is the Fourier transform of the surface function $S(\mathbf{r},t)$ defined by

$$S(\mathbf{r},t) = \delta[z-\sigma\zeta(\rho,t)] \tag{10.20}$$

which represents an impulse located at the rough surface. The transform is

$$\tilde{S}(\mathbf{k},\omega) = \int d^2\rho dt\,\exp[i\mathbf{k}_t.\rho - i\,\omega t + ik_z\,\sigma\zeta(\rho,t)] \tag{10.21}$$

which may be thought of as the Fourier transform of a phase-changing screen. Similar equations have been given by DeSanto (1979).

Before constructing solutions, it is useful to define

$$\mu_{a1} = \mu_a(1) = \left[\frac{\omega_1{}^2 n_a{}^2}{c^2} - k_{1t}^2\right]^{\frac{1}{2}} \tag{10.22}$$

$$N_0 = N(0) = \mu_{a0} - \mu_{b0} \tag{10.23}$$

$$D_1 = D(1) = \mu_{a1} + \mu_{b1} \tag{10.24}$$

$$\tilde{f}(1|0) = \left[\frac{1}{\mu_{a1}-i\epsilon-k_{1z}} + \frac{1}{\mu_{b1}-i\epsilon+k_{1z}}\right]\left[\frac{1}{\mu_{a0}-i\epsilon+k_{0z}} + \frac{1}{\mu_{b0}-i\epsilon-k_{0z}}\right] \tag{10.25}$$

in which $\epsilon \to 0+$. The form of the perturbation expansion of the Green's function points to the definition of a new wave function, $\tilde{F}(1|0)$, given by

$$\tilde{G}(1|0) = \frac{(2\pi)^4}{2}\, \delta(1-0) \left[\frac{1}{(\mu_{a1}-i\epsilon)^2 - k_{1z}^2} + \frac{1}{(\mu_{b1}-i\epsilon)^2 - k_{1z}^2} \right]$$

$$+ \frac{N_0}{D_1}\, \frac{\tilde{f}(1|0)}{i}\, P\!\left(\frac{1}{k_{1z}-k_{0z}}\right)\, \tilde{F}(1|0). \tag{10.26}$$

This wave function has the perturbation expansion

$$\tilde{F}(1|0) = \sum_{n=0}^{\infty} \sigma^n\, \tilde{F}_n(1|0) \tag{10.27}$$

$$\tilde{F}_0(1|0) = (2\pi)^3\, \delta(k_{1t}-k_{0t},\omega_1-\omega_0) \tag{10.28}$$

$$\tilde{F}_1(1|0) = i(k_{1z}-k_{0z})\, \tilde{\zeta}(1-0) \tag{10.29}$$

$$\tilde{F}_2(1|0) = \frac{i^2}{2}\, (k_{1z}-k_{0z}) \int \frac{d^2 k_{2t} d\omega_2}{(2\pi)^3}\, \tilde{\zeta}(1-2)\tilde{\zeta}(2-0)\, [2N_2-N_1-N_0+k_{1z}-k_{0z}]$$

$$\tag{10.30}$$

up to second order terms. This expansion is not subject to the Rayleigh hypothesis (Valenzuela 1978), but its inverse transform is so limited as term by term transformation is only allowed when the Rayleigh hypothesis is valid.

A powerful Rytov expansion can be found for $\tilde{F}(1|0)$. It is a reorganisation of the perturbation expansion but is also valid for a very rough surface. Its leading term can be thought of as a form of physical optics (Beckmann & Spizzichino 1963). If the term in $2N_2-N_1-N_0$ in eqn (10.30) is small, then $\tilde{F}_0, \tilde{F}_1$, and $\tilde{F}_2$ form the first three terms of the perturbation expansion of the surface function in eqn (10.21). This implies the approximation $\tilde{F}(1|0) \sim \tilde{S}(1|0)$ and points to the Rytov expansion

$$\tilde{F}(1|0) = \int d^2\rho dt \exp\Big[i(k_{1t}-k_{0t}).\rho - i(\omega_1-\omega_0)t + i(k_{1z}-k_{0z})\sigma\zeta\,(\rho,t)$$

$$+ \frac{(k_{1z}-k_{0z})\sigma^2}{4} \int \frac{d^2 k_{2t} d\omega_2 d^2 k_{3t} d\omega_3}{(2\pi)^6}\, \tilde{\zeta}(2)\tilde{\zeta}(3) \exp\,[-i(k_{2t}+k_{3t}).\rho+i\,(\omega_2+\omega_2).$$

$$\times \{N(0+1+2+3) + N(0+1-2-3) -N(0+1-2+3) -N(0+1+2-3)\}+0(\sigma^3)\Big]. \tag{10.31}$$

This agrees with eqn (10.27) upon expansion of the exponent. The symmetry of eqn (10.31) is a powerful argument in favour of this form of expansion.

10.4 BACKSCATTERING FROM THE SEA SURFACE

The Green's function is expanded in a perturbation expansion in eqns (10.26) and (10.27). This is for an arbitrary surface. The sea surface can be expanded as in eqn (10.2). If these expansions are combined, one obtains a perturbation expansion for scattering from the sea surface. This can be written

$$\tilde{G}(1|0) = \sum_{n=0}^{\infty} \sigma^n \, \tilde{g}_n(1|0). \tag{10.32}$$

Obviously $\tilde{g}_0(1|0)$ is the half-space scalar Green's function given by substituting $\tilde{F}_0(1|0)$ in eqn (10.28) for $\tilde{F}(1|0)$ in eqn (10.26) whilst $\tilde{g}_1$ and $\tilde{g}_2$ are easily found to be

$$\tilde{g}_1(1|0) = \frac{N_0}{D_1} \, \tilde{f}(1|0) \, \tilde{\zeta}_1(1|0) \tag{10.33}$$

$$\tilde{g}_2(1|0) = \frac{N_0}{D_1} \, \tilde{f}(1|0) \int \frac{\mathrm{d}^2 \mathbf{k}_{2t} \mathrm{d}\omega_2}{(2\pi)^3} \, \tilde{\zeta}_1(1-2)\tilde{\zeta}_1(2-0)\left[\frac{i}{2}(2N_2 - N_1 - N_0 + k_{1z} - k_{0z})\right.$$

$$\left. + \, \tilde{T}(1-2,2-0)\right] . \tag{10.34}$$

This expansion combines both electromagnetic and hydrodynamic interactions in the scattering process (Valenzuela 1978).

In the same way combination of the Green's function expansion in eqn (10.26) and (10.27) with the two scale expansion of the sea surface in eqn (10.10) leads to a two-scale expansion of the Green's function given by

$$\tilde{G}(1|0) = \tilde{g}_{00} + \sigma_s \, \tilde{g}_{10} + \sigma_l \, \tilde{g}_{01} + \sigma_s^2 \, \tilde{g}_{20} + \sigma_s \sigma_l \, \tilde{g}_{11} + \sigma_l^2 \, \tilde{g}_{02} + \ldots \tag{10.35}$$

These terms will not be given explicitly here but are simply obtained from eqns (10.32) and (10.34) by splitting the linear wave field $\tilde{\zeta}_1(.)$ into long and short wave parts using eqns (10.8) and (10.9). For backscatter not too near vertical incidence some of these terms are small or zero. The zero[th] order field $\tilde{g}_{00}$ is dominant in the specular direction but zero for backscatter. The first order short-wave field $\tilde{g}_{10}$ is the Bragg scattered field. The first and second order long-wave fields $\tilde{g}_{01}$ and $\tilde{g}_{02}$ are zero or small for backscatter not too near vertical incidence because the wavevector $\mathbf{k}_{1t} - \mathbf{k}_{0t}$ lies in the short-wave region. The second order short-wave field $\tilde{g}_{20}$ is the second order Bragg scattered field.

The second order field $\tilde{g}_{11}$ is a Bragg scattered field linearly modulated by the long waves and leads to linear imaging mechanisms for the long waves. Thus the significant first and second order fields for backscatter are

$$\tilde{G}(1|0) \sim \sigma_s \tilde{g}_{10} + \sigma_s \sigma_l \tilde{g}_{11} + \varrho_s{}^2 \tilde{g}_{20} + \dots \tag{10.36}$$

If one groups terms in eqn (10.35) with the same power of σ_s in infinite subseries, the result can be written

$$\tilde{G}(1|0) = \sum_{n=0}^{\infty} \sigma_s{}^n \tilde{g}_n{}'(1|0) . \tag{10.37}$$

The near specular field is mostly contained in $\tilde{g}_0{}'(1|0)$. The important term for backscattering away from vertical incidence is $\tilde{g}_1'(1|0)$. It is composed of the terms

$$\tilde{g}_1' = \tilde{g}_{10} + \sigma_l\tilde{g}_{11} + \sigma_l{}^2\tilde{g}_{12} + \dots \quad . \tag{10.38}$$

This series converges slowly because of the Doppler shifting by the long-wave orbital motions, but it can be recast as the rapidly convergent Rytov series

$$\tilde{g}_1'(1|0) = \frac{N_0}{D_1} \tilde{f}(1|0) \int d^2\rho dt \, \frac{d^2k_{2t}d\omega_2}{(2\pi)^3} \, \tilde{\zeta}_s(2) \exp\left[i(k_{1t}-k_{0t}-k_{2t}).\rho - i(\omega_1-\omega_0-\right.$$

$$\left. + \sigma_l \int \frac{d^2k_{3t}d\omega_3}{(2\pi)^3} \, \tilde{\zeta}_1(3) \, \exp(-ik_{3t}.\rho+i\omega_3 t)\tilde{V}(0,1,2,3) + 0(\sigma_l{}^2) \right] . \tag{10.39}$$

By expansion and comparison with eqn (10.38), one readily finds

$$\tilde{V}(0,1,2,3) = 2\tilde{T}(2,3) - \frac{i}{2} \{N(0+1+2+3)+N(0+1-2-3)-N(0+1-2+3)-N(0+1+2-$$

$$+ k_{1z} - k_{0z} . \tag{10.40}$$

The first term in this expression is hydrodynamic in origin, whilst the other terms are electromagnetic in origin.

The expression for $\tilde{V}$ in eqn (10.40) can be further developed by expanding it in powers of ω_3/ω_2 or equivalent. This generates a WKB expansion. It is not valid near grazing incidence, unlike eqn (10.40). Using eqn (10.15) one readily finds

$$\tilde{V}(0,1,2,3) = \left[\frac{\mathbf{k}_{2t}.\mathbf{k}_{3t}}{k_{3t}} + i(k_{1z} - k_{0z})\right] - \left[\frac{\omega_3 k_{2t}}{2\omega_2} \sin^2\theta_{23}\right]$$

$$+ \left[k_{3t}\{1 - \tfrac{1}{2}\cos^2(\theta_{23}/2)\sin^2\theta_{23}\} + 2i\,\mathbf{k}_{2t}.\mathbf{k}_{3t}\left\{\frac{1}{\mu_a(k_{2t},\omega_0 + \omega_1)} - \frac{1}{\mu_b(k_{2t},\omega_0 + \omega_1)}\right\}\right] + \dots$$

$$(10.41)$$

in which θ_{23} is the angle between $\mathbf{k}_{2t}$ and $\mathbf{k}_{3t}$. Terms of successively smaller order in ω_3/ω_2 have been grouped in the square brackets. The first term represents the Doppler shift of the radar wave by the orbital motions of the long waves. The second originates in the short-wave dispersion relation in which both the length of the wavenumber and the gravitational acceleration are perturbed by the long waves. The third term contains both hydrodynamic and electromagnetic amplitude modulations. Eqn (10.39) with eqn (10.41) are the actual forms used in the imaging theory.

10.5 SIMPLE RADAR SYSTEM

The next step is to combine the scattering theory with a simple radar system model. The geometry is shown in Fig. 10.1. The radar is travelling in the azimuth or x-direction with velocity v and coordinates $(vt,0,z_s)$. The centre of the beam intersects the earth's surface at $(vt,y_s,0)$ with angle of incidence θ_s and slant range $R_s = z_s \cos\theta_s$. It will be assumed that resolution in the range or y-direction is pulse rather than beam limited with the centre of the resolution cell at $(vt,y,0)$ with angle of incidence θ and slant range R.

It is helpful to assume a Gaussian aperture distribution which, at $t = 0$, is

$$\alpha(x,y,z) = \delta\left[y\sin\theta_s - (z-z_s)\cos\theta_s\right]\exp\left[-\frac{2\pi x^2}{X^2} - \frac{2\pi\{y\cos\theta_s + (z-z_s)\sin\theta_s\}^2}{Y^2}\right]$$

$$(10.42)$$

so X and Y are the effective antenna dimensions. The transmitted signal has a carrier frequency ω_s and a Gaussian pulse shape centred on t_s, so the time variation is

$$\eta(t,t_s) = \exp\left[i\,\omega_s t - \frac{2\pi(t-t_s)}{\tau^2}\right]$$

$$(10.43)$$

where τ is the effective pulse length. Pulse compression is used in real systems, but the result is as if short pulses had been used. Corresponding to ω_s is the free space wavelength $\lambda_s = 2\pi c/\omega_s$ and wavenumber $k_s = \omega_s/c$.

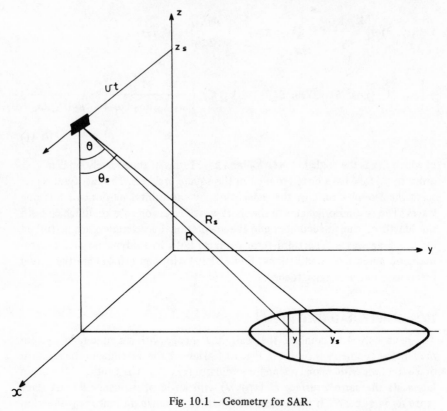

Fig. 10.1 – Geometry for SAR.

The received signal for a transmitted pulse centred on t_s and correlated with a reference pulse centred on t_f is

$U(x,y) =$

$$\int d^3 \mathbf{r}_1 dt_1 d^3 \mathbf{r}_0 dt_0 \, \eta^*(t_1,t_f) \, \alpha^*(x_1 - vt_1, y_1, z_1) \, G(\mathbf{r}_1, t_1 \, \mathbf{r}_0, t_0) \, \alpha(x_0 - vt_0, y_0, z_0) \, \eta(t_0, t_s)$$

(10.44)

$$x = \frac{v}{2} \, (t_f + t_s) \, , \; y = \frac{c \sin \theta}{2} \, (t_f - t_s)$$

(10.45)

in terms of the coordinates $\rho = (x,y)$ of the centre of a resolution cell. In a SAR a further stage of processing is carried out. Here this will be written

$$W(x,y) = \int dt_1 \, U(vt_1, y) \, \exp\left[-\frac{4\pi}{X_1{}^2} \, (x - vt_1)^2\right]$$

(10.46)

with $X_1{}^2$ chosen below.

The next stage is to combine the approximation for the Green's function in eqns (10.37), (10.39) and (10.41) with eqns (10.42) to (10.46). An extensive reduction must then be carried out involving saddle point integration and a number of minor approximations. The result is

$$W(x,y) = C \int d^2\rho_1 dt_1 \frac{d^2 k_{2t} d\omega_2}{(2\pi)^3} \tilde{\zeta}_s(2) \exp \left[-i k_{2t}.\rho_1 + i\omega_2 t_1 - 2i k_s \sin\theta \, y_1 \right.$$

$$- \frac{4\pi}{X_0^2} (x_1 - vt_1)^2 - \frac{4\pi}{X_1^2} (x - vt_1)^2 - \frac{4\pi}{Y_0^2} (y - y_1)^2$$

$$\left. + \sigma_1 \int \frac{d^2 k_{3t} d\omega_3}{(2\pi)^3} \tilde{\zeta}_1(3) \exp (-i k_{3t}.\rho_1 + i\omega_3 t_1) \, \tilde{V}_1(2,3) \right]. \tag{10.47}$$

in which C includes some constants, unimportant phase terms, and the beam shape in range. The other quantities are

$$X_0^2 = X^2 - 2iR\lambda_s , \quad Y_0^2 = \frac{c^2 \tau^2}{\sin^2\theta} \tag{10.48}$$

$$V_1(2,3) = \frac{k_{2t}.k_{3t}}{k_{3t}} - i k_{2y} \cos\theta - \frac{\omega_3 k_{2t}}{2\omega_2} \sin^2\theta_{23} + k_{3t} \{1 - \tfrac{1}{2} \cos^2(\theta_{23}/2)\sin^2\theta_{23} \}$$

$$-i k_{3y} \cos\theta + i k_{2t}.k_{3t} \left(\frac{1}{\mu_{as}} - \frac{1}{\mu_{bs}} \right) + \dots \tag{10.49}$$

in which $\mu_{as} = \mu_a (k_s \sin\theta, \omega_s)$ and $\mu_{bs} = \mu_b (k_s \sin\theta, \omega_s)$. With $n_a = 1$ and n_b the complex refractive index of the sea, these are $\mu_{as} = k_s \cos\theta$ and $\mu_{bs} = k_s (n_b^2 - \sin^2\theta)^{1/2}$.

The object of SAR processing is to choose X_1^2 so as to remove the phase part of the term X_0^2 in eqn (10.47). The simplest choice is

$$X_1^2 = 2iR\lambda_s \tag{10.50}$$

which is a focused one-look SAR processor. It is focused because it matches exactly the imaginary part of X_0^2 in eqn (10.48) and one-look because it does not modify the real part. A more general choice is

$$X_1^2 = (N^2 - 1) X^2 + 2iR\lambda_s (1 - \frac{2\delta v}{v}) \tag{10.51}$$

which gives one look of an N-look defocused SAR processor with δv the focus mismatch velocity which originates in replacing v by $v + \delta v$ in eqn (10.46).

In some treatments (Alpers & Rufenach 1979, Swift & Wilson 1979) the t_1 integration in eqn (10.47) is made Gaussian by Taylor expanding the exponent about $t_1 = x/v$ to second order. Although this expansion converges well for $\omega_3 T \ll 1$, where $T = R\lambda_s/vX$ is the one-look integration time, the first neglected term is $0(\omega_3^3 T^3 \sigma_l/\lambda_s)$. In microwave remote sensing one usually has $\sigma_l \gg \lambda_s$, and the first neglected term may not be small. This expansion is derived from point scatterer ideas (Raney 1971) but is often invalid for ocean wave imaging. Its use leads to incorrect results for the effect of orbital accleration on resolution, and also for focus effects (Alpers & Rufenach 1980).

10.6 LINEAR IMAGING

The SAR image intensity is WW^*. This is a random variable and its assembly average is the expected image intensity $I(x,y)$.

$$I(x,y) = <W(x,y)\, W^*(x,y)> \quad . \tag{10.52}$$

It can be expanded as a long-wave perturbation expansion of the form

$$I(x,y) = \sum_{n=0}^{\infty} \sigma_l^n \, I_n(x,y) \quad . \tag{10.53}$$

The determination of the terms in this expansion involves the substitution of eqn (10.47) in eqn (10.52). The expectation is evaluated using eqns (10.16) and (10.18). In some treatments (Alpers & Rufenach 1979, Swift & Wilson 1979) the short-wave field is assumed to be spatially δ-correlated upon taking the expectation, but this is only an approximation. An extensive reduction is required to determine the I_n and only I_0 and I_1 have been determined for a general wave field.

The zero[th] order term I_0 is the pure Bragg scattered component given by

$$I_0 = |A|^2 \; \frac{T^2 XY_0}{8N} \; [E_s(\mathbf{k}_{bt}) + E_s(-\mathbf{k}_{bt})] \tag{10.54}$$

in which $\mathbf{k}_{bt} = (0,-2k_s \sin \theta)$ is the Bragg wavevector. The Bragg angular frequency is $\omega_b = g^{\frac{1}{2}} k_{bt}^{\frac{1}{2}}$.

The first order term $I_1(\rho)$ represents linear modulations and can be written

$$I_1(\rho) = I_0 \int \frac{d^2 \mathbf{k}_{lt}}{(2\pi)^2} \; \tilde{\xi}_1(\mathbf{k}_{lt}) \tilde{M}(\mathbf{k}_{lt}) \tilde{Q}(\mathbf{k}_{lt}) \exp(-i\mathbf{k}_{lt}.\rho + i\omega_1 x/v) + cc \tag{10.55}$$

where cc denotes complex conjugate and $\omega_1 = g^{1/2}k_1^{1/2}$. In this expression $\tilde{\zeta}_1(\mathbf{k}_{l'})$ is the long-wave spectral amplitude in eqn (10.4), $\tilde{M}(\mathbf{k}_{l'})$ is the linear modulation transfer function, and $\tilde{Q}(\mathbf{k}_{l'})$ is the linear system transfer function. There are described below.

The complex linear modulation transfer function $\tilde{M}(\mathbf{k}_{l'})$ is given by

$$\tilde{M}(\mathbf{k}_{l'}) = \frac{2}{[\tilde{E}_s(\mathbf{k}_{bt}) + \tilde{E}_s(-\mathbf{k}_{bt})]} \sum_{\pm} \tilde{E}_s(\pm\mathbf{k}_{bt}) \exp{(\mp i\omega_b\, t_c)}\ [\tilde{L}^{(\pm)}(\mathbf{k}_{l'})\cos{(\omega_1 t_c/2)}$$

$$+ i\,\tilde{A}_1(\mathbf{k}_{l'})\sin{(\omega_1 t_c/2)}] \tag{10.56}$$

in which

$$t_c = \frac{R\lambda_s \mathbf{k}_{1x}}{4\pi v} = \left(\frac{T}{N}\right)\left(\frac{NX}{2}\right)\left(\frac{k_{1x}}{2\pi}\right) \tag{10.57}$$

$$\tilde{A}(\mathbf{k}_{l'}) = -2\,k_s \sin\theta\ \frac{k_{1x}}{k_{1t}} + 2i\,k_s \cos\theta \tag{10.58}$$

$$\tilde{L}^{(\pm)}(\mathbf{k}_{l'}) = \frac{k_{1t}}{2} - \frac{k_{1x}^2}{4k_{1t}} - \frac{i\,k_{1x}}{2}\cot\theta - 2i\,k_{1y}\tan\theta\left[1 - \mathrm{Re}\left\{\frac{\theta}{n_b{}^2 - \sin^2\theta)^{\frac{1}{2}}}\right\}\right]$$

$$+ \frac{k_{1y}}{k_{1t}} - i\cot\theta\ \frac{k_{bt}}{2}\ k_{1t}\cdot\frac{\partial}{\partial\mathbf{k}_{bt}}\ln\tilde{E}_s(\pm\mathbf{k}_{bt}) \tag{10.59}$$

where Re denotes real part. In eqn (10.57) t_c is a characteristic time for a SAR when imaging a wave with azimuthal wavenumber k_{1x}. As indicated by the three groups of terms in eqn (10.57), it is the integration time (T/N) reduced by the ratio of the azimuthal resolution $(NX/2)$ to the azimuthal wavelength $(2\pi/k_{1x})$. The modulation transfer function includes tilting and straining, velocity bunching and Doppler splitting and these are now described.

In other treatments (Alpers *et al.* 1982) tilting and straining are treated separately, but here they come together as a single mechanism given by the term $2\tilde{L}^{(\pm)}(\mathbf{k}_{l'})\cos(\omega_1 t_c/2)$ in eqn (10.56). The cosine is a long integration time effect and for $\omega_1 t_c \ll \pi$ one can just consider $2\tilde{L}^{(\pm)}(\mathbf{k}_{l'})$. This becomes non-dimensional if divided by k_{1t}, so one defines

$$m(\psi_1) = \frac{2\tilde{L}^{(\pm)}(\mathbf{k}_{l'})}{k_{1t}} \tag{10.60}$$

$$= 1 - \frac{\cos^2 \psi_1}{2} - i \sin \psi_1 \cot \theta - 4i \sin \psi_1 \tan\theta \left[1 - \mathrm{Re}\left\{\frac{\cos^2\theta}{(n_b{}^2 - \sin^2\theta)^{1/2}}\right\}\right]$$

$$+ (\sin \psi_1 - i \cot \theta)\, k_{bt}\, (\cos \psi_1, \sin \psi_1) \cdot \frac{\partial}{\partial k_{bt}} \ln \tilde{E}_s(\pm k_{bt}) \qquad (10.61)$$

in which ψ_1 is the azimuth angle that the wavevector of the long-wave $k_{1t} = k_{1t}(\cos \psi_1, \sin \psi_1)$ makes with the azimuth or x-direction. A simplified form results if one assumes that $\tilde{E}_s(\pm k_{bt})$ is given by an isotropic Phillips spectrum (Phillips 1977) proportional to $k_{bt}{}^{-4}$. Also $|n_b| \gg 1$ so the term in n_b can be discarded in a first approximation to give

$$m(\psi_1) = \tfrac{1}{2} \cos^2 \psi_1 + 5 \sin^2 \psi_1 - i \sin \psi_1 (4 \tan \theta + 5 \cot\theta) \quad (10.62)$$

Fig. 10.2 shows $|m(\psi_1)|$ plotted against ψ_1 for a number of values of θ. It is small in the azimuth direction ($\psi_1 = 0$) but large in the range direction ($\psi_1 = 90°$) showing how tilting and straining provides a good imaging mechanism for range waves.

The orbital motion of the long waves provides a periodic Doppler shift and thus a periodic displacement of points in a SAR image. This leads to an apparent buncing of scatterers in an image known as velocity bunching (Larson et al. 1976). It is given by the term $2i\tilde{A}(k_{1t})\sin(\omega_1 t_c/2)$ in eqn (10.56). For $\omega_1 t_c \ll \pi$ this can be approximated by $i\omega_1 t_c A(k_{1t})$. With t_c and $A(k_{1t})$ given by eqns (10.57) and (10.58) and $\omega_1 = g^{1/2} k_{1t}{}^{1/2}$, this becomes non-dimensional if divided by $k_{1t}{}^{3/2}$, so one defines

$$n(\psi_1) = \frac{i\omega_1 t_c \tilde{A}(k_{1t})}{k_{1t}{}^{3/2}} \frac{v}{g^{1/2}R} = -\cos \psi_1 (\cos \theta + i \sin \theta \sin \psi_1). \qquad (10.63)$$

Fig. 10.3 shows $|n(\psi_1)|$ plotted against ψ_1 for a number of values of θ. It is zero in the range direction ($\psi_1 = 0$) and is large in the azimuth direction ($\psi_1 = 0$) except towards grazing incidence. Velocity bunching thus provides a imaging mechanism for azimuth waves.

The terms $\exp(\mp i\omega_b t_c)$ in eqn (10.56) give rise to a short-wave motion effect known as Doppler splitting (Alpers et al. 1982) which arises because there are two Bragg waves moving in opposite directions. If the two values $\tilde{E}(\pm k_{bt})$ are very different this has little effect, but if they are close together, the modulation transfer function will be proportional to $\cos(\omega_b t_c)$. This has zeros when $\omega_b t_c = (j + \tfrac{1}{2})\pi$, where j is an integer, or at azimuthal wavelengths $2\pi/k_{1x} = \omega_b TX/(2j + 1)\pi$. For SEASAT this is $2\pi/k_{1x} = \pm 120, \pm 40, \pm 24$ m. etc.

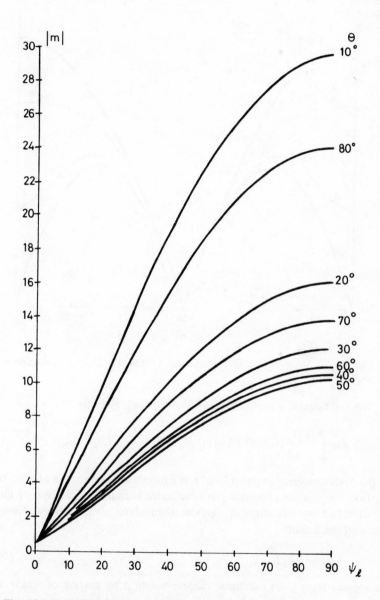

Fig. 10.2 – Modulus of the modulation transfer function |m| against azimuth angle ψ_1 for various angles of incidence θ. Tilting and straining.

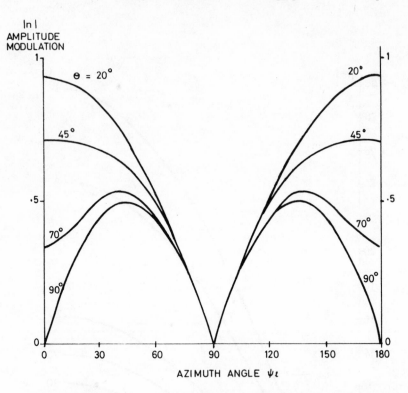

Fig. 10.3 — Velocity bunching dependence on azimuth angle ψ_1 and angle of incidence θ.

The real linear system transfer function $\tilde{Q}(k_{1t})$ is given by

$$\tilde{Q}(k_{1t}) = \exp\left[-\frac{1}{32}(k_{1x}^2 N^2 X^2 + k_{1y}^2 Y_0^2) - \frac{T^2}{8\pi N^2}(\omega_1 + 2k_{1x}\delta v)^2 \right] \quad (10.64)$$

for the SAR processor in eqn. (10.51). It expresses the resolution of the system. The first term in the exponent gives the static azimuthal resolution by cutting off azimuthal wavenumbers k_{1x} greater than about $2\pi/NX$, giving a nominal azimuthal resolution

$$\gamma_x = NX/2. \quad (10.65)$$

The second term gives the static range resolution by cutting off range wavenumbers k_{1y} greater than about $2\pi/Y_0$, giving a nominal range resolution, using eqn (10.48).

$$\gamma_y = \frac{Y_0}{2} = \frac{c\tau}{2\sin\theta}. \quad (10.66)$$

For a focused SAR with $\delta v = 0$, the third term in the exponent gives the temporal resolution of the system by cutting off angular frequencies ω_l greater than about $\pi N/T$, giving the nominal temporal resolution

$$t_r = T/N. \tag{10.67}$$

Using the dispersion relation $\omega_l^2 = g\,k_{lr}$ this can also be expressed as an isotropic spatial resolution

$$\gamma_r = \frac{g\,T^2}{\pi N} . \tag{10.68}$$

For SEASAT $\gamma_x = 6N$ m whilst $\gamma_r = 20/N^2$ m. It follows that for one look the resolution is about 20–25 m. For two looks it is about 10–15 m, and for three looks about 20 m.

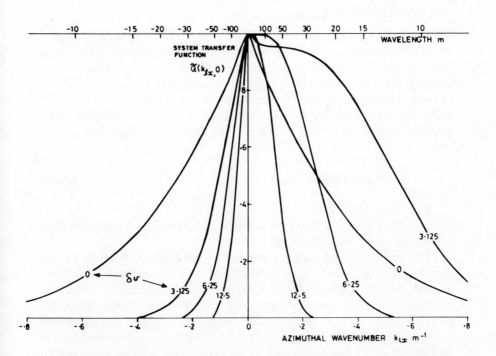

Fig. 10.4 – Resolution effect of changing the SAR focus 1-look SEASAT. Phase velocity of 100 m wave = 12.5 m/s.

If one changes the SAR focus, the temporal resolution is altered and can be completely cancelled if δv takes the value

$$\delta v = -\frac{\omega_1}{2k_{1x}} = -\frac{c_1}{2\cos\psi_1} \tag{10.69}$$

where $c_1 = \omega_1/k_{1t}$ is the long-wave phase velocity. To illustrate this, Fig. 10.4 shows $\tilde{Q}(k_{1x},0)$ plotted against k_{1x} for one-look SEASAT parameters ($N = 1$, $X = 12$ m, $T = 2.5$ s) for a series of values of δv which are 0, ¼, ½, and 1 times the phase velocity, 12.5 m/s, of a wave of wavelength 100 m. Similar conclusions about the SAR focus have been given by Ivanov (1982).

In some treatments (Alpers & Rufenach 1979, Swift & Wilson 1979) a modification of the resolution is given in terms of the orbital acceleration. For reasons given in section 10.5 this has limited validity, as have related conclusions about focus effects (Alpers & Rufenach 1982).

10.7 NONLINEAR IMAGING

The imaging mechanisms can only be considered linear for a restricted range of wave parameters which is described below. Outside of this range the mechanisms, especially velocity bunching, become nonlinear. Velocity bunching becomes nonlinear when the displacement of points in an image become comparable with the long wavelength. Tilting and straining can also become nonlinear for sufficiently large waves, but this aspect is not included properly in the formulation given below. Only the theory for a single long ocean wave with surface displacement.

$$z = a_1 \cos(k_{1t}.\rho - \omega_{1t}) \tag{10.70}$$

has been developed. The theory for a system of waves presents great difficulties which have not yet been overcome.

The expected image intensity is found by substituting eqn (10.47) in eqn (10.52) followed by an extensive reduction to a 2-dimensional diffraction integral. Related integrals (Alpers & Rufenach 1979, Valenzuela 1980) have been further reduced by saddle point integration. However, only a single saddle point has been used in these treatments, and it may be shown that this is valid only in the linear domain. In the nonlinear domain multiple saddle points exist. As many as nine saddle points must be included to give a complete asymptotic description of the integral. A simpler course is to develop the integral as a Fourier series, as the expected image intensity of a periodic scene must itself be periodic.

One finds

$$I(\rho) = |C|^2 \; \frac{T^2 \, XY_0}{8N} \sum_{\pm} \sum_{n=-\infty}^{\infty} \tilde{E}_s \, (\pm \mathbf{k}_{bt}) \, \tilde{Q}(\mathbf{k}_{1t})^{n^2} \, \exp \, [in(\mathbf{k}_{1t}.\rho - \omega_1 x/v \mp \omega_1 t_c + \delta)]$$

$$x \, [J_n(\alpha_n) + a_1 \cos(n\omega_1 t_c/2) \, \{\tilde{L}^{(\pm)}(\mathbf{k}_{1t}) \, e^{i\delta} J_{n+1}(\alpha_n) + \tilde{L}^{(\pm)}(\mathbf{k}_{1t})^* e^{-i\delta} J_{n-1}(\alpha_n)\}]$$

$$\tag{10.71}$$

$$\alpha_n = -2a_1 \, |\tilde{A}(\mathbf{k}_{1t})| \sin \, (n\omega_1 t_c/2) \tag{10.72}$$

$$\delta = 2/\pi - \arg \tilde{A}_1(\mathbf{k}_{1t}) \quad . \tag{10.73}$$

Convergence of this series is guaranteed by the term in $\tilde{Q}\,(\mathbf{k}_{1t})$, the system transfer function in eqn (10.63). If γ is the appropriate resolution measure of γ_x, γ_y or γ_r in eqns (10.64), (10.65), or (10.67), then of order λ_1/γ harmonics can be excited for a wave of wavelength λ_1.

Whether these harmonics are excited depends upon the other terms in eqn (10.71), especially upon $J_n(\alpha_n)$. For small $\omega_1 t_c$ one can approximate by

$$\alpha_n = -a_1 \, |A \, (\mathbf{k}_{1t})| \, \omega_1 \, t_c n = -\mu n \tag{10.74}$$

where, from eqn (10.57) and (10.58), one finds

$$\mu = a_1 \, k_{1t}^{3/2} \, g^{\frac{1}{2}} \, \frac{R}{v} \, \cos \psi_1 \, (\sin^2 \theta \sin^2 \psi_1 + \cos^2 \theta)^{\frac{1}{2}}. \tag{10.75}$$

This is essentially the same quantity described in section 10.6 on linear velocity bunching. Fig. 10.5 shows the function $J_n(\mu n)$ plotted against n for a series of values of μ. For $\mu = 0.1$ it decreases very rapidly, and only the $n = 0$ and 1 values are significant, giving linear imaging. For $\mu = 0.5$ harmonics up to about $n = 3$ are significant. For $\mu = 1$ all harmonics up to the resolution limit are significant. For $\mu > 1$, $J_n(\mu n)$ becomes oscillatory, but all harmonics are excited except those near the zeros of $J_n(\mu n)$. The value $\mu = 1$ marks a convenient dividing line between linear and nonlinear imaging. For given R/v and θ, this gives a relation between long-wave amplitude a_1 and wavelength $\lambda_1 = 2\pi/k_{1t}$ which can be plotted for a series of values of azimuth angle ψ_1. This is shown in Fig. 10.6 for $R/v = 100$ s and $\theta = 20°$. Note that the transition takes palce at relatively moderate values of a_1. It is worst for azimuth waves, and disappears for range waves. Similar results have been given by Alpers *et al.* (1982). For $\mu \gg 1$ the ability of a SAR to image waves becomes severely degraded because of the excitation of harmonics which confuse the image. For a system of waves the SAR spectrum will look as if it has been put through a filter which one may conjecture will have the shape of the wavenumber and azimuth dependence in eqn (10.75). This is given by $k_{1t}^{3/2} \cos \psi_1 \, (\sin^2 \theta \sin^2 \psi_1 + \cos^2 \theta)^{1/2}$. A contour of this function is plotted in Fig. 10.7 for $\theta = 20°$ and Fig. 10.8 for $\theta = 70°$. A spectrum like Fig. 10.7 is described in section 10.10.

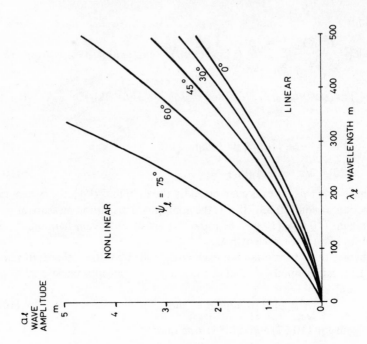

Fig. 10.6 – Linear – nonlinear transition in velocity bunching $R/v = 100$ s, $\theta = 20°$

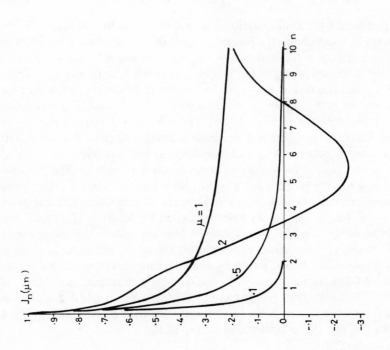

Fig. 10.5 – Envelope of spectral response to a single wave.

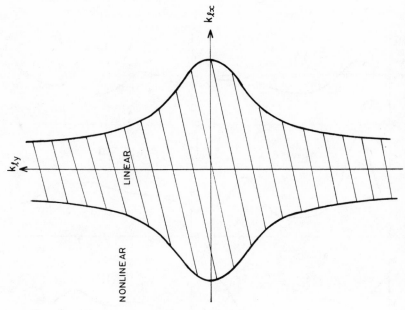

Fig. 10.8 — Spectral response of a SAR as a consequence of nonlinear imaging $\theta = 70°$

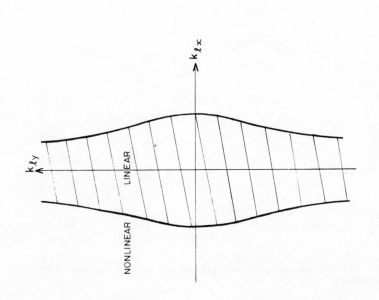

Fig. 10.7 — Spectral response of a SAR as a consequence of nonlinear imaging $\theta = 20°$

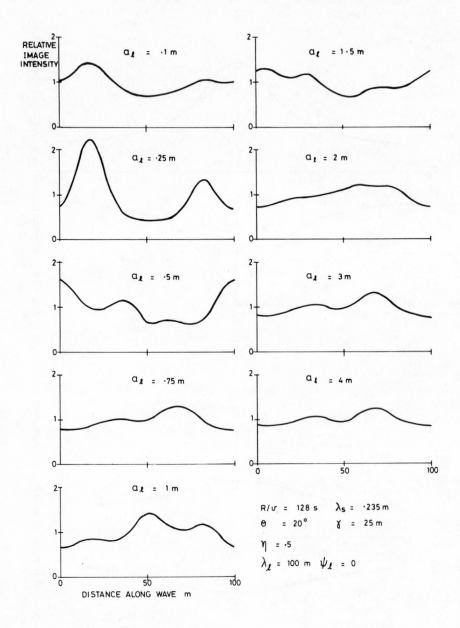

Fig. 10.9 – Expected image intensities for 100 m azimuth waves of different amplitudes.

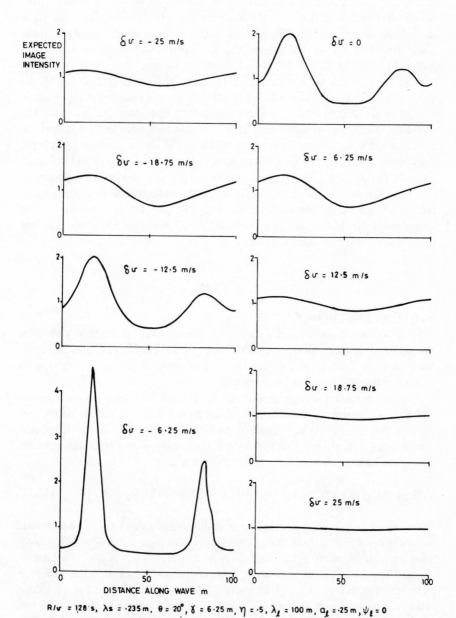

$R/v = 128$ s, $\lambda_s = \cdot235$ m, $\theta = 20°$, $\delta = 6\cdot25$ m, $\eta = \cdot5$, $\lambda_l = 100$ m, $a_l = \cdot25$ m, $\psi_l = 0$

Fig. 10.10 —Expected image intensities for 100 m azimuth wave of amplitude 0.25 m and different focuses.

Some aspects of nonlinear imaging are illustrated in Fig. 10.9 and 10.10 which show the results of calculations with eqn (10.71). The diagrams show the relative image intensity I/I_0 plotted against distance along a wave which has a wavelength of 100 m. The calculations are for SEASAT parameters of $R/v = 128$ s, $\lambda_s = 0.235$ m, and $\theta = 20°$. Fig. 10.9 is for four looks with resolution $\gamma = \gamma_x = \gamma_y = 25$ m. The extent of Doppler splitting is determined by the ratio $\eta = E_s(-k_{1t})/\tilde{E}_s(k_{1t}) = 0.5$. Fig. 10.9 shows the expected relative image intensities for 100 m azimuth waves of increasing amplitude. As the amplitude increases, the image modulations increase but then fade away as the nonlinearity increases and many harmonics are excited. Fig. 10.10 shows the expected relative image intensities for 100 m azimuth waves of amplitude 0.25 m and different focuses. The phase velocity of such a wave is 12.5 m/s, and strong modulations occur in the vicinity of $\delta v = -6.25$ m/s. The modulations fade away quite rapidly as one moves away from this value. The possibility of modifying the image modulations by changing the SAR focus has been advocated by Jain (1978), and these results support this view.

10.8 INVERSE THEORY

One of the main goals of imaging theory is to construct an inverse theory whereby the shape of the sea surface, on scales larger than the resolution, can be reconstructed from SAR data. The development of such an inverse theory is in its infancy, but a few steps have been taken.

An approximate inverse theory can be based on the linear relation in eqn (10.55). An interesting point is that, although it is linear, it cannot be inverted to give the function $\zeta_1(k_{1t})$. This can be seen by taking the Fourier transform. To simplify this assume that the SAR is spaceborne so v is large and the term $\omega_1 x/v$ can be neglected. The Fourier transform is then

$$\tilde{I}_1(k_{1t}) = I_0 [\tilde{\zeta}_1(k_{1t}) \tilde{M}(k_{1t}) \tilde{Q}(k_{1t}) + \tilde{\zeta}_1^*(-k_{1t}) \tilde{M}^*(-k_{1t}) \tilde{Q}(-k_{1t})] \qquad (10.76)$$

form which one cannot determine the combination $\tilde{\zeta}_1(k_{1t}) + \tilde{\zeta}_1^*(-k_{1t})$ needed to reconstruct the surface. This is because the complex function $\tilde{\zeta}_1(k_t)$ needs two real functions of k_{1t} to determine it, and this requires two independent images. It may be traced back to the existence of two solutions to the dispersion relation. The construction of alternative SAR images is considered in section 10.8. An approximate inversion is achieved if real, symmetric approximations to $\tilde{M}$ and $\tilde{Q}$ are used. From eqn (10.64) $\tilde{Q}$ is real and symmetric for $\delta v = 0$. A real symmetric approximation to $\tilde{M}$ is obtained if Doppler splitting is ignored, tilting and straining are given by $|2\tilde{L}^{(\pm)}(k_{1t})|$ using the approximation in eqn (10.62), and velocity bunching is given by $|i\omega_1 t_c \tilde{A}(k_{1t})|$. The result is

$$\tilde{M}(\mathbf{k}_{1t}) = \left[\frac{(0.5k_{1x}^2 + 5k_{1y}{}^2)^2}{k_{1t}{}^2} + (4\tan\theta + 5\cot\theta)^2 k_{1y}{}^2\right]^{\frac{1}{2}}$$

$$+ g^{\frac{1}{2}}\frac{R}{v} k_{1t}{}^{\frac{1}{2}} |k_{1x}| \left[\sin^2\theta \frac{k_{1y}{}^{\frac{3}{2}}}{k_{1t}{}^2} + \cos^2\theta\right]^{\frac{1}{2}}. \quad (10.77)$$

The reconstructed sea surface is then

$$z = \frac{\sigma_1}{I_0} \int \frac{d^2\mathbf{k}_{1t}}{(2\pi)^2} \frac{\tilde{I}_1(\mathbf{k}_{1t})}{\tilde{M}(\mathbf{k}_{1t})} \frac{\exp(-i\mathbf{k}_{1t}.\rho)}{\tilde{Q}(\mathbf{k}_{1t})} \quad (10.78)$$

The important contributions to the integral come from k_{1t} sufficiently small for $\tilde{Q}(\mathbf{k}_{1t})$ to be close to unity so that $\tilde{M}(\mathbf{k}_{1t})^{-1}$ is the important function in the inversion. Fig. 10.11 shows a contour map of this function using eqn (10.77) and SEASAT values of $\theta = 20°$ and $R/v = 128$ s. This function blows up as $k_{1t} \to 0$ and some modification is necessary. Pending a better theory, $\tilde{M}(\mathbf{k}_{1t})^{-1}$ has been given a maximum value of 4. An example of such an inversion is given in section 10.10.

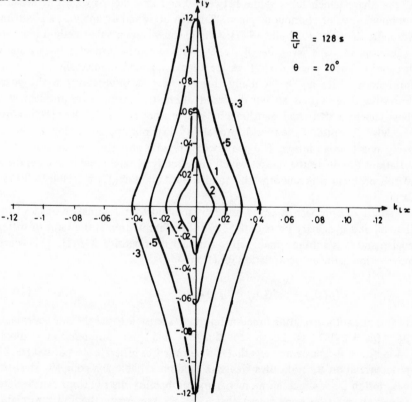

Fig. 10.11 – Lines of constant $|M^{-1}|$ in wavenumber plane (k_{1x}, k_{1y}).

In many ways this inverse theory is unsatisfactory. No consideration is given to the stochastic nature of the SAR image due to speckle and noise, and non-linear aspects are completely ignored. Further developments of the theory are being explored using the methods of Bacchus & Gilbert (1970).

10.9 ALTERNATIVE IMAGES

A conventional SAR image is constructed from the squared modulus WW^* after SAR processing to form W. This is not the only way to construct a SAR image, and almost certainly does not yield the full information content of the data. Alternatives can be formed in a number of ways. Changing the SAR focus represents one approach, and an example was given in Fig. 10.10.

Another approach is the concept of phase imaging. The complex SAR image W has both modulus $|W|$ and phase arg W, but only $|W|^2$ is displayed and arg W is discarded. An alternative image is formed by arg W which is given by Im ln W where Im denotes imaginary part. Unfortunately this only gives the principal part of the phase which takes values between 0 and 2π, whereas what one needs is the unwrapped or continuous phase. This can be found by applying a phase unwrapping algorithm (Tribolet 1977). A simple phase unwrapping method amounts to forming Im $\partial lnW/\partial x = \partial argW/\partial x$. Derivatives can be formed in both x and y directions. These phase derivatives are then integrated numerically to give the continuous phase arg W. Actually this last step is unnecessary as the phase derivative $\partial argW/\partial x$ is an equally valid alternative image. One problem with these images is that the logarithmic derivative amounts to $(1/WW^*)(W^*\partial W/\partial x)$ and WW^*, because of the speckle probability distribution, passes close to zero at many points in an image. The phase varies rapidly near these points, and these variations dominate the images which are effectively randomised. One approach to this problem is to smooth WW^* to give $\overline{WW^*}$ and then form $(1/\overline{WW^*})(W^*\partial W/\partial x)$.

Another approach is to go another step and form alternative images from the real and imaginary parts of $W^*\partial W/\partial x$. Differentiation is the limit of differencing and it is another small step to consider the quantity $W(\rho_1)W^*(\rho_2)$ whose expectation is the autocorrelation function.

$$C(\rho_1,\rho_2) = < W(\rho_1)\, W^*(\rho_2) >. \tag{10.79}$$

In forming autocorrelation function images it is sensible to use the sum coordinates $\rho_p = (\rho_1 + \rho_2)/2$ as image coordinates, and the difference coordinates $\rho_m = (\rho_1 - \rho_2)$ as parameters that can be varied to generate a set of images. The autocorrelation function of a Gaussian random variable is a complete statistical description from which all moments and probability distributions can be determined. It may be conjectured that $C(\rho_1,\rho_2)$ represents the full information content of a SAR signal. It is complex quantity so that it gives a pair of images for each value of ρ_m. Fig. 10.12 shows a series of relative image intensities, $C(\rho_1,\rho_2)/I_0$, plotted against the sum coordinate along a wave for a series of values of x_m, the azimuthal difference coordinate. The curves are for 1 − look SEASAT parameters and 100 m azimuth waves of amplitude 0.25 m. The letters R and I refer to the real and imaginary parts. The waves have been calculated from a generalisation of eqn (10.71).

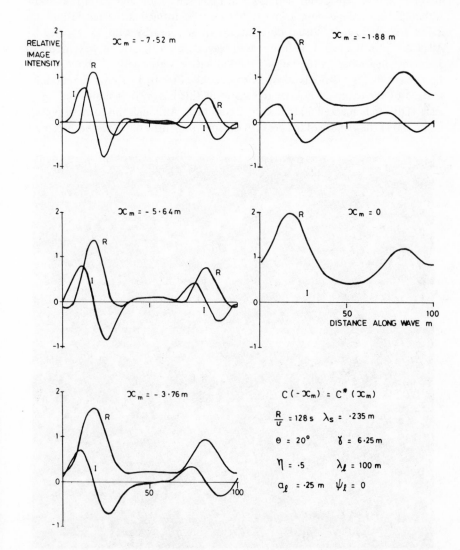

Fig. 10.12 − Autocorrelation function for 100 m azimuth waves of amplitude 0.25 m and different separations.

10.10 EXAMPLE

In this section a SEASAT image is analysed to exemplify many aspects described in this chapter. The image and five derived images are shown in Figs 10.13 to 10.18. The image is from orbit 762 with coordinates 60° 11' N and 6° 41' W at 0649 GMT on 19/8/78. It was processed by RAE with image reference 762298. In these photographs range is horizontal and azimuth is vertical. The original image is 256 × 256 pixels with 12.5 m pixel separation and 25 m resolution. Although four independent azimuth looks can be formed at this resolution, this is just one such look. Visual observations from a nearby ship, as reported by Allan & Guymer (1981), indicate swell waves with a significant waveheight of 5 m travelling close to the range direction with a wavelength of about 200 m. The image in Fig. 10.13 is consistent with this. Fig. 10.14 shows the modulus squared of the complex Fourier transform of this image. It is shown on a grey-scale. The circle marked 50 is at $k_{1f} = 2\pi/50 \text{ m}^{-1}$. Although this form of presentation shows the vicinity of the peaks in the spectrum, little detail can be discerned.

Fig. 10.13 – SEASAT SAR image.

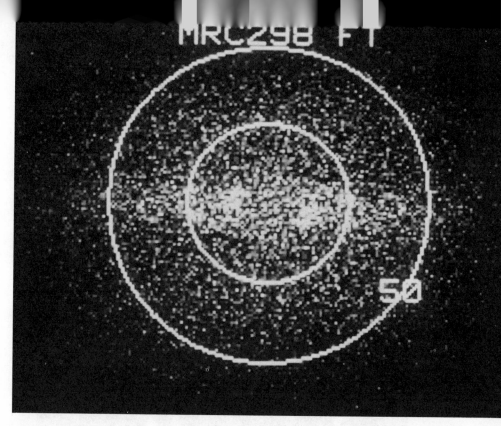

Fig. 10.14 – SAR image spectrum.

Col. Fig. 10.15 shows the same data as in Fig. 10.14 but smoothed and intensity coded. The intensity code is in Col. Fig. 10.17. Col. Fig. 10.15 shows some remarkable features. There is a marked radial decrease which is due to the system transfer function. The swell waves are seen to consist of two systems at angles of about $30°$ to each other. The central part of the spectrum is barrel shaped as in Fig. 10.7. The most remarkable feature is that the harmonics of the two swell wave systems are clearly visible. The system transfer function $\tilde{Q}(k_{1t})$ is easily determined using a method given by Beal *et al.* (1981) in which the spectrum of a featureless random image is formed. This is smoothed and gives a good approximation to $\tilde{Q}(k_{1t})$. The form determined was $\exp(-33.5\ k^2_{1t})$. This factor was removed from the Fourier transform and the modulus squared is shown smoothed and intensity coded in Fig. 10.16. Another step is to divide by the modulation transfer function $\tilde{M}(k_{1t})$. The form in eqn (10.77) was used, and the modulus squared of the result is in Col. Fig. 10.17. The result is not entirely satisfactory because the demodulation greatly magnifies the values in the centre of the spectrum. Finally the demodulated complex spectrum is inverse transformed as in eqn (10.78) to give the reconstructed sea surface in Fig. 10.18. The results of this reconstruction should not be taken too seriously. This is just the first tentative step in the construction of an inverse theory.

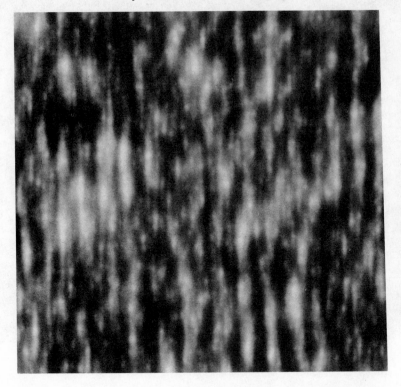

Fig. 10.18 – Demodulated SAR image.

10.11 FUTURE WORK

Almost every aspect of sea imaging theory requires development, but two aspects seem crucial. The first is to develop the nonlinear imaging theory for a system of waves. To fully understand the effects of nonlinearity one must be able to calculate all the interaction products between waves as well as the harmonics that appear in the theory for a single wave. This is a difficult problem, but the determination of the term I_2 in eqn (10.53) is feasible. It is quadratic functional of the long wave field. A generalisation is to compute the corresponding terms for the autocorrelation function in eqn (10.79) as this is likely to play a part in the inverse theory. Another important quantity in the construction of an inverse theory is the variance of the image intensity or autocorrelation function which involves fourth moments of the SAR signal W. Armed with these results, the second aspect can be developed which is to construct a general inverse theory within the framework described by Bacchus & Gilbert (1970).

ACKNOWLEDGEMENT

Part of the work was carried out with ESA support.

REFERENCES

Allan, T. D. & Guymer, T. H. (1981). A preliminary evaluation of SEASAT's performance over the area of JASIN and its relevance to ERS-1. *ESA SP-167.*

Alpers, W. R. & Hasselmann, K. (1978). The two frequency microwave technique for measuring ocean wave spectra from an airplane or satellite. *Boundary-Layer Meteorology* **13** 215-230.

Alpers, W. R. & Rufenach, C. L. (1979). The effect of orbital motions on synthetic aperture radar imagery of ocean waves. *IEEE Trans. Ant. Prop.* **AP-27** 685-690.

Alpers, W. R. & Rufenach, C. L. (1980). Image contract enhancement by applying focus adjustment in synthetic aperture radar imagery of moving ocean waves. *Proc. 2nd SEASAT SAR processing workshop,* Frascati, Italy. ESA Scientific Report.

Alpers, W. R., Ross, D. B., & Rufenach, C. L. (1982). On the detectability of ocean surface waves by real and synthetic aperture radar. *J. Geophys. Res.* To be published.

Bacchus, G. & Gilbert, F. (1970) Uniquences in the inversion of inaccurate gross earth data. *Phil. Trans. Roy. Soc.* **A266** 123-192.

Barrick, D. E. (1977). The ocean waveheight non-directional spectrum from inversion of the HF sea-echo Doppler spectrum. *Remote Sensing of the Environment* **6** 201-227.

Beal, R., Goldfinger, A., Tilley, D., & Geckle, W. (1981) System calibration strategies for spaceborne synthetic aperture radar. *JHU/APL CP 084.*

Beckmann, P. & Spizzichino, A. (1963). *The scattering of electromagnetic waves from rough surfaces.* Pergamon Press.

DeSanto, J. A. (Ed.) (1979). Ocean Acoustics. *Topics in Current Physics* **8** Springer-Verlag.

Jain, A. (1978) Focusing effects in the synthetic aperture radar imagery of ocean waves. *Appl. Phys.* **15** 323-333.

Larson, R., Moskowitz, L. I., & Wright, J. W. (1976). A note on SAR imagery of the ocean. *IEEE Trans. Ant. Prop.* **AP-24** 393-394.

Phillips, O. M. (1977) *The dynamics of the upper ocean.* C.U.P.

Raney, R. K. (1971). Synthetic aperture imaging radar and moving targets. *IEEE Trans. Aerosp. Electron. Syst.* **AES-7** 499-505.

Swift, C. T. & Wilson, L. R. (1979) Synthetic aperture radar imagery of moving ocean waves. *IEEE Trans. Ant. Prop.* **AP-27** 725-729.

Tribolet, J. M. (1977) A new phase unwrapping algorithm IEEE Trans. *Acoustics Speech Sig. Proc.* **ASSP-25** 170-177.

Valenzuela, G. R. (1978) Theories for the interaction of electromagnetic and oceanic waves – a review. *Boundary-Layer Meteorology* **13** 61–85.

Valenzuela, G. R. & Wright, J. W. (1979). The modulation of short gravity – capillary waves by longer scale periodic flows. A higher order theory. *Radio Science* **14** 1099–1110.

Valenzuela, G. R. (1980). An asymptotic formulation for SAR images of the dynamical ocean surface. *Radio Science* **15** 105–114.

Chapter 11

An assessment of SEASAT-SAR image quality

G. E. KEYTE & M. J. PEARSON
Space Department, RAE Farnborough

11.1 INTRODUCTION

The majority of European users of SEASAT Synthetic Aperture Radar (SAR) images will have acquired their data through ESA -- EARTHNET. In most cases these images were prepared from raw data received and recorded at the Oakhanger, UK, ground station, operated by RAE on behalf of ESA. The raw data were processed when required into radar images. In the main this was done using one of three processors: an optical processor at ERIM, Michigan, USA, a digital processor at DFVLR, Oberpaffenhofen, West Germany, and a further digital processor at RAE, Farnborough, UK.

Both the format of the SAR image and its quality vary considerably, depending on which processor was used. In some cases there are variations in image quality even for the same processor. The SEASAT SAR images have been widely used for studying both oceanographic and land applications; for example the application of SEASAT data to geological and land use was extensively studied by Hunting (1981). In many cases, the results of these studies will depend greatly on the quality of the SAR images used, and it is therefore important to be able to monitor and, if practicable, control the quality of SAR images in an objective manner.

The measurement of SAR image quality (or for that matter, of any image quality) is not without difficulty. In the first place it is necessary to define the factors which determine image quality; secondly, it is necessary to devise methods of measuring these factors in quantitative terms. Given that all such image quality parameters can be measured, then it should be possible to evaluate images produced by different processors and even to arrange that they are produced to a common standard.

It should be noted that the assessment of image quality is often made in subjective terms, and this will of course vary greatly from one observer to another. It is important, therefore, for image quality measurements to be based on the physical properties of the image (e.g. photographic density, pixel values, width of impulse response, etc.) since only then will different images be comparable. It should also be noted that some definitions of image quality have been devised as a basis for SAR systems or for SAR processor design. Many of these latter definitions may prove to be unpractical as a means of checking the quality of real SAR images.

In this chapter we describe the image quality parameters which can be used to evaluate the quality of SAR images. Some results are given for measurements on optically processed images, but the main emphasis is on the geometric fidelity and spatial resolution of digitally processed images.

11.2 DEFINITIONS OF SAR IMAGE QUALITY

The quality of any monochromatic image (whether produced by optical or non-optical sensors) may be considered to depend on the following four factors:

(a) geometric fidelity,
(b) spatial resolution,
(c) radiometric accuracy,
(d) radiometric resolution.

These factors also control the quality of SEASAT SAR images, although there are many independent parameters in the overall radar–data handling-processor system which influence these factors. The definition of these factors and the methods by which they may be measured are considered in turn:

11.2.1 Geometric fidelity

This aspect of image quality is concerned with the cartographic aspect of SAR images, i.e. the accuracy with which the radar backscatter characteristics of the ground are mapped. There are two parts to be considered; one is the accuracy with which the overall image is located in terms of latitude and longitude – this is particularly important for those images where there are no recognisable land forms. The other is the accuracy with which different points in the same image are mapped relative to each other for a particular projection of the Earth's surface.

Earlier examples of radar imagery have tended to suffer in terms of geometric fidelity. Distortions have been introduced by slant range projection and (in the case of airborne radar) by radar platform roll and yaw movements. However, the use of digital processors and spaceborne radars has now made it possible to produce images to a given projection with a high degree of accuracy.

The assessment of geometric fidelity of SAR images may be carried out by comparing the relative location of selected features in the image with the same features on the ground or on an accurate map. The absolute location of the overall image can be measured in the same way. There are, however, two significant difficulties; one is the correction for changes in apparent positon of the feature in the image as a consequence of its elevation above the projection plane (this requires accurate contour information); the other is in precise identification of the feature concerned on both map and image.

11.2.2 Spatial resolution

Normally, the spatial resolution of an imaging system is evaluated by measuring the contrast of periodic features of varying spacing (i.e. measuring the modulation transfer function). In radar systems it is usual to describe the spatial resolution by reference to the half-power (−3 dB) width of the impulse response of the radar system (in effect measuring the width of a point spread function, as in optical systems). This, however, introduces a single number to represent a many-valued function, and so the half-power width does not fully describe the spatial resolution properties of the system. Nevertheless, the half-power width may be used as a means of comparing images from different systems and processors.

In a SAR imaging system the spatial resolution (as defined by the half-power width of the impulse response) is controlled mainly by the effective system bandwidth, analogous to aperture width in optics. In the across track direction the bandwidth is determined by the pulse modulation frequency (or chirp) and in the along track or azimuth direction it is determined by the synthetic aperture length (Doppler bandwidth). Hence the spatial resolution of the SAR image in range and azimuth directions may be different, although in SEASAT both were designed to be around 20 m. There is usually some weighting of these frequency bandwidths (to suppress sidelobes), and this will have the effect of slightly increasing the half-power width of the impulse response, hence reducing the spatial resolution.

The spatial resolution can be evaluated by measuring the image of a bright point-like target, e.g. a corner reflector. Several such targets were set up by JPL at Goldstone, California, but the early failure of SEASAT prevented similar targets from being deployed in the UK. It was established later that some naturally occurring point targets could also be used with fair accuracy.

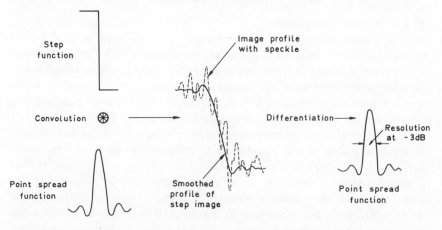

Fig. 11.1 – Derivation of point spread function from image of sharp edge or step.

An alternative method is to measure the amplitude profile of a linear feature in the image and to derive the point spread function or impulse response from the gradient of such a profile (see Fig. 11.1). It was considered that this method might be of use in optically processed images where the limited dynamic range of the photographic emulsion prevented point targets from being accurately recorded.

11.2.3 Radiometric accuracy

This aspect of image quality is concerned with the precision with which the different levels of a radar image represent the corresponding changes in radar backscatter of the ground at the particular wavelength, polarisation and incidence angle being used. In a conventional radar image (e.g. SLAR) the brightness (or intensity) of a point is determined by the radar equation and when corrections are made for such variables as antenna response and range, the brightness should vary linearly with radar backscatter. In a SAR system the image brightness is produced by convolving the radar signal with the pulse and Doppler replicas, but this should not affect the linearity of the system response. In addition to having a linear response it is desirable for different SAR images to be produced such that a given brightness in any image indicates the same radar backscatter cross-section, i.e. for the system to be calibrated.

To measure the radiometric accuracy it is necessary to use calibrated targets of different radar cross-section. Ideally, distributed targets (i.e. large uniform areas such as fields) should be used since calibrated corner reflector targets will always include some radiometric uncertainty from the surrounding terrain. In the case of SEASAT it was not possible in the UK to measure the backscatter of distributed targets or to provide calibrated point targets. The Goldstone corner reflector array included several reflectors of different size and hence of different radar cross-section, and was used to provided a rough check on processor linearity.

11.2.4 Radiometric resolution

The radiometric resolution is determined mainly by the random signal fluctuations present in the image; it is, of course, also determined (in digital processors) by the word size used in processing stages and in the final image. The noise may be introduced by (1) signal fluctuations in the radar system, caused by atmospheric turbulence, thermionic effects, etc., and (2) by coherent noise resulting from interference between signals from random scatterers. This latter noise – or 'speckle' – can to some extent be suppressed by means of 'multilook' techniques in which several images of the same scene are produced from different sets of data (in the one pass) each having a reduced azimuthal spatial resolution. These images are added incoherently to produce a single image in which the speckle noise is reduced by a factor $\sqrt{N}$ (where N is the number of images used, or 'looks'). Although the spatial resolution is reduced (as measured by half-power impulse response width, the *apparent resolution* may be improved as a consequence of speckle noise reduction.

SEASAT was designed to produce 'four-look' images with an azimuth resolution of 25 m. Even with this number of looks the speckle noise is still obtrusive and will probably need to be greatly reduced in future systems. The assessment of radiometric resolution can be made by performing statistical measurements on SAR images of uniform areas, or, for non-uniform areas, by filtering out the image content of the data leaving only the speckle noise. Examples of both kinds of measurement have been reported by Miller & Belcham (1981).

In this chapter the measurements are concerned mainly with geometric fidelity and spatial resolution, although some additional results on radiometric accuracy (using Goldstone data) are given. The quality of optically processed and digitally processed images are each discussed in turn.

11.3 OPTICALLY PROCESSED IMAGES

11.3.1 Image format

As part of the EARTHNET programme the raw SEASAT data recorded at Oakhanger were replayed through the ERIM film camera to generate optical signal film. These were subsequently used by ERIM on their optical correlator to produce survey mode and precision mode image transparencies of virtually all the SAR swaths received at Oakhanger. The transparancies were produced in strip form, each strip corresponding to a subswath 850 km long and 33 km wide, hence four strips were needed to cover the full SEASAT swath of 100 km. The nominal scale of the image in both across track (range) and along track directions was 1:685000 (note: the projection in the range direction was in ground range, not slant range). The nominal resolution (for EARTHNET products) was 20 m (in range) and 20 m (in azimuth) decreasing to 20 m and 9 m respectively for precision mode products (both modes were single 'look' only).

The measurements described here were carried out on precision mode images of Rev 762, acquired independently by RAE but otherwise identical to EARTH-NET products. The measurements have been reported more fully by Keyte & Pearson (1979).

11.3.2 Geometric fidelity

There were no markings on the as received transparencies to indicate latitude or longitude, and this aspect of image quality was not investigated. Subsequently, however, as much as possible of the optically processed imagery was overlaid with a latitude-longitude grid using landmarks and coastal features to identify the image. The accuracy of this gridding is of the order of 1.5 km where landmarks exist and about 15 km where such marks are absent.

The accuracy of projection of the optical images was assessed by measuring the relative position of recognisable features (ground control points) in the image and comparing these locations with those on Ordnance Survey maps. Table (11.1) shows the variations in scales obtained. It can be seen that there is a considerable variation in the across track scales from subswath 4 (near range) to subswath 2 (far range) of about 13%, although the along track variations are only 0.13%. These measurements seem to be in general agreement with those measured by ERIM, and are the result of changes in ground range scale with changes in angle of incidence of the radar beam.

Table 11.1

Optically processed image scale variation

Direction of measurement	Sub-swath	Location of measurement	Measured scale	Accuracy of measurement
	4	North West of swath	683396	
Along azimuth	4	Centre of swath	682892	0.5%
	4	South East of swath	683832	
	4	North West	743865	
	4	South East	729021	
Across track	3	North West	697791	3%
	3	South East	696927	
	2	North West	655120	
	2	South East	657258	

11.3.3 Spatial Resolution

Measurements of spatial resolution were confined mainly to measurements of the edge profiles of linear features in the image, as described in section 11.2.2. It was not possible to measure the half-power (-3 dB) width of point-like targets because the majority of such targets were clipped by the limited response of the photographic emulsion, making any estimate of peak intensity virtually impossible. Where point-like targets were not saturated they were indistinguishable from the background clutter. Some estimate of the system bandwidth could be made from measurements of the sidelobe spacing of bright point-like targets. Since the nominal impulse response is represented by a $\sin^2 x/x^2$ function, the spacing of the successive nulls is $1/\Delta F$ where ΔF is the frequency bandwidth. The half-power width is then $0.885/\Delta F$, assuming that no sidelobe reduction has been used. For the optically processed images (Rev 762) the -3 dB width was found to be 16 m in azimuth and 22 m in range. This suggests that the full resolution capability of SEASAT SAR was not used, but this cannot be proven as measurements are very approximate.

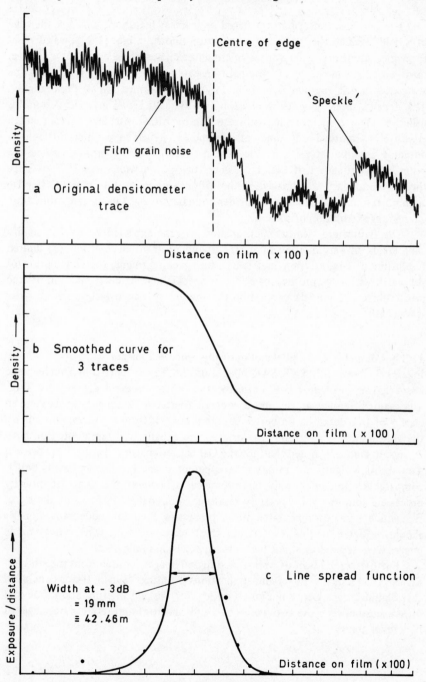

Fig. 11.2 – Resolution measurement.

The resolution measurement based on a linear feature (in this case the edge of a field) yielded the 'point spread function' shown in Fig. 11.2. It was derived from measurements made with a microdensitometer of slit width 5×10^{-6} m made on enlargements of the original transparency to a scale of 1:227000. The sharpness of the original field boundary was confirmed by air photography. The density measurements were expressed as radar signal intensity variations using the step wedge provided with the film by ERIM. Although the slit length effectively smoothed out some of the speckle it was in practice difficult to obtain a smooth intensity profile. The differential of the profile yielded the point spread function of Fig. 11.2 where half-power width was 23.5 m. Since the field boundary was imaged in the azimuth direction this gave an effective azimuth resolution of 23.5 m. A further boundary, imaged in the range direction, gave a range resolution of 42.5 m.

The figures are clearly much greater than the expected values, but significant errors in this method of measurement are introduced from: (a) limited resolution of original film and losses during enlargement, (b) convolution of slit width with edge profile, (c) effect of speckle making determination of true profile difficult, and (d) assumption that boundary was in fact 'sharp' at radar wavelength.

11.3.4 Comparison of digital and optically processed images

Fig. 11.3 shows SEASAT SAR images of the Tay Estuary and Dundee, one image having been processed optically, the other processed digitally by RAE. The main differnces are in image contrast (better in the digital image) and in clarity of fine detail (again better in digital image). Other features, particularly the appearance of river and sea surfaces, are equally well represented on both images. It should be noted that, for the Oakhanger images the optically processed data shows a significant variation in apparent image quality, and at its best is comparable with digital products. In some cases, however, the clarity of optically processed data was poor, possibly resulting from changes in focus of the signal film which were uncompensated during processing. Precision mode transparencies should be better in this respect, but there were occasions when survey mode images were as good as, if not better than, precision mode data.

Further comparison of optical and digital images resulted from the study of SEASAT SAR data for overland applications Huntings (1981). It was concluded that digital images were superior to optical for flatland interpretation, but that differences between the two were not significant for interpretation of images of high relief areas.

Fig. 11.3b — Digitally processed (RAE) image of Tay Estuary.

Fig. 11.3a — Optically processed image of Tay Estuary, Scotland.

11.4 DIGITALLY PROCESSED IMAGES

11.4.1 Image format

The measurements described herein were carried out mainly on RAE processed images although some DFVLR images were obtained for comparison. The RAE processor, the experimental SAR processing facility (or ESPF), was designed as a flexible processor using standard computing hardware aimed at investigating the effect of different processing parameters on SAR image quality. The ESPF has been described more fully in earlier papers, e.g. Corr & Haskell (1978), Keyte (1980).

The SAR image is mapped on to a rectangular grid whose orientation, dimensions, and pixel size are all selectable. The azimuth resolution, sidelobe levels, and gain are also controllable, and the output image is available in amplitude or in complex form. Speckle reduction by 'multilooking' is carried out as a separate process. An additional feature is the SAR simulator which enables raw data for simple targets to be synthesised and later processed as if they were SAR data.

11.4.2 Geometric fidelity

The accuracy of location and of projection of the digital images was measured by selecting ground control points in the image which could also be readily identified on Ordnance Survey maps. During the processing the radar image is built up on to a rectangular grid, tangential to the Earth's surface at the origin, whose latitude and longitude are specified. Hence it is straightforward to calculate the location (in terms of map coordinates or grid references) of any feature in the image, and to compare this with the actual map coordinates. Corrections were made when necessary for any shift in target location caused by elevation above the image plane.

Fig. 11.4 shows the relative errors in target position when compared with map coordinates for a number of different images taken from different SEASAT passes. Although there is a substantial amount of scatter in the measurements there is a distinct trend, centred on a mean error of +465 m in azimuth and −129 m in range (after neglecting some extreme, and possibly spurious, errors). The scatter within a given image is much less (as shown by Fig. 11.5) and indicates that there is a distinct trend for the image to be produced at a shorter range and slightly ahead of that specified. It is considered that these errors result from differences in timing between the orbit data and SAR recorded data (based on ground station clock times). The accuracy of location within a given image is good to within two or three pixels (i.e. 40–60 m) and does not show any significant trends (i.e. errors are randomly distributed, as shown in Fig. 11.5).

11.4.3 Spatial resolution

The spatial resolution of digitally processed images was mainly evaluated using point-like targets although some work was also done using linear targets. Measure-

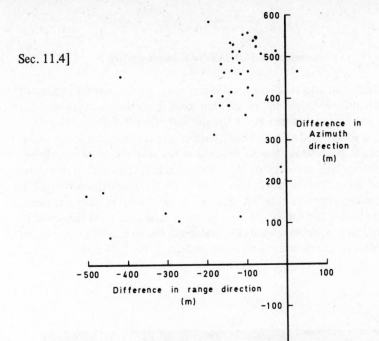

Fig. 11.4 – Target positions compared with target ordnance survey grid references

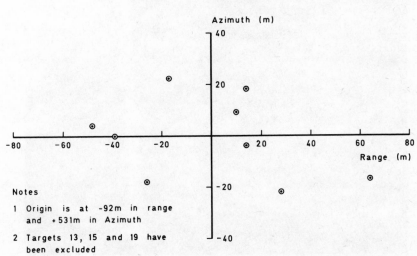

Fig. 11.5 – Image 791045: Actual target positions compared with mean target positions.

ments were made on images of corner reflectors using the raw SAR data collected by JPL at Goldstone, California. In addition some measurements were made on naturally occurring point targets in the UK and also of simulated point targets.

Fig. 11.6 shows an RAE processed image of the Goldstone reflector array. The long bright band (from top to bottom) is the sidelobe of the Goldstone receiving antenna, and the array of ten reflectors is to the one side of this band. The sizes of the reflectors varied from 2.44 m (8 ft) square down to 1.22 m (4 ft) triangular, although this last reflector is not easily perceived amongst background clutter. The two bright 2.44 m square reflectors are at either end of the array and were oriented towards the spacecraft; the remainder were oriented vertically. Most of the measurements were made with the 2.44 m square reflectors.

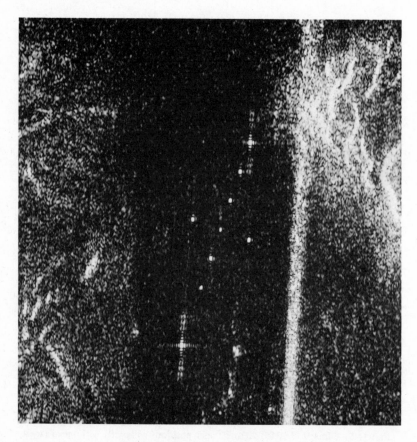

Fig. 11.6 – Goldstone corner reflector array.

Fig. 11.7 shows an image of one of the 2.44 m square reflectors and, for comparison, an image produced from simulated data. Similar processing parameters have been used for both images, although the Goldstone image has the benefit of interpolation to remove range migration effects (i.e. the faint 'ghosts' seen on the simulated image). The amplitude profiles of these images were measured and are shown in Figs 11.8 and 11.9 for the across track (range) and along track (azimuth) response respectively. The pixel spacing used for these images was 3 m hence greatly oversampling the spatial resolution (25 m) but enabling clear profiles of the impulse response to be obtained. The two profiles are identical for the along track response and given an azimuth resolution of 24.9 m and 25.02 respectively for simulated and Goldstone images.

Fig. 11.7a – Digitally processed image of a simulated point target.

Fig. 11.7b – Digitally processed image of real (Goldstone) point target.

The across track response (Fig. 11.8) contains a slight unevenness in the sidelobe levels for the Goldstone image as compared with the simulated data. In fact some minor changes were required to modify the expression used for the radar chirp frequency before an image of the Goldstone reflectors of maximum clarity could be obtained – this modification was also found to improve significantly clarity and detail of other images. The measured half-power widths were 18.44 m and 21.95 m respectively for simulated and Goldstone images, the difference being due to the variation in ground range resolution as a function of target position across the radar swath.

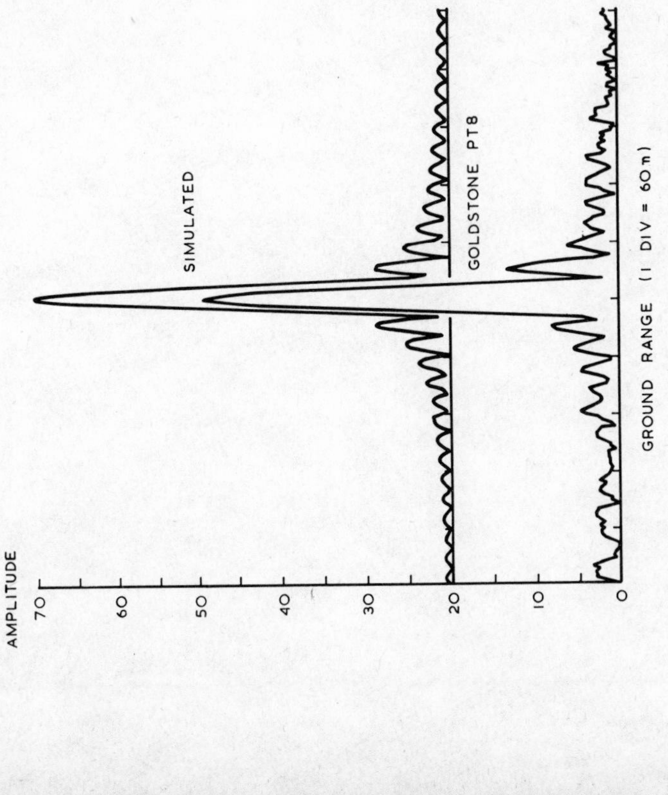

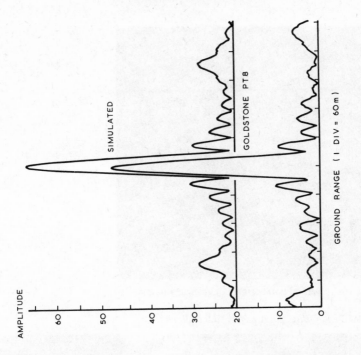

Fig. 11.9 – Along track response.

Fig. 11.8 – Across track response.

Fig. 11.10a – Natural point target: Oakhanger antenna.

Fig. 11.10b – Natural point target: Hull region.

Some further measurements were made to see if naturally occurring point targets could yield similar results. Fig. 11.10 shows images of the Oakhanger antenna (in effect a point-like target) and of an unknown target in the region near Hull, UK. Both images contain characteristic sidelobe structures associated with bright point targets, and they yielded the along track profiles shown in Fig. 11.11. Similar profiles were obtained for the across track direction except that the sidelobe level unevenness observed earlier was again present. The azimuth resolution was measured as 25.0 m for the Oakhanger target and 26.5 m for the Hull target (the nominal resolution was 25.0 m). There were similar results obtained for other point-like targets, but there were occasions when such targets failed to give sensible results. In these cases the targets are perhaps extended targets (e.g. an array of roof tops) reflecting specularly and of sufficient size to distort the shape of the SAR impulse response.

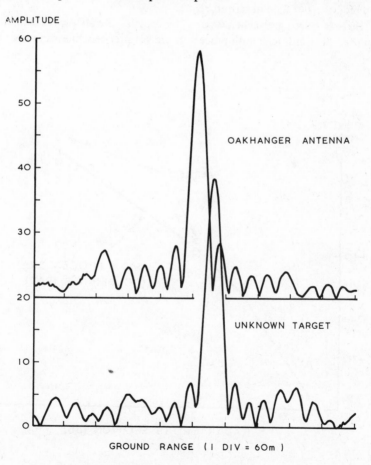

Fig. 11.11 – Along track response.

11.4.4 Radiometric accuracy

Only a limited assessment of radiometric accuracy was possible, because of the absence of calibrated targets. Some measurements were made on the Goldstone image using the known dimensions of the reflectors as a means of calculating their radar cross-section. These values were compared with the peak impulse response measurements obtained for each reflector, and the results were expressed in terms of power relative to the 2.44 m reflector. Fig. 11.12 shows the variation in relative power as a function of radar cross-section in dB. The solid line shows the results obtained by Goldfinger for the MDA processor (used at CCRS, Ottawa, Canada, and at DFVLR) and the circled points are those obtained with the ESPF. The scatter in ESPF results is attributed to variation in the actual cross-section of the reflectors and to statistical contributions from the surrounding terrain (estimated mean backscatter of terrain was −27 dB relative to maximum brightness of 2.44 m reflector).

There is close agreement between the data for the MDA processor and that from the ESPF, although both processors are not precisely linear in response.

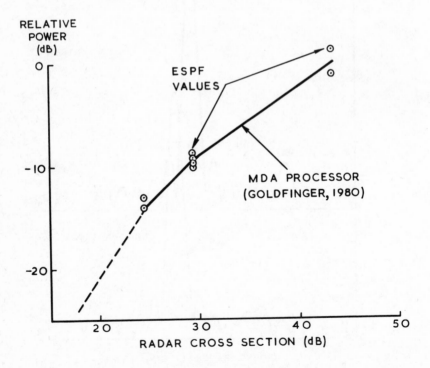

Fig. 11.12 – Power response of RAE (ESPF) processor.

RAE
MEAN=104.5 SD=31.9
DENSITY RANGE 0-207

MDA
MEAN=101.7 SD=30.2
DENSITY RANGE 0-207

Fig. 11.13 – Comparison of RAE and MDA (DFLVR) digitally processed images.

11.4.5 Comparison of ESPF and MDA (DFVLR) images

A comparison of the photographic products derived from the digital images
produced by the ESPF (RAE) and MDA (DFVLR) processors normally shows
substantial differences. These are mainly caused by differences in the way the
digital data are compressed before being written on film. The RAE use a simple
linear contrast stretch, whilst DFVLR use a transformation which optimises the
mean grey level (to the photographic density — exposure curve) and controls
the standard deviation of the pixel values in the image.

The digital images themselves do not differ greatly, however, given that similar processing parameters are used. It should be noted that although the principles of processing are the same for each processor, the actual computations are carried out by different methods. Fig. 11.13 shows digital images produced by both processors of the Tay Estuary scene; in both cases the mean grey level and standard deviation were arranged to be similar before writing on film. The figure shows that the images are greatly alike; the RAE image suffers a little from lack of interpolation, and this causes the some smearing of detail in very bright areas (e.g. in Dundee region) on the north bank between the railway and road bridge.. [Interpolation could be carried out on the ESPF but was relatively time-consuming; however, after recent modifications to the ESPF, interpolation is now performed as a standard processor option.]

Fig. 11.14 shows examples of amplitude profiles for both digitally processed images; the profiles are of the Tay (railway) Bridge seen (left) in Fig. 11.13 and their appearance suggests that both processors exhibit similar image quality where spatial resolution is concerned.

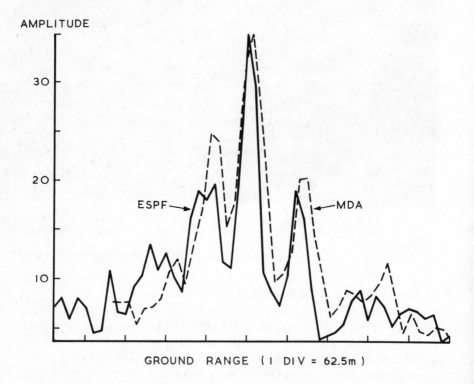

Fig. 11.14 – Comparison of MDA (DFVLR) and ESPF (RAE) processors, amplitude profile across Tay bridge.

The result of this comparison indicates that a complete assessment of image quality should take account of the smearing or ghosting caused by not interpolating during the azimuth compression process. Ideally, this would involve measuring the energy distribution scattered around the image of a point target, but this would be difficult to carry out on real data owing to the presence of clutter in the terrain surrounding the target. It is considered that the only practical method of assessing this aspect of image quality is to use simulated data, where the distribution of energy can be measured for a point target on a perfectly absorbing background.

11.5 CONCLUSION

Despite the early failure of SEASAT and the fact that few targets were deployed to assist in the evaluation of SAR image quality, it has been possible to measure some image quality parameters and to verify system performance.

It was found that digital images gave measurements which were very close to those expected, especially for spatial resolution and geometric fidelity. For optically processed data the main problem was to devise accurate methods for measuring spatial resolution, and only indicative measurements were obtained. It is considered that spatial resolution measurements for optical data are best carried out in the output plane of the optical processor.

The value of corner reflectors, both as a means of assessing image quality and for calibration purposes, has been demonstrated. Such reflectors are also invaluable as a means of 'tuning' the processor where some uncertainties exist, e.g. in radar chirp parameters. It is concluded that future spaceborne SAR remote sensing programmes should include provision for the deployment of numerous corner reflectors of various sizes, and also for the measurement of backscatter of distributed targets. The latter would be of especial value in calibrating the radar system, particularly at low backscatter values.

REFERENCES

Hunting Geology and Geophysics Ltd. *The evaluation of the data content of overland SEASAT SAR imagery*, (Feb 1981) (unpublished report).

Miller & Belcham, *Image Quality: SEASAT SAR evaluation*, MRL Contract Report No.MTR 81/39, (Apr. 1981).

Keyte & Pearson, *A preliminary evaluation of optically processed SEASAT SAR imagery*, ESA SP-154, (Dec. 1979).

Corr & Haskell, A digital SAR processor for SEASAT A, *Proc. Int. Conf. on spacecraft on-board data management*, Nice, (1978) ESA SP-141.

Keyte, *Preliminary results from the RAE digital processor*, SEASAT SAR Workshop, ESRIN, Frascati, (Dec 1980).

A. D. Goldfinger, *SEASAT SAR processor signatures: point targets*, John Hopkins University Report JHU/APL CP078 (Apr. 1980).

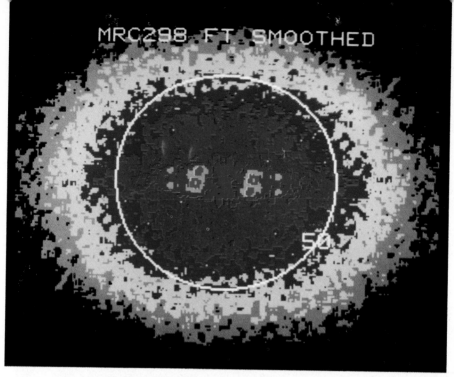

Above: Col. Fig. 10.15 – Smoothed spectrum.

Below: Col. Fig. 10.16 – Spectrum with system transfer function removed.

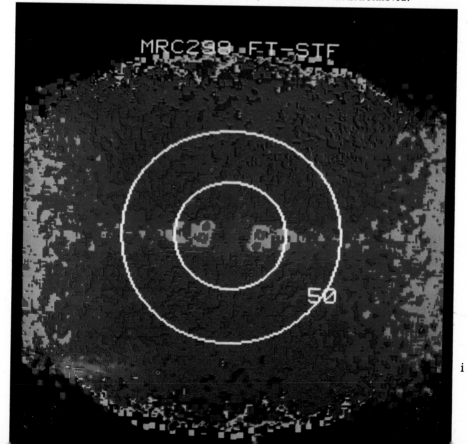

i

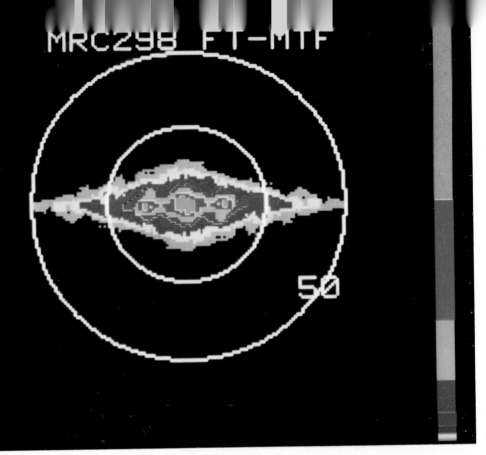

Above: Col. Fig. 10.17 — Spectrum with modulation transfer function removed.

Below: Col. Fig. 17.5 — (*See page iv for legend*)

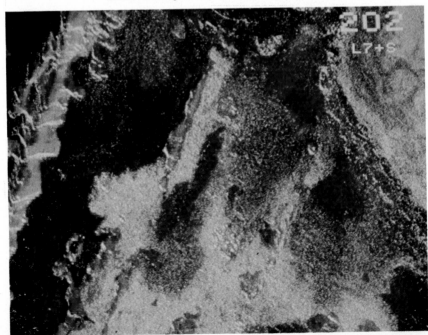

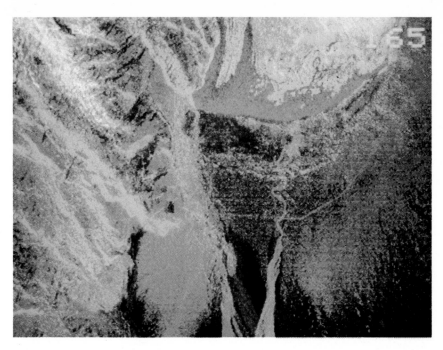

Above: Col. Fig. 17.6 – (*See page iv for legend*)

Below: Col. Fig. 17.9 – (*See page iv for legend*)

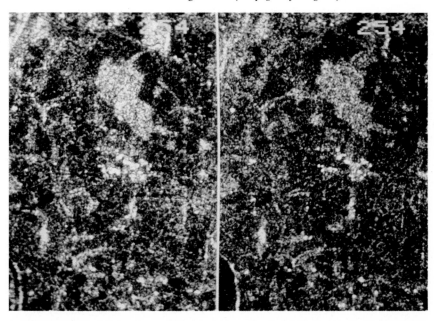

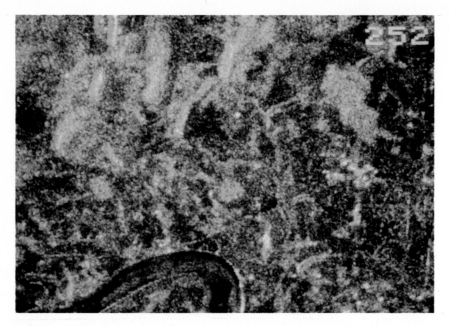

Col. Fig. 17.10 — (*See below for legend*)

Col. Fig. 17.5 — Combination of SEASAT and LANDSAT on the IDP 3000 monitor screen covering a 10 × 7 km area in Iceland. LANDSAT Band 7 is shown in read and SEASAT in green, an ice cap is shown on the top right.

Although there is lack of registration in the hill ridges, this digital combination of images is quite effective. Detail is shown in the ice and the fringing moranies are well depicted. Rock outcrops tend to be well outlined, principally by LANDSAT. The rougher areas of the lava fields, in green, are mainly registered by SEASAT.

Col. Fig. 17.6 — Combination of SEASAT and LANDSAT covering a 10 × 7 km area of Sleidarajokull glacier, Iceland. LANDSAT Band 4 is shown in green and SEASAT in red.

The bright red fringe to the tongue of the glacier stems from bouldery morainic accumulations, the roughness of which produces strong radar response. On the left hand of the scene there is little or no registration of LANDSAT and SEASAT owing to severe layover distortion.

Col. Fig. 17.9 — Comparison of SAR imagery from two Orbits (963 and 834) adjacent to the River Severen Estuary, UK. Changes in radar reflectance can be seen in fields near to the large block of woodland (bright signature) in the upper central part of these images. Each image covers approximately 4 km × 6 km. OS Grid reference of the southwest corner is SO823 120.

Col. Fig. 17.10 — Superimposition of SAR imagery from two Orbits. The images partly shown in Col. Fig. 17.9 have been displaced on the IDP 3000: Orbit 834 in green and 963 in red. Red and green areas on this picture are areas of change. Where the picture is mainly black or yellow the two images are the same.

Chapter 12

Effect of defocusing on the images of ocean waves

K. OUCHI

Blackett Laboratory, Physics Department, Imperial College, London, UK

12.1 INTRODUCTION

The effects of defocusing on the images of dynamic sea surfaces produced by SAR are of considerable current interest (Shuchman & Zelenka 1978, Jain 1978, Valenzuela 1980, Alpers, & Rufenach 1979, Raney 1981). Two main theories have been postulated to explain the experimental observations in which the images of ocean waves are enhanced by adjusting the azimuth focus of a SAR processor. One of them is that the azimuth component of the wave phase velocity changes the relative velocity of the radar platform (Shuchman & Zelenka 1978, Jain 1978, Valenzuela 1980), and the other is that defocusing is caused by the slant-range component of the acceleration associated with the wave orbital motion (Alpers & Rufenach 1979).

In this chapter we investigate the defocusing effect by applying a different approach to the formulation of SAR imagery of dynamic objects from the conventional phase perturbation method. The main advantage of the present approach is that it is rigorous and could lead to simple analytical results (Rufenach & Alpers 1981). We consider that the main contribution to the final image is backscattered amplitude (power) modulation arising from surface roughness and/or surface tilt (Elachi & Brown 1977), and the velocity bunching effect (Valenzuela 1980, Alpers & Rufenach 1979, Rufenach & Alpers 1981, Ouchi 1982a) is not included. The cases where the two effects contribute to the images of azimuth waves have been previously discussed by Ouchi 1982b. We also consider the spatial and temporal random fluctuations of backscattered amplitude which in effect degrade image quality (Ouchi 1982c, Ouchi 1981, Raney 1980, Raney & Shuchman 1978, Raney 1981). We then derive and discuss expressions for the local mean intensity and contrast of the images of ocean waves propagating in an arbitrary direction. It is shown that if the reference signal of the SAR processor is matched to a stationary point target, the images of waves are always defocused, irrespective of their propagation direction. The images can be enhanced by applying a defocused reference signal, and the amount of defocus depends on the phase velocity and propagation direction of ocean waves.

12.2 IMAGING SYSTEM

A side-looking radar antenna sends out a sequence of chirped (FM) pulses

$$E_i(\tau) = W_r(\tau) \exp(i\omega\tau) \exp\left(-i\frac{\alpha}{2}\tau^2\right) \tag{12.1}$$

as it propagates in the azimuth direction, where τ is a time variable, W_r is the amplitude weighting of the pulse, α is the linear FM rate, and ω is the pulse centre frequency.

Let us consider a local single scatterer at a position (x,y) on the ground and denote the backscattered complex amplitude from the scatterer by $U(x,y;t)$, where x and y are the spatial variables in the ground azimuth and range direction respectively, and the time variable t defines the azimuth position of the radar when a chirped pulse is emitted. Note that t is discrete and τ is the range time variable relative to each t. In general, t can be considered to be continuous if an appropriate sampling interval defined by the pulse repetition frequency is taken to satisfy the Nyquist sampling theorem. The return signal from the scatterer is

$$E_r(t,\tau;x,y) = U(x,y;t) W_a(t-x/V) W_r(\tau-2r/c) \exp\left[i\omega(\tau-2r/c)\right]$$
$$\exp\left[-i\frac{\alpha}{2}(\tau-2r/c)^2\right] \tag{12.2}$$

In this expression W_a is the amplitude weighting (far-field antenna pattern) of the illuminating beam in the azimuth direction, V is the radar platform velocity, $2r/c$ is the time delay of the return signal relative to the transmitted pulse, c is the velocity of the radio wave, and r is the slant-range distance from the radar to the scatterer at time t. To a good approximation, r is given by

$$r = (R^2 + 2R y \sin\theta + y^2 + (Vt-x)^2)^{\frac{1}{2}}$$

$$\cong . R + y \sin\theta + \frac{(Vt-x)^2}{2(R+y\sin\theta)} \tag{12.3}$$

where R is the slant-range distance between the radar at time $t = 0$ and the origin of the ground coordinate system, and θ is the radar look angle as shown in Fig. 12.1.

At this point it is useful to replace the time variables by the equivalent spatial variables. We define

$$\xi = Vt$$

$$\eta = c\,\tau/2 - R \tag{12.4}$$

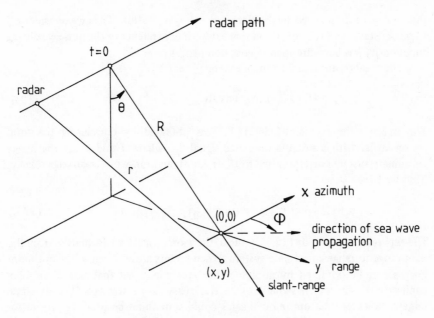

Fig. 12.1 – Geometry of SAR system.

Then, retaining the terms up to the quadratic, the return signal can be written as

$$E_r(\xi,\eta;x,y) = E_0\; U(x,y;\xi)\; W_a(\xi-x)\; W_r(\eta-y\sin\theta)$$
$$.\exp(-i2ky\sin\theta)\exp\left[-i\;\frac{k}{R+y\sin\theta}\;(\xi-x)^2\right] \qquad (12.5)$$
$$.\exp\left[-i\;\frac{2\alpha}{c^2}(\eta-y\sin\theta)^2\right]$$

where

$$E_0 = \exp\left\{i\,2k\eta + i\;\frac{2\alpha\,(\xi-x)^2\,(\eta-y\sin\theta)}{c^2(R+y\sin\theta)}\right\}$$

$$k = 2\pi/\lambda$$

and λ is the radar wavelength.

In the above formulae, the linear phase term in E_0 gives a constant shift of the origin of the image coordinate but the amount of the shift is known as *priori* and the image coordinate could be rescaled accordingly. The azimuth-range cross term in E_0 may result in ghost images which could be corrected (Report, Systems Designers Ltd 1979). In general applications, the contribution of this cross-product term is small compared with that of the doppler and chirp phase quadrature and is often ignored in the data processing operation. Thus we consider E_0 to be constant, and the corresponding azimuth dependence of the pulse weighting function W_r has been dropped in equation (12.5).

The total return signal from an extended target is

$$E(\xi,\eta) = \int\int\limits_{-\infty}^{+\infty} E_r(\xi,\eta ; x,y)\, dx\, dy. \tag{12.6}$$

The image complex amplitude $A(X,Y)$ is obtained by correlating the total return signal with a suitable reference signal E_p, where X and Y are the image coordinate system corresponding to the azimuth and range direction respectively. Thus we have

$$A(X,Y) = \int\int\limits_{-\infty}^{+\infty} E(X + \xi, Y\sin\theta + \eta)\, E_p(\xi,\eta)\, d\xi\, d\eta. \tag{12.7}$$

The sine term in equation (12.7) is due to the geometrical transformation from the slant-range to range coordinate system. When targets are stationary, the optimum images can be obtained by applying the reference signal that is the complex conjugate of the return signal from a stationary single scatterer. For dynamic targets, however, the optimum images could sometimes be produced by introducing a defocus term in the azimuth component of the reference signal (Shuchman & Zelenka 1978, Jain 1978). Let this defocus parameter be v_p and write the reference signal as

$$E_p(\xi,\eta) = W_x(\xi)\, W_y(\eta)\, \exp\left(i\,\frac{k\,(1 + v_p/V)^2\,\xi^2}{R + y\sin\theta}\right)\exp\left(i\,\frac{2\alpha}{c^2}\,\eta^2\right) \tag{12.8}$$

where W_x and W_y are the amplitude weighting of the reference signal in the azimuth and range direction respectively.

If the azimuth extent of the physical antenna is rectangular, the amplitude weighting W_a of the illuminating beam is given by a sinc function. But this weighting can be made rectangular in a pre-processing stage (Report, System Designers Ltd 1979). We denote the effective extent of the weighting by L_a and the range extent of the weighting W_r by $L_r = c\tau_0$ for a rectangular pulse envelope of duration τ_0. We also define the amplitude weightings of the reference signal as

$$W_x(\xi) = \quad 1 \qquad |\xi| \leqslant L_x/2$$
$$0 \qquad \text{otherwise}$$

$$W_y(\eta) = \quad 1 \qquad |\eta| \leqslant L_y/2$$
$$0 \qquad \text{otherwise} \tag{12.9}$$

where L_x and L_y are the azimuth and range extent of the reference signal respectively. In general applications of SAR processing, L_x and L_y are smaller or equal to L_a and L_r respectively, i.e., the resolution is limited by the extent of the reference signal of the processor. Then the integrals with respect to ξ and η in equation (12.7) are determined only by W_x and W_y. Thus, from equations 12.5–8, and assuming

$$(v_p/V)^2 \ll 1 \tag{12.10}$$

the image complex amplitude is given by

$$A(X,Y) = E_0 \iint_{-\infty}^{+\infty} dx \, dy \iint_{-\infty}^{+\infty} d\xi \, d\eta \, E_1(x,y;X,Y) \, U(x,y;\xi+X) \, W_x(\xi) \, W_y(\eta)$$

$$\cdot \exp\left[i \frac{2k(v_p/V)\xi^2}{R + y \sin \theta}\right] \exp\left[-i \frac{2k(X-x)\xi}{R + y \sin \theta}\right] \cdot \exp\left[-i \frac{4\alpha}{c^2} \sin \theta \, (Y - y)\eta\right] \tag{12.11}$$

where

$$E_1(x,y;X,Y) = \exp(-i2ky \sin \theta) \exp\left[-i \frac{k(X-x)^2}{R + y \sin \theta} -i \frac{2\alpha \sin^2\theta \, (Y-y)^2}{c^2}\right]$$

and other constants have been absorbed in E_0.

12.3 IMAGING OF OCEAN WAVES

Let us consider that the backscattered complex amplitude U is given by

$$U(x,y;t) = u(x,y;t) \, U_r(x,y;t) \tag{12.12}$$

where u is the backscattered complex amplitude corresponding to large scale undulations that is deterministic, and u_r is the backscattered random complex amplitude due to small-scale capillary waves. The deterministic amplitude modulation could be a result of surface roughness and/or surface tilt, the former being dominant for azimuth waves (Elachi & Brown 1977). The random complex amplitude could be considered as the product of two uncorrelated terms u_s and u_t, and we write

$$U(x,y;t) = u(x,y;t) \, u_s(x,y) \, u_t(t/x,y) \tag{12.13}$$

in which u_s is the spatial variation of the random component of the backscattered complex amplitude and u_t represents the temporal change of this component at a position (x,y) on the surface (Ouchi 1981, Raney 1980).

Owing mainly to the presence of the random complex amplitude u_r, the resultant images are immersed in speckle, and no simple results for image complex amplitude can be obtained by substituting equation (12.13) into (12.11). Such image structure may best be described by the ensemble average of local image intensity. To do this we must take the spatial and temporal average of the backscattered complex amplitudes. For SAR imagery, the spatial random process is often modelled as a Gaussian white noise because a single resolution cell contains many ($\geqslant 8$) scatterers (capillary waves). Representing the spatial and temporal average by ensemble average and denoting it by $\langle\ \rangle$ we obtain

$$\langle U(x_1,y_1;t_1)U^*(x_2,y_2;t_2)\rangle = u(x_1,y_1;t_1)\,u^*(x_2,y_2;t_2)$$

$$.\,\delta\,(x_1{-}x_2,y_1{-}y_2)\,\Gamma\,(t_1{-}t_2) \tag{12.14}$$

where δ is a Dirac delta function arising from the white noise approximation, the asterisk denotes the complex conjugate, and Γ is a temporal correlation function which we assume to have a Gaussian form:

$$\Gamma(t_1 - t_2) = \exp\left[-(\frac{t_1 - t_2}{t_0})^2 \right] \tag{12.15}$$

with t_0 being a decorrelation time of the temporal random fluctuations. The validity of the present surface model has been shown (Ouchi, 1982b). We also assume that there are many temporal fluctuations within the integration time T required to construct a synthetic aperture. Thus we put

$$t_0/T \ll 1 \tag{12.16}$$

Substituting equation (13.14) and (13.15) into (13.11), integrating over x_1 and y_1, and putting $x = x_2$ and $y = y_2$, we have for the local mean image intensity

$$\langle I(X,Y)\rangle = |E_0|^2 \underset{-\infty}{\overset{+\infty}{\iint}} dx\ dy \underset{-\infty}{\overset{+\infty}{\iiiint}} d\xi_1\,d\eta_1\,d\xi_2\,d\eta_2\ W_x(\xi_1)\,W_x^*(\xi_2)\,W_y(\eta_1)\,W_y^*(\eta_2)$$

$$.\,u(x,y;\,\xi_1 + X)\,u^*(x,y;\,\xi_2 + X)\,\exp\left[-(\frac{\xi_1 - \xi_2}{Vt_0})^2 \right] \tag{12.17}$$

$$.\,\exp\left[i\frac{2k(v_\mathrm{p}/V)}{R + y\sin\theta}(\xi_1^2 - \xi_2^2) \right]\,\exp\left[-i\frac{2k(X - x)}{R + y\sin\theta}(\xi_1 - \xi_2) \right]$$

$$.\,\exp\left[-i\,(4\alpha/c^2)\sin\theta\,(Y - y)\,(\eta_1 - \eta_2) \right].$$

We now assume that the major contribution to the image of a monochromatic ocean wave is backscattered amplitude with the degree of modulation m, wavenumber k_0, and phase velocity v, which propagate at an angle φ from the azimuth axis as shown in Fig. 12.1. This modulation arises because there are more fully developed Bragg scatterers near the crests of the wave than the troughs. Then the backscattered deterministic amplitude can be written as

$$u(x,y,t) = 1 + m \cos [k_0(x \cos \varphi + y \sin \varphi - vt)]. \tag{12.18}$$

We note in passing that inclusion of deterministic waveheight in equation (12.18) would lead to complicated expressions for the local mean image intensity in which the contribution comes from both amplitude modulation and velocity bunching. Such cases have beeen discussed in detail for azimuth waves (Ouchi 1982b, c), in which the general conclusion is that the image contrast, at the stationary focus ($v_p = 0$) of the processor, is greater for waves propagating against the radar ($\varphi = 180°$) than those travelling with the radar ($\varphi = 0°$) and that the optimum focus position is the same as that when the main contribution is amplitude modulation. Similar results would be expected in the two-dimensional imagery.

Our final assumptions are

$$\exp [i \, \frac{Rk_0^2 \cos \varphi}{kV} (v + 2v_{p_1} \cos \varphi)] \cong 1$$

and

$$\exp \left\{ - \left[\frac{k_0 \cos \varphi}{kVt_0} \left(R + \frac{y_{max}}{2} \sin \theta \right) \right]^2 \right\} \cong \exp \left\{ - \left[\frac{Rk_0 \cos \varphi}{kVt_0} \right]^2 \right\}$$

$$\tag{12.19}$$

The first approximation in equation (14.19) is valid for general parameters of SAR and ocean waves, and the second approximation implies that the effect of temporal random motion of capillary waves at $y = 0$ is rougly the same as that at $y = y_{max}/2$, where y_{max} is the range extent of the ocean wave. As shown previously (Ouchi 1982, 1982c), Ouchi 1981, Raney 1980, Raney & Schuchman 1978, Raney 1981), the contrast of local mean image intensity depends on the slant-range distance $R + y \sin \theta$, but the y-dependence can be neglected if y_{max} and/or θ are sufficiently small. If this condition is not satisfied, the image would be degraded with increasing y, and the evaluation of image structure becomes very complicated.

The local mean intensity of the ocean wave can be obtained by substituting equation (12.18) into (13.17) and using the condition given by equation (12.16) and (12.19). We then obtain

$$\langle I(X,Y) \rangle = 1 + m^2/2 + 2m \, \gamma_1 \, C_1 \, D_1 \cos (k_x X + k_y Y)$$

$$+ (m^2/2) \, \gamma_2 \, D_2 \cos [2(k_x X + k_y Y)] \tag{12.20}$$

where

$$k_x = k_0 \left(\cos \varphi - v/V \right), k_y = k_0 \sin \varphi$$

$$\gamma_n = \exp \left[- \left(nRk_0 \cos \varphi /2kVt_0 \right)^2 \right]$$

$$C_n = \cos(nRk_0^2 v \cos \varphi /4kV) \qquad\qquad (12.21)$$

$$D_n = \operatorname{sinc} \left[n\pi \left(T/T_0 \right) (1 + (2v_p/v) \cos \varphi) \right]$$

$$T = L_x/V \text{ and } T_0 = \lambda_0/v.$$

In the above formulae the normalising factor E_0 has been dropped, T and T_0 are the SAR integration time and the ocean wave period respectively, and λ_0 is the ocean wavelength.

12.4 DEFOCUSING EFFECTS

Equations (12.20) and (12.21) constitute a general expression for the local mean intensity of SAR images when an ocean wave introduces amplitude modulation to backscattered complex amplitude. We now discuss the main consequences of our result.

As mentioned by several workers (Valenzuela 1980, Ouchi 1982b, Raney 1978), the wavenumber of the SAR images of ocean waves differs from the true value if the radar platform velocity is comparable with the wave phase velocity. This is seen in the azimuth wavenumber k_x of equation (12.21). The φ-dependence on k_x shows that the wavenumber distortion also occurs for range waves ($\tilde\varphi = 90°$). For azimuth waves where $\varphi = 0°$ and $180°$, we recover the previous result (Ouchi 1982b).

The term γ_n arises owing to temporal random fluctuations of the return signals. As noted earlier, the value of γ_n, and hence the image contrast, decrease with increasing slant-range distance. It also decreases as the azimuth component of ocean wavelength and/or the decorrelation time t_0 of temporal fluctuations decrease (Ouchi 1982c). The n-dependence on γ_n (γ_n decreases with increasing n) indicates that the images of sharp-crested backscattered amplitude are smoothed: this is also true for the images produced by the velocity bunching effect.

The term C_n is a result of interference between two backscattered amplitudes at time t_1 and t_2. This term takes a value of unity for general SAR and wave parameters and can be ignored.

The term D_n represents defocusing, which is our main interest. Let us first consider cases where the focusing parameter v_p of the reference signal is set to zero, i.e., stationary focus. D_n is then given by

$$D_n = \operatorname{sinc}(n\pi T/T_0). \qquad\qquad (12.22)$$

This expression is independent of the radar wavelength, SAR geometry, and the direction of ocean wave propagation. The amount of defocusing is thus a function of the ratio of the SAR integration time T to the wave period T_0. It also depends on the shape of the backscattered amplitude modulation because of the n-dependence on D_n: the amount of defocusing is large for sharp-creasted waves (Ouchi 1982a). As in equations (12.20) and (12.22), the image contrast decreases with increasing T/T_0 ratio at the stationary focus, and the images completely disappear when $T = T_0$. In the presence of speckle noises, however, they may become invisible even when $T > T_0$. The disappearance of images can be understood by considering the point spread function $P(X, Y)$ from a local scatterer introducing a sinusoidal backscattered temporal amplitude change with random fluctuations at a position $x = x_0$. For simplicity we consider the point spread function in the azimuth direction only. Let the backscattered complex amplitude from the scatterer be

$$U(x;t) = (1 + m \cos [k_0(x - vt)]) u_t(t) \delta(x - x_0). \qquad (12.23)$$

Because this signal contains temporal random fluctuations represented by u_t, the resultant point spread function is a speckle pattern (Ouchi 1981, Raney 1980), and no simple expression can be obtained for the detailed speckle pattern. However, we can deduce the mean intensity of the point spread function that is a speckle envelope. We refer to this envelope as the effective point spread function. Substituting equation (12.23) into (12.11), putting $v_p = 0$, and assuming

$$t_0/T_0 \ll 1 \qquad (12.24)$$

the effective point spread function is

$$\langle |P(X)|^2 \rangle = M \exp \left\{ -\left[\frac{kVt_0}{R} (X - x_0) \right]^2 \right\} \qquad (12.25)$$

where

$$M = |E_0|^2 \{ 1 + m^2/2 + 2m \ \text{sinc} \ (\pi T/T_0) \cos [k_0(x_0 - vX/V)]$$
$$+ (m^2/2) \ \text{sinc} \ (2\pi T/T_0) \cos [2k_0(x_0 - vX/V)] \} \qquad (12.26)$$

The mean image intensity of the ocean wave is the incoherent addition of equation (12.25) from each x_0. It can easily be seen that the image contrast is maximum if the effective point spread function from a crest of the wave ($x_0 = n\lambda_0$) is optimum and there is no point spread function from a trough ($x_0 = (n + \frac{1}{2})\lambda_0$). Such cases can be realised in stationary waves, i.e., $T/T_0 = 0$, as shown in Fig. 12.2a and 12.2b where the effective point spread function from (a) wave crest and (b) wave trough are illustrated for different T/T_0 ratio with $m = 1$. With

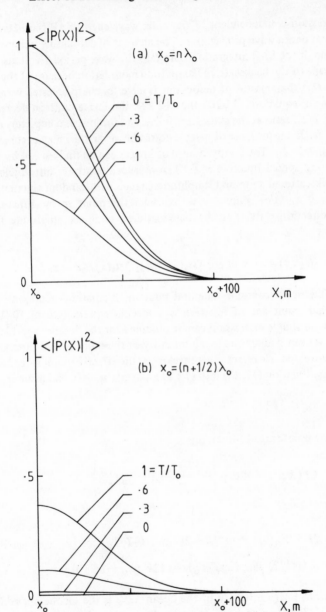

Fig. 12.2 – Effective point spread function $\langle |P(X)|^2 \rangle$ from a scatterer at (a) crest and (b) trough of a sinusoidal wave. The effective point spread function from the crest is degraded with increasing T/T_0 ratio; while that from the trough is upgraded, resulting in a decrease in contrast, where T and T_0 are the SAR integration time and wave period. When $T=T_0$ they become identical and no image structure can be observed.

increasing ratio the effective point spread function from the crest is degraded, while that from the trough is upgraded and the consequence is a decrease in the image contrast. In the limit $T = T_0$, the effective point spread function from any scatterer on the surface becomes identical since in this limit M, given by equation (12.26), is independent of x_0. The resultant mean intensity of the image is constant with no observable image structure. Thus defocusing of the images of ocean waves is due to this systematic degradation and enhancement of the point spread function, resulting from the periodic spatial and temporal change of the return signal. Since we have not included the temporal wave height variation, defocusing cannot be due to the acceleration in the slant-range direction; nor can it be because of the change in the relative radar platform velocity as D_n is independent of radar wavelength.

Defocusing of the images of ocean waves can be compensated by introducing the focal adjustment parameter v_p to the reference signal as seen in the expression D_n of equation (12.21). The optimum images can be produced if

$$v_p = \frac{v}{(2\cos\varphi)} \tag{12.27}$$

To illustrate the effect we define the image contrast as

$$\text{Contrast} = \frac{\langle I \rangle_{\max} - \langle I \rangle_{\min}}{\langle I \rangle_{\max} + \langle I \rangle_{\min}} \tag{12.28}$$

where $\langle I \rangle_{\max}$ and $\langle I \rangle_{\min}$ are the maximum and minimum local mean intensity of the image respectively. Examples are shown in Fig. 12.3 where the image contrast is plotted as a function of v_p for various direction of wave propagation.

For azimuth waves the optimum images can be obtained when $v_p = \pm v/2$, where the positive and negative sign indicate waves travelling against and with the radar respectively: this is in agreement with the previous result (Ouchi 1982b). As the direction of wave propagation deviates from the aximuth axis, v_p for optimum images increases, and when $\varphi = 60°$ the best images can be produced by setting $v_p = -v$. The focal setting v_p increases further as $\varphi \rightarrow 90°$, and simultaneously the contrast curves tend to become constant. When $\varphi = 90°$ (range waves) the optimum images can be found at $v_p = -\infty$ and, of course, the images cannot be optimised by adjusting v_p. Thus for range waves the defocus term D_n is given by equation (12.22) and the images are always defocused.

The present theory seems to be in good agreement with the experimental observations (Shuchman & Zelenka 1978, Jain 1978) of which the main results indicate that the effect of defocusing is small for (a) range waves, (b) long waves, and (c) short radar wavelengths. We have already explained the observation (a). In addition, there seems to be a tendency that the optimisation of images is significant for waves with $\varphi \backsim \pm 60°$ at $v_p \backsim \pm v$ (Shuchman & Zelenka 1978).

This is in agreement with our result. Defocusing is small for long waves because their long periods (Lighthill 1978) reduce the T/T_o ratio. As to the observation (c), the defocus term D_n is independent of the radar wavelength. But to obtain the same resolution the SAR integration time T becomes shorter for shorter radar wavelength; this in turn reduces the T/T_o ratio.

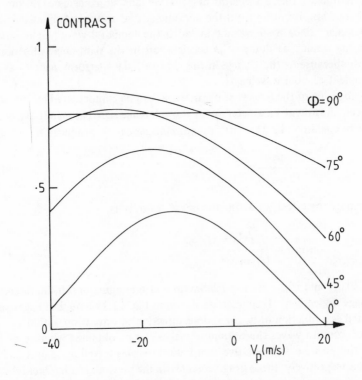

Fig. 12.3 – Image contrast as a function of the focal adjustment parameter v_p for different values of wave propagation direction φ. The SAR parameters are taken for SEASAT as $L = 16$ km, $R = 850$ km, $T = 2.3$ s, $V = 7$ km/s, and $\lambda = 0.235$ m. The parameters of the ocean wave are taken as $m = 0.6$, $\lambda_o = 150$ m, $v = 25$ m/s, and $T_o = 6$ s.

12.5 SUMMARY

We have examined the effect of defocusing on the images of ocean waves propagating in an arbitrary direction, by taking account of the spatial and temporal dependence of small-scale and large-scale ocean waves. Our main results are

(1) defocusing is caused by the systematic degradation and enhancement of the point spread function resulting from the periodic spatial and temporal change of the return signal,

(2) if no focal adjustment is applied to the reference signal, the images of ocean waves are always defocused, irrespective of their direction of propagation,

(3) the amount of defocusing at the stationary focus is a function of the SAR integration time and wave period, and

(4) the optimum images can be produced by adjusting the focal parameter by an amount $v_p = -v/(2 \cos \varphi.)$

REFERENCES

Alpers, W. & Rufenach, C. L. (1979) Image contrast enhancement by applying focus adjustment in synthetic aperture radar imagery of moving ocean waves *Proc. 2nd SEASAT-SAR Workshop,* Frascati (Italy) pp, 25–31.

Elachi, C. & Brown, W. E. (1977) Models of radar imaging of the ocean surface waves, *IEEE Trans. Antennas Propagat.,* **AP-25** 84–95.

Lighthill, J. (1978) *Waves in fluids,* Cambridge University Press, Cambridge.

Jain, A. (1978) Focusing effects in the synthetic aperture radar imaging of ocean waves, *Apl. Phys.,* **15** 323–333.

Ouchi, K. (1982a) Defocus dependence on ocean wave shape in synthetic aperture radar imagery *Opt. Quant. Electron.* (in press).

Ouchi, K. (1982b) Imagery of ocean waves by synthetic aperture radar, (*Apl. Phys.* **B-29** pp. 1–11.

Ouchi, K. (1981) Statistics of speckle in synthetic aperture radar imagery from targets in random motion, *Opt. Quant. Electron.,* **13** 165–173.

Ouchi, K. (1982c) Effect of random motion on synthetic aperture radar imagery, *Opt. Quant. Electron.* **14** pp 263–275.

Raney, R. K. (1978) The occurrence and correction of wave spectra distortions through scanning sensors, *Proc 5th Canadian Symp. Remote Sensing, Victoria,* 356–362.

Raney, R. K. & Shuchman, R. A. (1978) SAR mechanisms for imaging waves, *Proc. 5th Canadian Symp. Remote Sensing, Victoria,* 495–505.

Raney, R. K. (1980) SAR processing of partially coherent phenomena, *IEEE Trans. Antennas Propagat,* **AP-28** 777–787.

Raney, R. K. (1981) Wave orbital velocity, Fade and SAR response to azimuth waves, *IEEE Ocean Engneer.,* **OE-6** 140–146.

Rufenach, C. L. & Alpers, W. R. (1981) Imaging ocean waves by synthetic aperture radar with long integration times, *IEEE Trans. Antennas Propagat,* **AP-29** 422–428.

Shuchman, R. A. & Zelenka, (1978) J. S. Processing of ocean wave data from a synthetic aperture radar, *Boundary-Layer Meteorol,* **13** 181–191.

System Designers Ltd., UK, (1979) *Report on ESPF Study and Syst. Impl. Manual.*
Valenzuela, G. R. (1980) An asymptotic formulation for SAR images of the dynamic ocean surface, *Radio Science*, **15** 105-114.

Chapter 13

The Canadian SAR experience

R. K. RANEY

Radarsat Project Office Suite 200, 110 O'Connor St. Ottawa, Ontario. KIP 5M9 Canada

13.1 INTRODUCTION

Since 1976 a serious effort has been made in Canada to concentrate on synthetic aperture radar (SAR) as the primary sensor to provide information over the frozen and open oceans bordering Canada, and in support of information needs in forestry, geology, hydrology, and agriculture. During the period 1977 to 1979 through its SURSAT project, Canada participated in the NASA SEASAT program. Despite the early failure of SEASAT, 35 SAR orbits were recorded at Shoe Cove and 80 SAR orbits obtained over Western Canada from USA recording stations. The project supported the development of the MDA digital SAR processor, which continues to set the standard of excellence for SEASAT SAR imagery world-wide.

In the same timeframe, CCRS acquired the ERIM multi-channel SAR, modi-fied it, and installed it in their Convair 580 aircraft. The aircraft flew over 530 hours in fulfilling the data acqustion needs for approximately 100 experiments. The majority of the aircraft SAR data were optically processed under contract to ERIM, of Ann Arbor, Michigan.

The experience of both aircraft and spacecraft SAR in serving Canadian needs as demonstrated in the SEASAT and SURSAT projects was sufficiently persuasive that both the airborne and the national satellite SAR interests of Canada are being extended. The SAR 580 was fitted with a third frequency — C-band ($\lambda = 5$ cm) — in 1981, and there has been a commitment to Phase A of the RADARSAT program (1981–1990), which is outlined below. Of course these national activities are in parallel with, and complementary to, the participation of Canada in the ERS-1 microwave remote sensing satellite program.

This chapter is intended to give a brief description of this experience, touching on highlights of historical or technical interest. These fall naturally in the areas of the SEASAT satellite and key applications results in Canada, development of digital SAR data processing, and an outlook for the future.

13.2 SEASAT, SURSAT, AND APPLICATIONS

Canada has diverse requirements for surveillance of human activities and environ-mental phenomena in ocean and remote areas. Such requirements include location and identification of ocean traffic, data on the type and extent of ice coverage, information for preparation of weather and sea-state reports and forecasts, and location and identification of ocean pollution. The increase in exploration and development activities offshore and in the Arctic and the extension of jurisdictional limits to 370 kilometers offshore created an increased need for surveillance data. The Canadian government thus initiated an interdepartmental program in April 1977 to explore the extent to which satellites might contribute to these needs for information. This program, the Surveillance Satellite (SURSAT) Program, served as the basis for Canadian participation in the US SEASAT Project.

In general, surveillance data is required on a regular, frequent basis, and consequently for the Canadian North only sensors which can penetrate cloud and operate in both darkness and daylight are applicable. The SURSAT Program thus considered only microwave sensors, and gave particular emphasis to synthetic aperture radar (SAR) since this is the only microwave sensor capable of providing high-resolution imagery from space.

The major effort of the SURSAT Program was directed toward performing a set of experiments to assess the feasibility of satisfying surveillance requirements via satellite and gaining experience in surveillance satellite technology. Principal activities were participation in the NASA SEASAT satellite experiment and a complementary research and development program based upon the 'SAR-580' facility − a Convair 580 aircraft equipped with a four channel X- and L-band synthetic aperture radar.

13.2.1 SEASAT data reception and processing

The NASA SEASAT satellite was launched on 26 June 1978, and functioned until 10 October 1978, when a short-circuit in the satellite power system prematurely terminated its operation. The satellite carried four microwave sensors including a SAR.

Through the SURSAT Program, Canada proceeded to prepare for reception and processing of SEASAT SAR data, and hence to distribute it to a network of users. The Canada Centre for Remote Sensing (CCRS) undertook modification of Shoe Cove Satellite Station near St John's Newfoundland, for reception and recording of SEASAT data. (This work was largely carried out through contracts with Nordco and Applied Physics Laboratories.) The National Research Council of Canada, with support of the departments of Supply and Services, and Energy Mines and Resources, contracted with MacDonald, Dettwiler and Associates for development of software for digital conversion of the raw SAR data recorded at Shoe Cove to images.

In addition, CCRS, with project financial support, contracted with Intera Environmental Consultants Ltd for lease of the ERIM four-channel X- and L-band SAR (which was subsequently purchased by CCRS); installation of the radar in the CCRS Convair 580 aircraft; collection of radar data; and support of experimenters and data analysis. The radar was modified to provide digital recording of two of the four channels of data and digitally process one channel of the data in real time.

13.2.2 Results of experiments

Results of SURSAT experiments (as well as other aspects of the Program) are described in Discussion Paper DOC-6-79DP. Jan. 1980; Report of the ITFSS, August 1977; Experiment Plan Part 1, October 1978; Final Report of the Airborne SAR Project, March 1980; SURSAR Program Part I, Sept. 1980, and Van Koughnett *et al.* 1980. The following principal results were obtained.

Experiments dealing with surveillance of human activity were largely concerned with detection of ships on the ocean using SEASAT and airborne SAR data, detection of ocean pollution, detection of selected human activities on land and ice, and applications of SAR data to search and rescue tasks. For activities that involve objects on the surface, such as seismic lines or fishing vessels, experiments suggested that finer resolution, shallower incidence angle, and a shorter wavelength were required, as compared to the 25 metres resolution, 20° incidence, and 23 centimetres wavelength of SEASAT. Detection of oil spills was verified using radar, but subtle distinctions of petroleum slicks from natural phenomena resembling oil slicks required further work.

Land experiments spanned many fields, including cartography, agriculture geology, forestry, wetlands, renewable resources, and hydrography. Selected highlights of this work are mentioned here. Both SEASAT and aircraft SAR data were used.

Work on the cartographic potential of SAR imagery was very encouraging. Normally not considered an accurate mapping tool, SAR, as flown on SEASAT and as processed by the MDA digital method, produces imagery that meets high standards for cartographic accuracy (discussed below). Optically processed SAR data and aircraft data do not yield such good results.

Agricultural experiments were based on six Canadian test sites. Using only manual interpretation, classification levels for major crop groups were consistently about 90% — approaching the performance of LANDSAT systems.

SAR proved to be of excellent value in various wetland experiments, including waterfowl breeding at Humboldt, Saskatchewan and an ecological test site near Neepawa, Manitoba.

For flood mapping, it was found from the Red River near Winnipeg that SAR imagery was superior to air photography in several regards. In open areas the boundary of the flood was easy to identify by its consistent tone contrast, whereas the aerial photography had varying tonal contrast due to varying sun angle. In wooded areas the tone variation in the SAR image was still sufficient to delineate the flood. Aerial photographs overestimated the flooded area because wet soil has a similar appearance to standing water.

One of the better-known land applications of radar is in surficial and structural geology. Project results in this area tended to agree with known work in the field. One of the more interesting results was the discovery in Nova Scotia of evidence of volcanic activity through structural features not previously known.

Ice and ice-related applications are the most important information needs to be addressed by a SAR system for Canada. The project grouped nineteen ice-related proposals into three major experiment areas, located in the Beaufort Sea, the Baffin-Labrador Seas, and the Gulf of St Lawrence. Many individuals and organisations contributed to this major effort.

Data sets available to the project included 30 SEASAT passes over the western Arctic between mid-July and early October of 1978, and airborne X-band SLAR missions in the Beaufort Sea during late August and in Viscount Melville Sound in September. Analysis of SEASAT SAR data was therefore concentrated on passes closely associated, geographically and temporally, with the airborne SLAR data and related visual observations. General conclusions of the analysis were as follows:

- Under summer melt conditions, SEASAT L-band imagery of sea ice is relatively featureless, while airborne X-band SLAR imagery over identical ice fields shows topographical features clearly. The effect of the low SAR incidence angle combined with presence of surface melt water is presumed to be responsible for this difference.

- During freeze-up conditions, information available from SEASAT SAR is much improved. Identifiable ice types included old, first-year, and young ice. Smooth floes of second year, often not identifiable on SLAR because of the high incidence angles involved, also showed a unique signature. Newly formed ice showed on open-water areas as regions of no sea clutter.

- Open-water areas provided strong sea clutter returns, often more so than adjacent ice surfaces, resulting in a reverse image from that normally observed in SLAR. In the vicinity of ice floes, open water often showed a margin of low sea clutter along one side of the floes. This effect was confirmed in several cases to be associated with wind directions.

From the examination of other optically processed data, it is quite evident that the dynamic range of first-year and multi-year ice equals the dynamic range of new ice. This makes it virtually impossible to classify sea ice at L-band frequencies based on backscatter alone. Using shape and texture however, does, permit the interpretation of a large number of ice types and features.

12.2.3 Conclusions of the SURSAT/SEASAT Program

The program has permitted the following conclusions to be drawn:

a) The technical feasibility of a spaceborne synthetic aperture radar and ground reception and processing system capable of providing high-resolution images from space was demonstrated.

b) Spaceborne SAR data can significantly assist meeting Canada's needs for surveillance information by providing information on ice coverage, type, and drift.
c) Digital techniques for processing SAR data offer significantly higher quality data and processing flexibility than optical techniques.
d) Sufficient promise was shown by SAR to meet Canadian needs that major investment in this technology is justified.
The consequence of these conclusions is described in section 13.4.

13.3 DIGITAL PROCESSING

From the standpoint of signal processing, SAR signals need only to be correlated in range and azimuth in order to form an image, a procedure representable by a single equation in two dimensions. However, the amount of calculation required to accomplish this using digital techniques is far from trivial. SEASAT, for example, required approximately 400 000 data points to be processed in the signal domain for each output image data point. Therefore, practical use of digital techniques for SAR has been paced by the development of large memories, fast input/output devices, and processing hardware speed. In Canada, where there was not a substantial investment in optical data processing hardware, SEASAT served as a real challenge and motivation for digital SAR processing.

The world's first digitally processed radar satellite image (of Trois Rivières, Quebec (Cumming & Bennett 1979) was produced by MacDonald, Dettwiler and Associates of Richmond, British Columbia, in November 1978.

13.3.1 Trois Rivières bridge
The Trois Rivières image (Fig. 13.1) contains a suspension bridge that was used to verify the radar's range resolution and differential range distance measurement precision.

The La Voiliette bridge over the St Lawrence River at Trois Rivières, Quebec, is a rigid inverse suspension bridge (Fig. 13.2). The bridge image consists of three distinct reflectivity loci, shown in Fig. 13.3. After some thought, one can rationalise the occurence of these three signals as follows:
(1) Referring to Fig. 13.4 and the preceding figures, the first signal component corresponds to energy directly reflected from the bridge structure back to the radar, hence the arch and the roadway curvature are visible in the image.
(2) The second signal component is the sum of all of the energy reflected from the side of the bridge structure facing the radar and the river surface. This path is equivalent in radar propagation delay to the point on the river surface that is in the plane of the bridge side, hence this locus is linear, as it should be.
(3) The third component is intriguing − it corresponds to the bottom of the bridge as seen in (microwave) reflection at the river surface. This locus shows the curvature of the suspended roadway, as it should.

Fig. 13.1 – Detail of the first SEASAT digital SAR image of Trois Revieres, Ouebec.

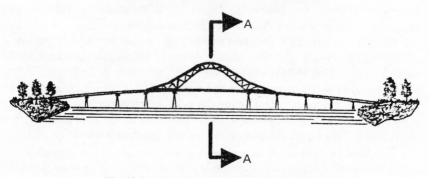

Fig. 13.2 – Sketch of the Trois Rivieres bridge

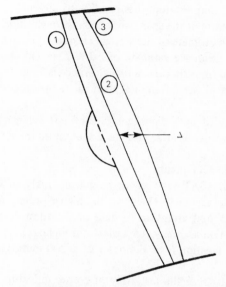

Fig. 13.3 – SEASAT radar image loci of Trois Rivieres bridge

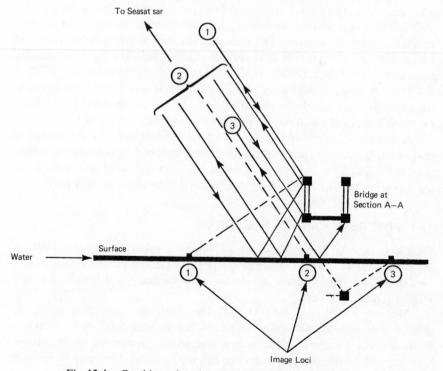

Fig. 13.4 – Graphic explanation of source of image constituents

Having made this interpretation, it is possible to measure the separation of traces 2 and 3 at the centre of the span, work back through the governing trigonometry, and arrive at an estimate of the vertical clearance beneath the bridge.

This process yields a clearance estimate of 50 ± 7 m. On checking with the Seaway Authority, one finds that the actual clearance is 55 metres. Hence, there is excellent verification of the differential slant range measurement precision of SEASAT.

(This work was aided by a digital image subframe enlargement and scaled interactive display capability at MDA, for the use of which the author is grateful).

13.3.2 SEASAT image data quality

Digitially processed SEASAT test scenes have been analysed with respect to spatial and radiometric accuracy. Owing to the higher order slant range/gound range conversion used, and attention to data organisation detail in the MDA approach, relative position accuracy is ± 2 pixels throughout a typical 40 × 50 km scene. Absolute scene position is of the order of ± 200 metres relative to Earth coordinates.

Based on the analysis of the test array of corner reflectors near the Goldstone receiving station in California, the combined performance of the SEASAT SAR and a variety of SAR data processing systems were evaluated (Goldfinger 1980). The major variable of interest was point object response, in terms of resolution, sidelobe structure, and linearity. The optical products from JPL and ERIM were used (from two different processings each), the digitised optically processed method from ERIM and JPL, and the all-digital techniques of JPL and and MDA. Principal results show that of all methods tested, the MDA example was superior in all regards. Second, the SEASAT SAR was shown to have met or exceeded design goals with respect to resolution in range and azimuth.

A secondary aspect of the SEASAT SAR performance of some interest is that example data were processed at the normal level of 5 bits of signal quantisation, and reduced levels down to only one bit. Although some additional noise was present in the one bit example, the imagery remained remarkably stable.

13.4 FUTURE OF CANADIAN PROGRAM

Canada is moving firmly ahead on its domestic airborne SAR development program, a domestic SAR satellite project, and cooperative international activities such as ERS-1. These are all seen to be complementary, and beneficial to the long-term needs in Canada.

The major project is RADARSAT, for which there are three elements: mission requirements definition supported by aircraft SAR experiments; a Phase A concept study which analyses spacecraft and ground system alternatives that respond to the user requirements; and research and development of radar technology.

The objectives of the RADARSAT Project are:

● To perform the Phase A technical and economic studies necessary to define a radar satellite implementation program that would provide a limited operational capability to supply timely ice information for selected Arctic or east coast operations, and provide research data and operational data to meet selected land and ocean requirements. The satellite payload consists of an imaging radar plus one or more of the following secondary sensors: scatterometer, optical imager, scanning microwave radiometer, and altimeter.

● To develop Canadian industrial expertise in spaceborne radar technology through a research and development program, so that a radar satellite including radar systems and major subsystems in space and ground processing can be built in Canada.

● To develop program options for the satellite radar and secondary sensors, including international cooperation with other space agencies or companies, and cost-sharing arrragements with domestic and foreign data users.

The overall concept for the Canadian radar satellite (RADARSAT) is shown in Fig. 13.5. It is an end-to-end system that starts with a satellite in an inclined

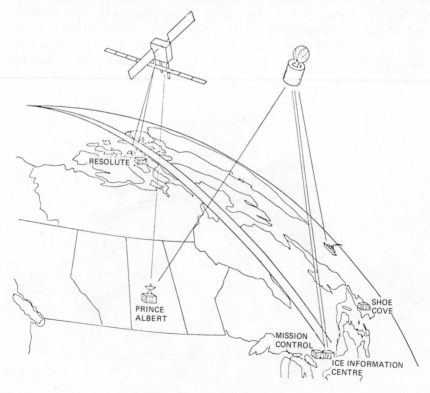

Fig. 13.5 – RADARSAT concept.

polar orbit transmitting signals from its wide swath SAR to ground receiving stations, where the data are converted into digital images. The image data are relayed via communications satellite (ANIK) to an ice-information centre, where other data such as from aircraft SAR or weather satellite are incorporated to provide a forecast of ice conditions. The forecast is again relayed by communications satellite to the tankers and other users in the form of annotated images.

The spacecraft will consist of a 3-axis stabilised platform with sufficient power and weight capacity to carry a C-band 6 SAR and complementary microwave or optical sensor. In addition a transponder for search and rescue mission might be carried. Preliminary studies of existing spacecraft buses have shown that there are several which could be adapted for this mission. The spacecraft would be designed for a 3-5 year life, with further spacecraft launched depending on the user needs and successes of the program. Three or four satellites would be needed to provide an operational service from 1990–2000. The baseline parameters for the radar payload are a swath of 150 km or greater, with a resolution of 25 to 30 metres at 4 looks and an incidence angle of $30°$ to $45°$. Preliminary orbit studies have shown that with this swath width it is possible to obtain twice-daily coverage over the primary Arctic transportation corridor, the North-West Passage. Fig. 13.6 shows coverage obtained from the preferred orbit. Coverage options may be extended through the use of a variable incidence angle antenna. Current design includes swath steering over an incidence angle range of $20°–45°$, to enable coverage of 150 km anywhere within a 500 km accessibility swath.

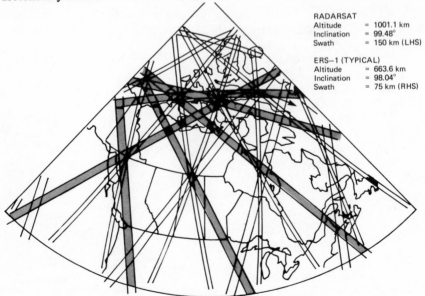

RADARSAT
Altitude   = 1001.1 km
Inclination = 99.48°
Swath   = 150 km (LHS)

ERS–1 (TYPICAL)
Altitude   = 663.6 km
Inclination = 98.04°
Swath   = 75 km (RHS)

Fig. 13.6 – One-day coverage from RADARSAT.

Direct telemetry will be sent to three Canadian ground stations. They will be equipped with hardware SAR processors, with sufficient throughput to maintain zero backlog for the areas requiring near real-time response plus a reserve capacity to deal with back-up load-sharing between station. The ice-information centre will house image analysis facilities that will enhance ice features in the images, and add annotation to show the location of ice-fields, the wind vectors and other information obtained from aircraft data. The total data handling time between acquisition of satellite data to delivery to ships must occur in 2 to 4 hours. A mission control facility will be used to monitor spacecraft performance, command orbit adjustments and schedule sensor coverage.

In order to develop Canadian industrial expertise in spaceborne SAR a substantial development program has been initiated. Selected SAR components, such as the antenna configuration and coherent microwave system technology, are being developed to establish confidence in building the complete flight SAR. An advanced digital aircraft SAR is being developed to provide SAR systems expertise, for use in the satellite support role, and for direct field use by user industries.

The current project is planning for a RADARSAT launch in 1990 with an estimated cost of the order of $300 M. This is an ambitious program and unlikely to reach fruition unless there are offsetting contributions from industry and other international initiatives. The project is being discussed with ESA, NASA, other nations, and the oil and gas industry in Canada to negotiate satisfatory cost-sharing. Early indications are promising. The results of these negotiations will be incorporated in the phase A documentation late in 1982, which will form the basis of Phase B development and final project definition in 1983 and following years.

13.5 CONCLUSIONS

Canada's interest in SAR has been increasing since 1976, and is now substantial, evidenced by national commitment to Phase A of a SAR satellite project in addition to domestic airborne and international SAR activites. The SEASAT experience was very positive, establishing a solid base of processing technology and informed users in Canada. There is a well-recognised need for SAR information products, particularly in the Arctic, where reliable and routine weatherproof surveillance is essential.

REFERENCES

Cumming, I. G. & Bennett, J. R., Digital processing of SEASAT SAR data *Proc. Int'l Conference on Acoustics, Speech, and Signal Processing* Washington, D.C., 2–4 April 1979.

Discussion Paper DOC-6-79DP, *The Canadian space program,* Five-Year Plan (80/81-84/85), January 1980, Government of Canada, Dept. of Communications.

Experiment Plan Part 1, *Surveillance satellite project,* SURSAT Project Office, October 1978.

Final report of the airborne SAR project, Intera Environment Consultants Ltd, Ottawa, Ontario, March 1980.

Goldfinger, A., *SEASAT SAR processor signatures: point targets,* Applied Physics Laboratories, John Hopkins Rd, Laurel, Maryland, April 1980.

Report of the interdepartmental task force on surveillance satellites, CCRS, Government of Canada, August 1977.

SURSAT Program Part 1: Executive Summary, Energy. Mines & Resources, September 1980.

Van Koughnett, A. L., Raney, R. K. & Langham, E. J., *The surveillance Satellite program and the future of microwave remote sensing,* Sixth Canadian Symposium on the Remote Sensing of the Environment, Halifax, Nova Scotia, May 1980.

Chapter 14

The use of SEASAT-SAR data in oceanography at the IFP

A. WADSWORTH, C. ROBERTSON and D. De STAERKE

Institut Français du Pétrole, 1 à 4 Avenue de Bois Prèau, 92506 Rueil Malmaison, France

14.1 INTRODUCTION

Since 1966 the Institut Français du Pétrole has been involved in different topics on remote sensing in meteo-oceanography. The first aim was to extract directional spectra from aerial photos. Then, in 1973, we began developing work on remote sensing of hydrocarbon pollution at sea. It soon became obvious that for those applications future operational sensors would be in the microwave range, and that further investigations of their performance had to be carried out. The creation of SURGE, and the launch of SEASAT, provided our Institute with useful tools for increasing research into imaging radar, particularly regarding its applications to the petroleum industry.

Numerous data (41 survey processed sub-scenes, 90 precision processed digital images, and 17 CCTs) have been used and proved to be useful in a variety of fields.

14.2 SEA-STATE MEASUREMENT

A major benefit from operational ocean-surveillance satellites will be the routine monitoring of wave parameters (direction, wavelengths) and wind information (speed and direction) from the SAR images on a daily basis.

SEASAT provided a unique opportunity to study the capacity of spaceborne synthetic aperture radar to measure such parameters. Unfortunately, SEASAT's mission was brought to a premature end, and it was not possible to carry out the measurements we intended from Noordwijk tower, simultaneously with the satellite passes. So, our research had to be divided into three parts:

- A study of a set of SEASAT SAR images over the North Sea to determine the ability of this sensor to observe ocean surface waves. This data set was supplemented by surface observations (waves and wind) from oil platforms (Frigg, MCPO1. . .) and synoptic meteorological charts. Waves with amplitude of 1 to 3 m travelling in directions other than azimuthal direction have been seen. The wavelengths of the waves observed on the images obtained from Fourier transforms (optical and digital) are generally about 20 per cent longer than waves observed on the surface. An example of optical and digital Fourier transforms is shown in Fig. 14.2 for a part of the image of Fig. 14.1 from orbit 1149 over the Shetland Islands. Unfortunately the lack of measurement of wind-generated ocean wave spectra prevented more accurate estimates.

Fig. 14.1 – Digitally processed image, part of orbit 1149.

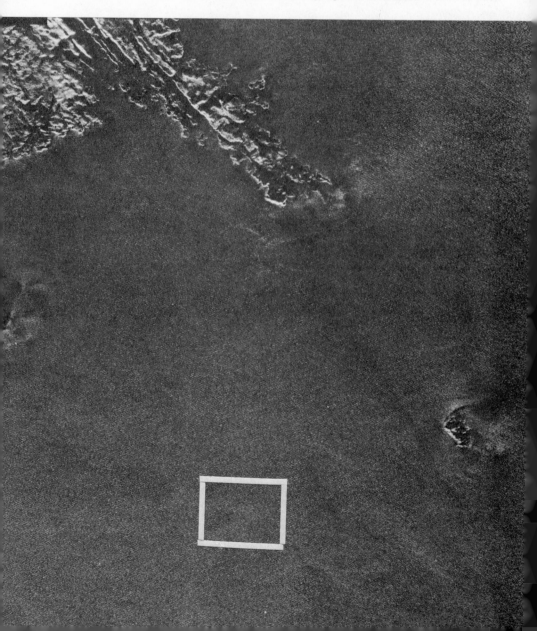

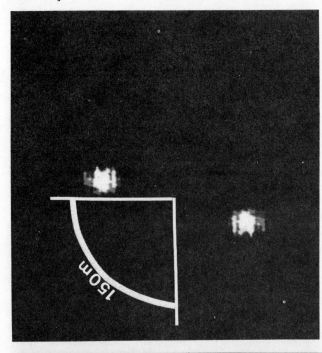

Fig. 14.2 – Fourier transforms. (Image 1149). a) Optically processed. b) Digitally processed.

● Another study was made with the data sets gathered from the Noordwijk tower with the scatterometer RAMSES. One of the objectives was to study the modulation of short waves by long waves and to calculate the modulation function for a large range of wave heights and wind speeds as a function of several radar parameters. The other objective was to measure ocean surface wind. These studies showed mainly that (i) the modulation transfer function cannot be described by a linear relationship for crosswind measurements, and (ii) that there is a difference of about 10 dB between the backscattering coefficient taken with ground based and with airborne (space) equipments. We cannot explain this difference on the basis of calibration error, since our equipment, the Dutch TNO, and Kansas University gave comparable results.

● Another interesting point is that the grey-level statistics of a number of images have been investigated as part of a study of the modulation transfer function and its application to SAR images. Results obtained with images from orbit 834 and 1149 indicate that it may be possible to detect the presence of waves on the image by using the shape of the grey-level histogram (Fig. 14.3). This could be significant in the future for the routine processing of SAR images.

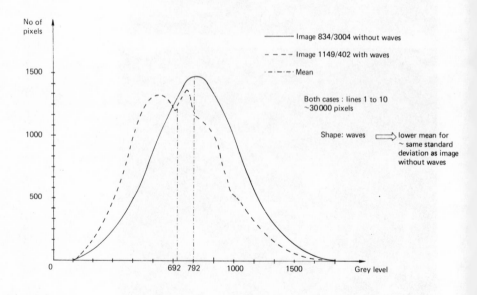

Fig. 14.3 – SEASAT histograms.

All those efforts to quantify the sea state by means of a SAR are leading to a better knowledge of the effects of waves on offshore structures and on oil tankers, to the future benefit of the oil industry.

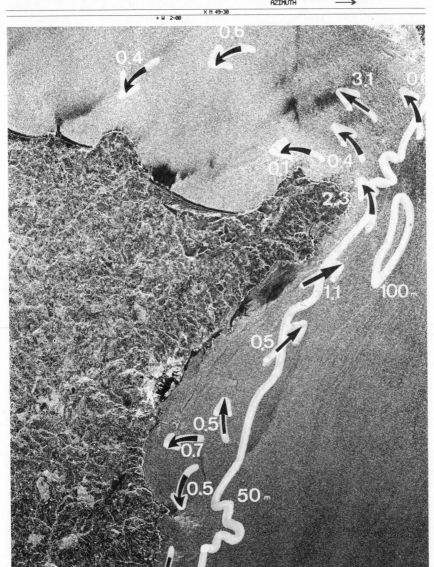

Fig. 14.4 – Currents Visualisation on a SAR image. ◄━ current direction, as
given by the French Oceanographic Survey (Service Hydrographique et Oceanographique de la Marine): *2.3* current speed in knots: *50 m* isobath. Time – 0732
(approx. 4 hours before high tide in Cherbourg).

14.3 CURRENT DETECTION

On a few occasions the SAR images indicated that local currents or the general ocean circulation may be detectable. Fig. 14.4 shows a case where directions seen on the image fit well with those from current tables. Nevertheless, no special use has yet been made at IFP of this current detection capability since the imaging phenomenon is not yet fully understood.

Horizontal turbulence as a part of the more general circulation patterns, and responsible for many features seen on SAR images, is currently under investigation in cooperation with a university research laboratory. The possibility of monitoring current directions in remote areas opens up an interesting potential application of SAR to the oil industry.

14.4 IMAGING OF SEA BOTTOM FEATURES

Our Dutch colleagues showed some years ago that sea-bottom features could be imaged with a SLAR under certain operating conditions. Later, with the French 'Vigie' SLAR we observed the same patterns on ocean radar images, but with the launch of SEASAT it became obvious that this aspect could become a standard product of radar imagery.

Every analogue or digital scene over the Channel and the southern North Sea was closely analysed. An experiment was conducted during the SAR 580 Campaign last year over the same region and the French oceanographic survey (SHOM, Service Hydrographique et Oceanographique de la Marine) provided a great number of original bathymetric data and charts. The results of these investigations have shown that many of the SAR images made over the region provided information about sea-bottom features (the quality of this information could range from excellent to almost nil, but something was to be seen on some 80 per cent or more of the orbits of SEASAT).

With the sea-truth we obtained after the overflights, we also found that the meteo-oceanographic conditions during the acquisition of those images were the following: wind speed ranging from 10 to 20 knots, wind from south, west, north, or even north-east, sea state from 2 to 4, current from less than 1 knot to 2.5 knots, direction between wind and current from 50 to 150 degrees, and direction between wind and track from almost 0 to 180 degrees.

In the only case in which we had no evidence of sea-bottom features (rev 719) the environmental parameters appeared to be little different from the other cases.

Theoretical investigations on the relationship between bottom features and surface ripples are proceeding, but as yet no generally accepted explanation has emerged. The mapping of shallow-water features by SAR — especially in remote areas — could be of obvious importance to the offshore industry.

Fig. 14.5 – Aspect of radar sea-bottom topography, in a case of medium quality.

arrows show a line corresponding to isobath–30 m, which is a slope change in that area. Time: 2215, wind: 340/18 kt (approx.), sea force: 2, currents: to the north; 2.1 kt off Boulogne, 1.2 kt S. Ridens.

SEASAT SAR PROCESSED BY DFVLR/GSOC FOR ESA/EARTHNET
ARCHNO S 0785 N 04737 W 00444 200878 7706 ORBIT 0785 AUG 20, 1978   4 LOOKS   25M RESOLUTION SCALE 1:250000
    FRAME CENTRE   N 047-37-34   W 004-44-06

N↙

AZIMUTH

0 1 2 3 4 5 6 7 8 9 10 KM
SATELLITE FLIGHT DIRECTION
→

Fig. 14.6 – Internal waves trains. South West Brittany. 30 km SW Pointe de Penmarc'H.

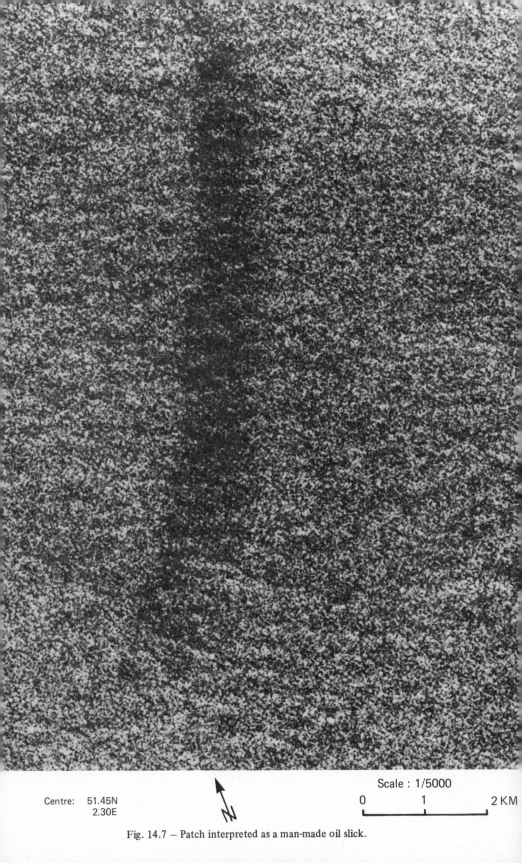

Centre: 51.45N
2.30E

Scale : 1/5000

0   1   2 KM

Fig. 14.7 – Patch interpreted as a man-made oil slick.

14.5 INTERNAL WAVES

This study is one of the most important we have made at IFP in remote sensing applied to oceanography. The creation of internal waves appears to be strongly correlated with sea-bottom topography (continental slopes, seamounts etc. . .) in most cases, and their detection might be a useful bathymetric indicator in deeper seas. The benefits to the offshore oil industry are evident: one should consider, for example, the strain on a 3000 feet drill pipe string, to appreciate the interest in early detection of a train of incoming internal waves.

Interesting examples of internal waves were imaged off the French coasts; one example on 20 August, extends from Brittany to Spain in the Bay of Biscay (orbit 785).

14.6 HYDROCARBON POLLUTION MONITORING

The SEASAT mission ended a few days before we began a series of experiments on the Noordwijk platform in the Netherlands. Our plans were to use the RAMSES scatterometer from CNES to monitor the sea state, and we also planned to make controlled oil spills at sea during overflights of the SAR. Unfortunately, the early demise of the satellite caused the experiment to be abandoned.

However, on some occasions a close look at SEASAT SAR scenes provides evidence of what appears to be oil pollution. Because of the lack of sea-truth we do not know if these spills are natural or not, but in one case the length and location (North Sea) lead us to believe that the slick was man-made.

Our interest in radar detection of oil pollution did not die with SEASAT, and the European ERS–1 project may provide opportunity for routine all-weather detection using SAR.

14.7 CONCLUSION

It has been proved, from our point of view, that the SEASAT Users' Research Group of Europe did a good job in triggering the interest of the research community in the use of the SAR.

In research concerning the applications to the oil industry, we found that the SAR was of great promise and could quickly become a part of the tools used on a routine basis in offshore operations.

BIBLIOGRAPHY

Alpers, W. & Jones, W. (1978) The modulation of the radar backscattering cross-section by long ocean waves. *Proc. 12. Int. Symp. on Remote Sens. of Env.* Manilla.

Alpers, W., *et al.* (1981) On the detectability of ocean surface waves by real and synthetic aperture radar. *J.G.R.* **86** no. C7.

Attema, E. & Hoogeboom, P. (1978) Microwave measurements over sea in *Netherlands Surveillance of Env. Poll. and Res. by E/M Waves* D. Reidel Publ. Co., Dordrecht.

Beal, R. *et al.* (1981) *Spaceborne synthetic aperture radar for oceanography,* John Hopkins Univ. Press, 7, Baltimore.

Brown, W. (1976) *et al.* Radar imaging of ocean surface patterns. J.G.R. **81**, no. 15.

De Staerke, D. & Wadsworth, A. (1981) Problemes d'interpretation sur des donnees SEASAT en mer. *Photo-Interp.* no. 81-5.

Fontanel, A. (1978) Détection de nappes d'huile par radars a antenne réelle (SLAR) ou antenne synthétique (SAR). *Photo-Interp.* Nov, Dec.

Kenyon, N. (1980) Bedforms of shelf sea viewed with SEASAT-SAR *Proc. Adv. in Hydrog. surveying* London.

De Loor, G. (1978) *Remote sensing of the sea by radar.* Report No. PHL, 178-53, TND, The Hague.

Piau *et al.* (1980) *Compte rendu des experiences RANO* 1979 Report No. 27944, IFP.

Schuchman *et al.* (1978) *SAR mechanism for imaging ocean waves.* EASCON'78, IEEE, New York.

Valenzuela, G. (1978) Theories for the interaction of E/M and oceanic waves (a review) *Boundary Layer Meteo.* **13**.

Chapter 15

Expressions of bathymetry on SEASAT synthetic radar images

D. W. S. LODGE
Space and New Concepts Department, RAE, Farnborough

15.1 INTRODUCTION

The depth of penetration achieved by the synthetic aperture radar on SEASAT into sea water was a small fraction of a millimetre. Therefore any features observed on the radar images of the sea relate only to variations in roughness at the immediate surface. However, the state of the surface can be greatly affected by processes occurring at considerable depth. This chapter describes features visible on SEASAT radar images that can be correlated with changes in depth. They fall into three cateogories, linked to wave refraction, shallow water, and deep water. Suggestions as to the mechanism whereby the surface effects manifest themselves are made where appropriate.

15.2 WAVE REFRACTION

One of the major aims for the SEASAT synthetic aperture radar was to produce images of ocean waves. The mechanism whereby it did so is not yet well understood (Alpers 1983). Nevertheless under certain circumstances waves are undoubtedly apparent. Fig. 15.1 shows such a case. The data were received at 0823 on 15 September 1978. The image shows an area 30 km square centred on Fair Isle, southwest of the Shetland Islands. Waves of length 200 to 220 m, corresponding to a period of about 11 seconds, may be seen travelling in an east-west direction. The direction would be ambiguous, but for the presence of the island which produces a well-defined refraction pattern showing that waves were travelling from west to east. The propagation of gravity waves such as these is unaffected by the topography of the sea bed, unless the depth of water is less than approximately half the dominant wavelength. Therefore the regular propagation of the waves on the image suggests that the water depth is not less than about 100 m, or at least constant, except in the immediate vicinity of Fair Isle.

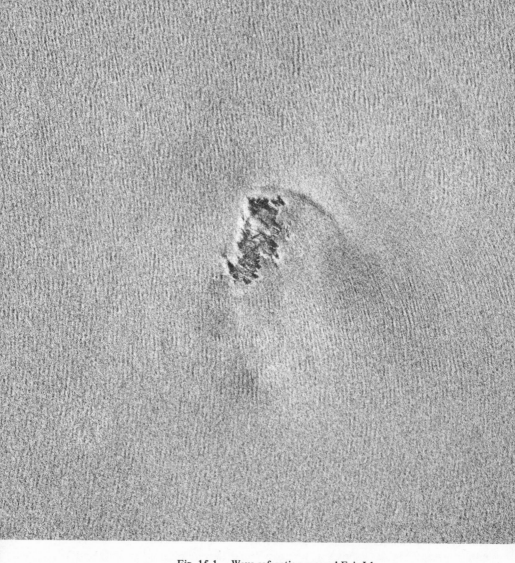

Fig. 15.1 — Wave refraction around Fair Isle.

Fig. 15.2 is an image of an area 30 km square around Foula, about 70 km north-west of Fair Isle, produced from data received a few seconds later during the same satellite pass. The overall conditions and effects are similar, but there is some indication of shallower water less than 100 m deep extending about 5 km eastwards from Foula. Also, there is evidence of refraction from an area of shallower water about 10 km north of the island coastline.

In principle there is a direct relationship between refraction patterns like those shown and depth of water. Therefore they can be used to provide quantitative bathymetric information. The other methods to be described only permit the drawing of qualitative conclusions about relief on the sea bed. They do not yet indicate any means to measure the vertical scale.

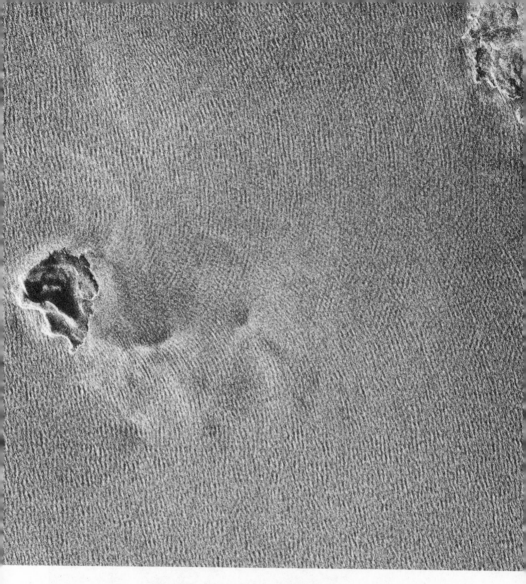

Fig. 15.2 – Wave refraction around Foula.

The concentration of energy by constructive interference between refracted wave trains has been suggested as an important mechanism in the deposition of shingle (Hardy 1964). The capability of an imaging radar to provide large-scale evidence of refraction patterns independently of the weather could be a most useful means to test that theory.

15.3 SHALLOW WATER FEATURES

The title for this section does not imply that there exists a known physical model linking features on synthetic aperture radar images with particular oceanographic parameters. Rather it refers to a type of feature so far observed only on images

East
Anglia

52°30'N

1°00'E

52°00'N

The
Englis
Chann

1°30'E

51°30'N

of areas where the water depth is less than 50 m or so. Fig. 15.3 shows an image made from data received at 0646 on 19 August 1978. The area is the English Channel and Thames Estuary Approaches north of Dunkerque. The striking linear features immediately suggest a link with sandbanks. Comparison with the appropriate hydrographic chart confirms that the correspondence is very close. Fig. 15.4 is a simplified version of that chart showing only the coastline and the 20 m depth contour. Although images like these caused some initial surprise, similar features had been observed on side-looking airborne radars since 1969 (de Loor 1981). The depth over the area rarely exceeds 40 m and may be less than 5 m over the tops of some of the banks. Apart from the Goodwin Sands, however, they are permanently submerged.

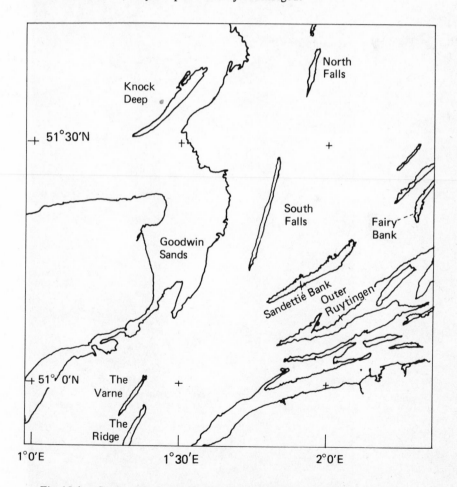

Fig. 15.4 — Straits of Dover and Thames Estuary Approaches, 20 m depth contour.

Facing page: Fig. 15.3 — English Channel 19 August 1978, 5 hours before high water at Dover (image produced by optical correlation at the Environmental Research Institute of Michigan, grid overlay by Hunting Surveys Ltd, Boreham Wood, on behalf of the European Space Agency).

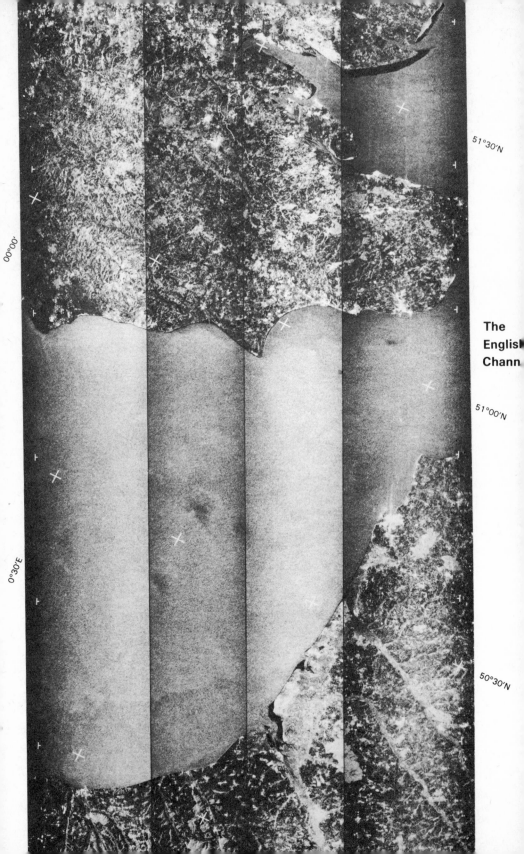

51°30'N

00°00'

The
Englis
Chann

51°00'N

0°30'E

50°30'N

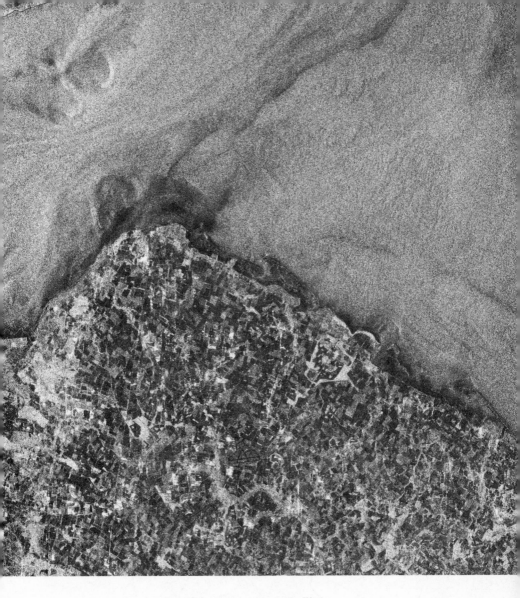

Fig. 15.6 – Approaches to the Wash.

Facing page: Fig. 15.5 – English Channel 16 August 1978, 2 hours before high water at Dover (image produced by optical correlation at the Environmental Research Institute of Michigan, grid overlay by Hunting Surveys Ltd, Boreham Wood, on behalf of the European Space Agency).

Further comparison between chart and image reveals that steep gradients are associated with a bright region on the radar image. That indicates a rougher surface with a higher prevalence of short, back-scattering waves. A possible mechanism to cause these effects could be that the tidal flow over the banks modulates the surface roughness. The image represents the area 5 hours before high water at Dover. At that time, according to the Admiralty *Tidal stream atlas,* the expected current would be about 1 to 2 knots along the long axes of the banks in question. The wind, which may also affect the surface roughness, was reported by coastal stations in the area to be light, south to south-easterly. That corresponds to the azimuthal or along-track direction for the radar.

Unfortunately SEASAT did not last long enough for much repeat coverage of any area to be obtained. In this case, however, at 0639 on 16 August 1978, that is three days before the data discussed above were received, an overlapping area was covered as shown Fig. 15.5. There are no surface features in the over-lapping and adjacent areas although from the chart The Varne and The Ridge might have been expected to produce similar effects to The Falls, Sandettié Bank, or Outer Ruytingen for example. The reported winds were similar in strength to those three days later, although westerly, that is blowing in the range direction. The most significant difference seems to be that on this occasion, with the time corresponding to 2 hours before high water at Dover, slack water would be expected over the two major banks in the area.

Fig. 15.6 shows an image of the approaches to The Wash. The data were received at 0646 on 19 August 1978, that is from the same satellite pass that produced Fig. 15.3. In general the water is shallower than in the English Channel.

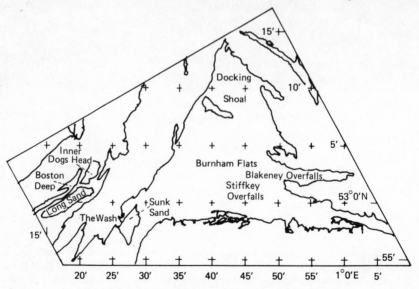

Fig. 15.7 – Approaches to the Wash, 10 m depth contour.

The depth is less than 25 m over most of the area. Fig. 15.7 shows a simplified version of the relevant hydrographic chart, this time with the 10 m depth contour marked. The meteorological conditions closely matched those reported for the corresponding Channel scene. Similarly, the tidal current would be expected to be strong, generally parallel to the coast away from the shore, and straight in through the mouth of The Wash, that is along the long axis of the linear banks. Once again the features on the image correlate strongly with the bathymetric detail. In addition, however, there are two nearly circular patterns around Sunk Sand and Inner Dogs Head that do not match the charted outline of those features. They may be affected by circulatory currents set up by the incoming tide.

Fig. 15.8 – Solent and Eastern Approaches.

The data used to produce the image of the Solent and part of the eastern approaches to it shown in Fig. 15.8 were received at 0800 on 2 September 1978. Fig. 15.9 shows the 10 m and 20 m depth contours for the area. Strong features on the image correspond with the edges of St Catherines Deep, the 20 m contour extending east from Dunnose which here delineates a relatively steep gradient between two more level regions, the Spit east of Shanklin and Culver Spit. The time of the image corresponds to 3 hours before high water at Dover. The difference between this scene and those discussed previously is that the tidal stream in this instance flows across the banks showing so prominently. In addition, there are similar looking features, for example south and south-east of Culver Spit, which show no obvious correlation with the local chart.

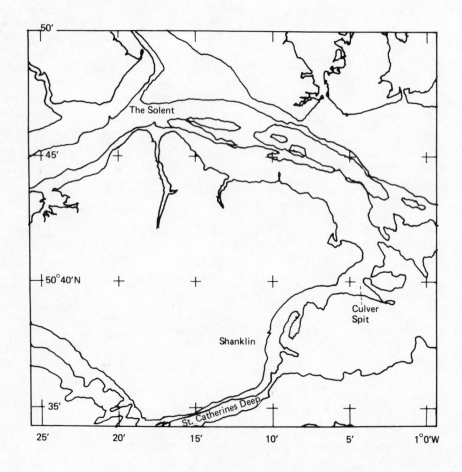

Fig. 15.9 – Solent and Eastern Approaches, 10 m and 20 m depth contours.

15.4 DEEP WATER FEATURES

Like the previous section, this one must be prefaced by a note that the appellation of deep water is a convenience to denote a particular type of observation. No quantifiable link has yet been uncovered which delimits the effect to prescribed water depths. However, many SEASAT synthetic aperture radar images show wavelike features, with wavelengths of several kilometres, propagating dispersively. They resemble internal waves. Moreover, they appear to be linked with changes of depth in water from 100 m to 1000 m or more deep. The first example, shown in Fig. 15.10, comes from data received at 0643 on 16 August 1978. The prominent V-shaped feature north-west of the Isle of Lewis corresponds approximately with the position of the 100 m depth contour, but is not obviously related to any major bathymetric changes. There are other features, extending from about 80 km north-west of Lewis or north-west from approximately 59°N, 7° 30'W. At first sight they appear to be confused perturbations of the sea surface. However, on closer inspection they show a very high degree of correlation with steep gradients on the sea bed, typically where the depth changes rapidly from about 500 m to 1000 or 1500 m. The onset of these features corresponds to the edge of the Hebrides Shelf. They appear to continue with close links with the positions of the Ymir Ridge and Wyville-Thomson Ridge. Three days later at 0649 on 19 August the adjacent area to the east was covered. The resulting image is included as Fig. 15.11. The structure on the surface seems less chaotic on first impression than in the previous case. It is apparent that the appearance of the internal wave-like features corresponds with the edge of the Hebrides and West Shetland Shelves. Going north-west similar patterns mark the crest of the Wyville-Thomson Ridge.

There were three other satellite passes from which images could be made of the area shown in Fig. 15.10. Data received on 5 September 1978 showed no comparable features. From 8 September 1978, when stronger winds were reported than had been the case for the August data, 300 m long swell waves were visible, and there was some structure in the vicinity of Lousy Bank. Finally, on 26 September 1978 with stronger winds still and swell visible on the image, a faint hint of a feature corresponding to the edge of the Hebrides Shelf could be discerned.

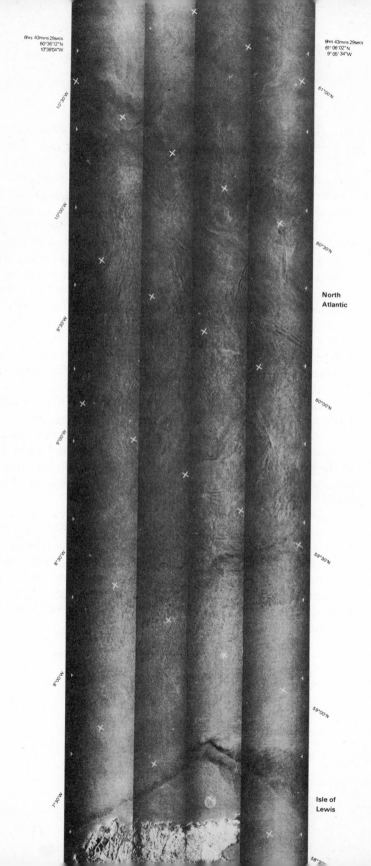

10°30'W

61°00'N

10°00'W

60°30'N

North Atlantic

9°30'W

9°00'W

60°00'N

8°30'W

59°30'N

8°00'W

59°00'N

7°30'W

Isle of Lewis

58°30'

Fig. 15.10 — Hebrides Shelf, Ymir Ridge and Wyville-Thomson Ridge 16 August 1978 (image produced by optical correlation at the Environmental Research Institute of Michigan, grid overlay by Hunting Surveys Ltd., Boreham Wood, on behalf of the European Space Agency).

North
Atlantic

Fig. 15.11 — West
Shetland and Heb-
rides Shelves and
Wyville - Thomson
Ridge 19 August
1978 (image pro-
duced by optical
correlation at the
Environmental Re-
search Institute of
Michigan, grid over-
lay by Hunting
Survey Ltd, Bore-
ham Wood, on be-
half of the Euro-
pean Space Agency).

15.5 CONCLUSIONS

Three distinct types of feature can be identified on SEASAT synthetic aperture radar images that can unequivocally be correlated with changes of depth. The refraction and diffraction of gravity waves provide in principle a quantitative method for making bathymetric measurements. However, such a technique requires a better understanding of the wave imaging mechanism. In shallow water, that is where the depth is less than 50 m or so, the tidal flow over the undulations of the sea bed can modulate the surface roughness in a way detectable by the radar. Other factors such as the wind and gravity wave fields will undoubtedly prevent the detection of bottom features under all conditions, but nevertheless the technique offers the useful prospect of monitoring the position of sandbanks. More work is needed to determine the environmental limits within which data may be useful and to investigate the possibility of drawing quantitative conclusions about depth. In deep water, internal waves propagating from discontinuities on the sea bed seem to modulate the surface roughness in a detectable way. The region used for demonstration in this paper is one where such internal waves may be expected, since the Wyville-Thomson Ridge separates the cold, less-saline water of the Arctic from the warm, more-saline water of the Atlantic. Nevertheless, similar effects have been observed very much more widely. As a technique for making bathymetric measurements, it does not have any obvious significance. However, it is more likely to aid the understanding of ocean dynamic processes. The immediate problem is to define the modulation process that causes the manifestation on the image.

REFERENCES

Alpers, W. (1983) Imaging the sea surface a review (Chapter 6 of this book).
Hardy, J. R. (1964) The movement of beach material and wave action near Blakeney Point, Norfolk. *Transactions and Papers of the Institute of British Geographers* **34** 1964.
de Loor, G. P. (1981) The observation of tidal patterns, currents and bathymetry with SLAR imagery of the sea. *IEEE Journal of Oceanic Engineering* **OE-6** 4, 124-129.

Chapter 16

Tidal current bedforms investigated by SEASAT

Neil H. Kenyon

Institute of Oceanographic Sciences, Wormley, Godalming, Surrey, UK

16.1 INTRODUCTION

For many years there has been considerable study of the tide-dominated bedforms of the seas of northwest Europe (summarised in Stride 1982). Valuable information on the plan view of bedforms has come from the extensive use of side-scan sonar, e.g. Belderson et al. (1972). Despite the large body of data acquired from ships, both airborne and satellite-borne radars have a useful role to play in geological studies. Images of bedforms obtained in 1979 with airborne radar in the southern North Sea were discussed by McLeish et al. (1981) and with SEASAT in 1978 by Kenyon et al. (1981). The purpose of this chapter is to detail some aspects of the contribution made by the SEASAT SAR (Synthetic Aperture Radar) images to the understanding of sedimentary processes and to the refinement of the geological models that are used to help interpret ancient rocks.

The side-looking radar on SEASAT produced the first available radar images from space. The main areas covered were North America and Western Europe, and most interesting images of bedforms were from the Southern Bight of the North Sea and the Bristol Channel off the UK and from the Nantucket Shoals off the USA. The 23 cm wavelength of the SAR enabled it to penetrate cloud, and of course to operate in darkness. It looked to one side at an angle of incidence of between 17° and 23° and covered a swath that was 100 km wide.

The mechanism whereby seafloor topography is registered on SAR images has been much discussed, e.g. Beal et al. (1981) and this volume. It will suffice to say here that the radar was sensitive to short gravity waves comparable in length to the radar wavelength of 23 cm.

Because the SAR does not penetrate more than a few centimetres through water, only ocean surface effects are seen. Seafloor topography can have an effect at the sea surface in various ways. For instance (1) turbulence over bedforms can reach the sea surface and can be seen visually from ships and aircraft. The great variety of turbulent structures associated with bedforms in rivers is familiar to sedimentologists, and their surface effects have been mapped from aerial photographs (Coleman 1969). Similar turbulent structures should be associated with the bedforms of tidal seas, as many of them are comparable in shape and size. (2) However, the SAR should register relative changes in surface current velocity, such as would occur over shoals, because it is expected that the currents would modulate the short gravity waves. On the occasions when the best images of topographic features were obtained there were light winds favourable for the formation of short gravity waves. It is variations of this latter mechanism, rather than direct observation of turbulence, that are considered to be the most likely cause of these images.

16.1.1 Tidal current parallel streaks

Wind-driven surface streaks in the ocean (windrows) are a well-known phenomenon. They have been much investigated and are associated with helical roll vortices (Langmuir circulations) aligned within a few degrees of the wind direction (reviewed by Pollard (1977)). Although ocean current parallel streaks have been seen on satellite images by, for instance, Mollo-Christensen (1981), observations of tidal current parallel streaks have rarely, if ever, been made before. Such features have, however, been predicted to account for the formation of the longitudinal tidal bedforms such as sand ribbons (Kenyon 1970) and erosional furrows (Stride *et al.* 1972) that are found in regions of strong tidal currents (Fig. 16.1).

Tidal flow parallel streaks are seen on several SEASAT passes of the seas around the UK. The best examples are in the Bristol Channel (Fig. 16.2), but they are also seen in the Southern Bight of the North Sea (Fig. 16.3), St Georges Channel and in the central English Channel. The current speed was on each occasion greater than 80 cm/sec. The areas of occurrence of the flow parallel streaks are in all cases within the zones of sand ribbons mapped by Kenyon (1970) and longitudinal furrows mapped by Stride *et al.* (1972). This supports a causal relationship between the longitudinal current streaks and the longitudinal bedforms. Because current streaks in the Bristol Channel are seen in areas of bare rock where sediment bedforms are not present, but would be expected to occur if sufficient sediment was available, it can be surmised that the pattern in the water occurs before the pattern in the sediment. The patterns presumably indicate that there are filaments of water moving at different speeds, but it is not known whether there are helical roll vortices within these filaments.

The row spacing of the current streaks at the time of observation was approximately 300 m in the inner Bristol Channel, 800 m in the outer Bristol Channel, >1300 m in St Georges Channel, and 2500 m in the Southern Bight of the North Sea. The near surface current speeds varied from about 170 cm/sec for the closest spaced streaks to about 80 cm/sec for the widest streaks. However, as there was only one observation of streaks in each area it is not known whether streak spacing changes over a tidal cycle. It may be that depth is a factor or that there is a smaller scale of streaks there than observed by the SAR.

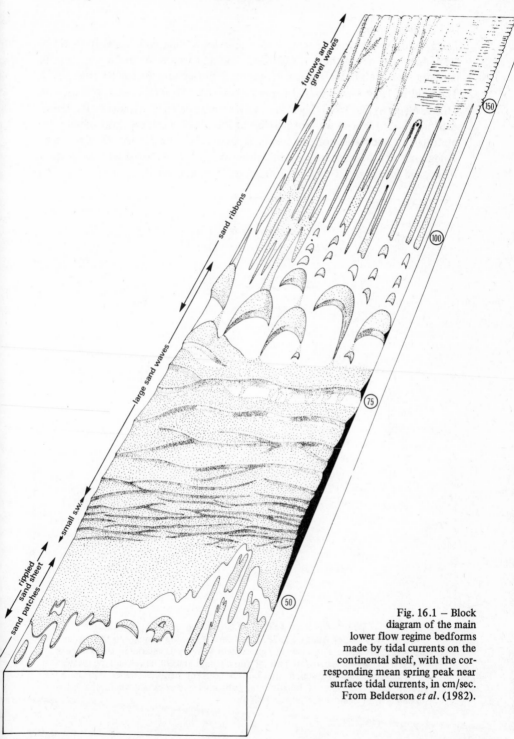

Fig. 16.1 – Block diagram of the main lower flow regime bedforms made by tidal currents on the continental shelf, with the corresponding mean spring peak near surface tidal currents, in cm/sec. From Belderson *et al*. (1982).

16.1.2 Sand waves

Large sand waves, often greater than 5 m in amplitude, are normally found in tidal shelf seas where there is sufficient sand and where the mean springs peak currents near the sea surface are between 65 cm/sec and 90 cm/sec (Fig. 16.1). The field of large sand waves at the mouth of the Bristol Channel is well seen on a SEASAT pass (Fig. 16.2). These sand waves were originally mapped by Belderson & Stride (1966), but, it was not possible at that time to achieve great accuracy in plotting the orientation of the crestlines, because of the distortions of side-scan sonographs and because of the difficulties of navigation. Although the crests of these sand waves are at depths of between 25 m and 50 m, the deepest seen on

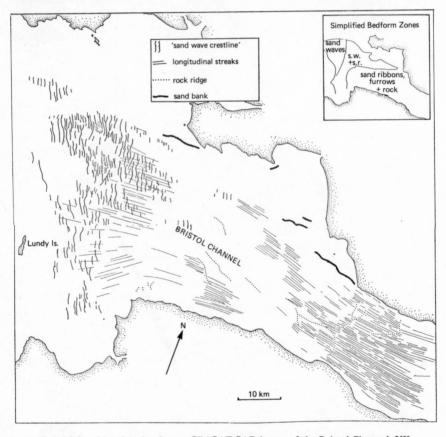

Fig. 16.2 – Line drawing from a SEASAT SAR image of the Bristol Channel, UK, taken at 0727 on 24 August, 1978. The tide was flowing strongly to the east and the wind was light. The near surface current velocity is estimated to have been 180 cm/sec in the narrowest part of the channel and 90 cm/sec at the western boundary of the longitudinal streaks. The current parallel streaks correspond closely with the zone of longitudinal bedforms and rock shown in the insert (updated from Belderson & Stride 1966)).

SEASAT SAR, the image is believed to give an accurate location and orientation of the crestlines. If, as discussed above, the streaks seen on the sea surface are parallel to the direction of peak current flow, and as at the time of the SEASAT pass the tide was near maximum (and flowing to the east), then the orientation of the sand wave crests vary from about normal to 15° to the peak tidal current flow. The sand wave crests appear to have a tendency to be rotated in an anti-clockwise sense relative to the currents. Whether this relationship is found in this area on other occasions remains to be seen, but predicted directions of bottom stress in this region (Uncles, in press) have a similar sense of obliquity to the large sand wave crests.

16.1.3 Tidal sand banks

Sheets of sand whose surface is partly covered by large and small sand waves are one of the main sand facies of tidal seas. The other main facies are the large sandbanks. Tidal sandbanks are the largest bedforms of shallow, strongly tidal seas. They tend to occur in groups or else are found as solitary banks tied to headlands, islands, or submerged rock shoals. They have generally been considered to extend parallel to the peak tidal current flow (Off 1963) but an investigation of all the major tidal sandbanks around the British Isles confirmed that they are oblique to the regional peak tidal flow by up to 20° and more (Kenyon *et al.* 1981). This is dramatically illustrated by the SAR image of part of the Southern Bight of the North Sea (Fig. 16.3). Streaks on the surface, believed to be parallel to the current flow, are at about 20° to the crestlines of the South Falls (SF) and Sandettie (S) Banks. This same configuration of streaks and bank crests was also seen on a near infrared Landsat image taken on 12 June 1975 (Viollier 1980). The SAR image of the Southern Bight (Fig. 16.3) and also the SAR image of the tidal sandbanks in the Nantucket Shoals region, off Cape Cod, USA (reprinted on p.22 of Beal *et al.* (1981)) confirm what had previously been demonstrated by side-scan sonar (Caston 1972), that sand waves as they approach the crest of a sandbank do not maintain their crestlines at approximately normal to the general direction of peak tidal flow but bend around to become tangential to the crest of the bank.

Next page: Fig. 16.3 − A portion of a SEASAT SAR image from the southwesternmost corner of the North Sea, taken at 0647 on 19 August, 1978. The sandbanks, named in Fig. 16.5, have a band of stronger reflectivity (white tone) just downstream from their crests. The tidal current was flowing to the south west at about 60% of maximum mean springs values. The near surface velocities were between about 60 and 100 cm/sec. The wind speed was about 25 km/hour. Longitudinal streaks on the water surface, believed to be flow parallel, are at an angle of about 20° to the sandbank crests.

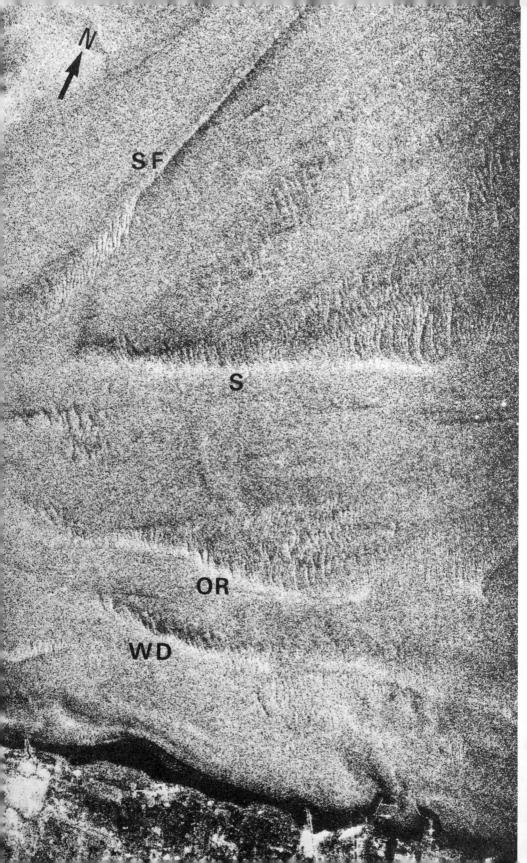

Two simplified models for the transport paths of sand near asymmetrical sandbanks can be envisaged (Fig. 16.4) In (A) the axes of the banks are offset in a clockwise sense from the peak tidal flow and the flow veers anticlockwise as it nears the bank crest. In the reciprocal case (B) the bank crest is offset anticlockwise and the flow veers clockwise as it nears the bank crest. Thus sandbanks can be considered as somewhat analogous to sand waves for the purpose of predicting regional sand transport paths. One needs to know only the plan view configuration of large sand wave crests in relation to the sandbank crest

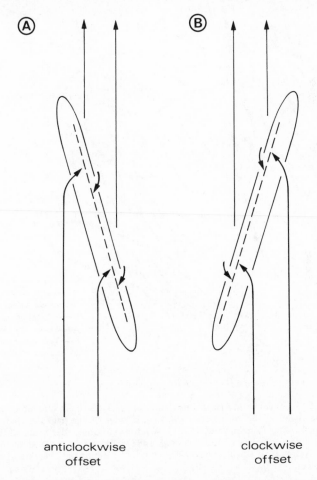

anticlockwise
offset

clockwise
offset

Fig. 16.4 − Two classes of active asymmetrical tidal sandbanks. The arrows indicate the direction of net sand transport which veers towards the right on approaching the bank crest in (A) and towards the left in (B). These models are confirmed by SEASAT SAR images of tidal sandbanks, such as Fig. 17.3. They have been used in Fig. 17.5 to predict net sand transport direction in regions of tidal sandbanks.

in order to decide which of the two models is applicable. This can be readily determined from SAR images obtained in the right conditions. An example of sand transport prediction for the Southern Bight of the North Sea is shown in Fig. 16.5, which uses all available data in addition to the SEASAT SAR data. It is fully discussed in Kenyon *et al.* (1981), as are predictions for less well known areas of tidal sand banks.

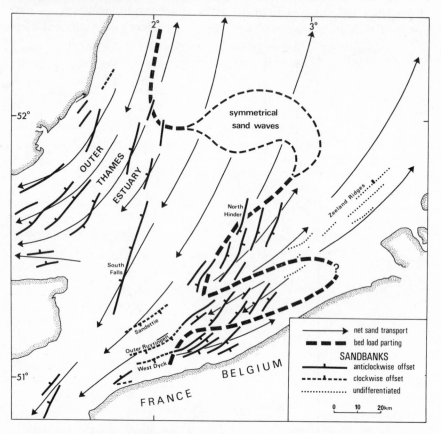

Fig. 16.5 – Offshore tidal sandbanks and net sand transport directions in the Southern Bight of the North Sea. The steeper sides of the banks are indicated with a tick. After Kenyon *et al.* (1981). A recent airborne radar image obtained after the figure was drawn (McLeish *et al.* 1981) indicates that at least one of the Zeeland Ridges is of the clockwise offset type, confirming the direction of transport shown.

16.2 CONCLUSIONS

There has been much necessary discussion of the mechanism responsible for the surface expression of bathymetry on radar, and there is a need for experimental quantification of the mechanism. A search for turbulence patterns at the sea surface during calm weather should help to determine whether or not they are part of the process. The ability of radars to accurately map the position of shoals is obviously of significance to hydrography even though depth measurements cannot be made. However, this chapter is concerned with pointing out the sedimentological value of the data already obtained. The first few SAR images of bedforms have already been of considerable value in testing hypotheses concerning tidal current sedimentation.

Perhaps the most significant new discovery is the presence of current parallel streaks. They not only help to explain the formation of longitudinal bedforms but also give a more accurate measurement of current orientation. Radar also offers a way of measuring bedform trends that is more accurate than shipborne acoustic methods. This new-found capability for accurate measurement should be useful in investigating the effects of Coriolis force on bedform trends, which Smirnov & Khramov (1975) consider to be significant, and in testing the ideas of Nio & Nelson (1982) who suggested that some large sand waves have a trend that is inherited from an early stage of sea level rise.

The confirmation of the obliquity of tidal sandbanks opens up a ready method for the prediction of sand transport paths in regions that are little known. In this respect it would be of interest to see radar images from regions such as the Yellow Sea, the White Sea, the straits and gulfs to the north of Australia, and the Gulf of San Matias, Argentina.

ACKNOWLEDGEMENTS

I would like to thank M. A. Johnson for making the calculations of current speeds.

REFERENCES

Beal, R. C., DeLeonibus, P. S. & Katz, I. (eds.) (1981). *Spaceborne synthetic aperture radar for oceanography.* Johns Hopkins Oceanographic Studies No. 7, 215 pp.

Belderson, R. H. & Stride, A. H. (1966). Tidal current fashioning of a basal bed. *Mar. Geol.* **4** 237-257.

Belderson, R. H., Kenyon, N. H., Stride, A. H. & Stubbs, A. R. (1972). *Sonographs of the sea floor.* Elsevier, 185 pp.

Belderson, R. H., Johnson, M. A. & Kenyon, N. H. (1982). Bedforms. In: Stride, A. H. (ed.), *Offshore tidal sands. Processes and deposits.* Chapman & Hall, London, 213 pp.

Caston, V. N. D. (1972). Linear sand banks in the southern North Sea. *Sedimentology,* **18** 63-78.

Coleman, J. M. (1969) Brahmaputra river: channel processes and sedimentation. *Sediment. geol.* **3** 129-239.

Kenyon, N. H. (1970). Sand ribbons of European tidal seas. *Mar. Geol.* **9** 25–39.

Kenyon, N. H., Belderson, R. H., Stride, A. H. & Johnson, M. A. (1981). Offshore tidal sand-banks as indicators of net sand transport and as potential deposits. pp. 257–268 in: Nio, S. D., Schuttenhelm, R. T. E. and van Weering, T. C. E. (eds.): Holocene marine sedimentation in the North Sea basin. *Int. Assoc. Sed. Spec. Publ. 5.*

McLeish, W., Swift D. J. P., Long, R. B., Ross, D., & Merrill, G. (1981). Ocean surface patterns above sea-floor bedforms as recorded by radar, Southern Bight of North Sea. *Mar. Geol.* **43** M1–M8.

Mollo-Christensen, E. (1981). Surface signs of internal ocean dynamics. pp. 140–154 in: Beal, R. C., DeLeonibus, P. S. & Katz, I. (eds.), *Spaceborne synthetic aperture radar for oceanography.* Johns Hopkins Oceanographic Studies No. 7, 215 pp.

Nio, S. D. & Nelson, C. H. (1982) The North Sea and northeastern Bering Sea: a comparative study of the occurrence and geometry of sand bodies of two shallow epicontinental shelves. *Geol. Mijnbouw* **61** 105–114.

Off, T., (1963). Rythmic linear sand bodies caused by tidal currents. *Bull. Am. Ass. Petrol. Geol.* **47** 324–341.

Pollard, R. T. (1977). Observations and theories of Langmuir circulations and their role in near surface mixing. pp. 235–252 in: Angel, M. (ed.), *A Voyage of Discovery: George Deacon 70th anniversay volume.* Pergamon, Oxford.

Smirnov, L. S. and Khramov, A. N. (1975). Coriolis force and the texture of the sandstone-siltstone rocks vis-a-vis the palaeomagnetic latitudes. *Izv. Earth Phys.* **3** 66–79 (translation).

Stride, A. H., Belderson, R. H. and Kenyon, N. H. (1972) Longitudinal furrows and depositional sand bodies of the English Channel. *Memoir Bureau Recherches Geologiques et Minieres* **79** 233–240.

Stride, A. H. (ed.) (1982). *Offshore tidal sands: Processes and deposits.* Chapman and Hall. London. 213 pp.

Uncles, R. J. (in press). Modelling tidal stress, circulation and mixing in the Bristol Channel as a pre-requisite for ecosystem studies. *Canadian J. Fisheries and Aquatic Sci.*

Viollier, M. (1980) Remote sensing of ocean colour in the Straits of Dover. In: M. B. Sorensen (ed.), Workshop of the EURASEP ocean colour scanner experiments 1977. *Proceedings, Joint Research Centre Ispra,* 30–31 Oct 1979.

Chapter 17

SEASAT over land

P. H. MARTIN-KAYE and M. McDONOUGH

Hunting Geology and Geophysics Ltd., Borehamwood, Herts, England

and G. C. DEANE, Hunting Technical Services Ltd., Borehamwood, Herts, England

17.1 INTRODUCTION

17.1.1 Background

The SEASAT satellite, launched by NASA in 1978, presented the first opportunity of evaluating some potentially important land applications of a satellite-borne Synthetic Aperture Radar (SAR) system; such possible applications include the monitoring of soil moisture and crop state for crop yield forecasting. The results of such an evaluation are particularly significant in Europe where the frequent cloudy conditions are a hindrance to using other forms of remotely sensed imagery for monitoring purposes.

The SEASAT receiving station at Oakhanger, England, operated by the Royal Aircraft Establishment (Farnborough) recorded some 272 minutes of data from 53 passes of SEASAT. These data represent imaging by the SAR of approximately 11 million km^2 between Greenland and North Africa, of which about 20 per cent are overland images. Almost complete coverage was accumulated for the United Kingdom and Iceland.

The evaluation of the overland image data from the SEASAT SAR outlined in this paper was commissioned by the Royal Aircraft Establishment prior to the launch of the satellite. This work was also partly financed under a wider contract held by the Royal Aircraft Establishment from the European Space Agency (ESA). A programme of work that would have assessed the geologic and land use data content of the SAR imagery over the United Kingdom was devised. Seasonal ground observations were to be made at the time of satellite overpass. In the event SEASAT failed on 10 October 1978, after 1503 revolutions; at which time the original programme had barely commenced. Fortunately SEASAT had already acquired ample SAR imagery of the United Kingdom for a retrospective evaluation to be carried out. These studies were eventually extended to cover selected areas in other parts of Europe.

SEASAT had an almost circular orbit with an inclination to the plane of the equator of 108° and its altitude at the equator was 791 kilometres. The SAR sensor which was L-band (23 cm) operating in HH polarisation, imaged a swath

100 kilometres wide. Prior to SEASAT there had been no satellite radar imagery, and airborne radar made comparatively little use of L-band imagery for natural resource studies; the majority of SLAR (Side-Looking Airborne Radar) imagery having been acquired in X-band (3.3 cm).

17.1.2 The imagery

Much of the SEASAT SAR imagery covering Europe is available in both optically and digitally correlated formats. The optically correlated imagery (processed by ERIM) was issued in both preliminary format (survey mode) and, at a later date, precision processed format, both to approximately 1:680 000 scale. Digital correlation was undertaken by both the Royal Aircraft Establishment and the DFVLR (at Oberpfaffenhofen, West Germany). The digitally correlated imagery was produced using a range of processing options. The results of some of these processing options were compared using the IDP 3000 interactive image processor of the RAE Farnborough.

Both optically and digitally correlated imagery have been used in the following largely subjective assessments but optically correlated 'survey mode' imagery was the most widely used for the geological studies. Digitally correlated imagery was used for many of the land use studies.

Fig. 17.1 – Glaciers, Greenland. Optically correlated image. Owing to the steep inspection angle of the SEASAT SAR, lay-over effects are pronounced in mountainous terrain. In this image, foreshortening and layover produce the sawtooth appearance of the mountain ranges and overprinting of valley-side features onto the glacier surfaces.

Some defects in the imagery due to radar system effects were observed but as these occurred infrequently they were not considered significant for the evaluation discussed here. The optically corrected imagery lacks ground range correction and often shows inconsistent along-track scaling. Digitally correlated imagery is corrected in these respects. Owing to the steep inspection angle of the SAR, foreshortening and lay-over effects are severe, as is indicated on the image from Greenland shown in Fig. 17.1.

17.3 Topographic features

Features that are well expressed on SEASAT SAR imagery include major physiographic units (e.g. mountain ranges, plateaux, coastal plains), general morphology (e.g. flat, rolling, irregular, hummocky, mountainous), principal drainage networks, coastlines, major industrialised centres, and agricultural patterns. Fig. 17.2 shows how the morphology of the drumlin field of central Ireland is clearly expressed on the imagery.

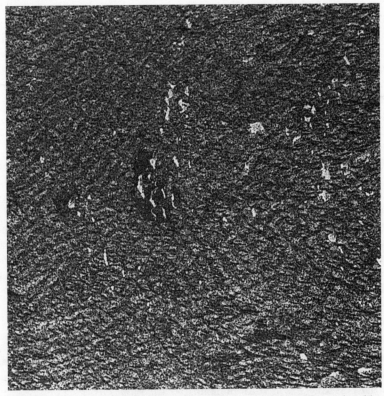

Fig. 17.2 — Drumlins, Central Ireland, Drumlins are rounded hillocks produced by retreating ice sheets. The direction of ice flow can be determined by drumlin orientation.

Certain man-made features can usually be clearly identified either as very dark or very bright signatures on the imagery. These include airfield runways, harbour installations, and large ships. Other features such as roads, railways, canals, bridges, electricity pylons, and stone walls can only be identified in certain conditions, particularly when the features are aligned parallel with the along-track direction of the satellite. Urban centres can usually be identified but the images contain little interpretable information about urban structure. Small villages are usually difficult to identify with certainty.

Some features such as the route of a stream or road may be inferred from adjacent patterns e.g. a particular vegetation type or a particular agricultural pattern, even if the feature is not directly expressed on the image. Low, water-logged shores can be difficult to differentiate from adjacent open water areas because they have similar dark radar signatures.

17.2 GEOLOGICAL STUDIES

17.2.1 Range of studies

Geological evaluation comprised orientation studies, systematic interpretations at various scales, lineament analysis, and interactive computer processing for image enhancement and LANDSAT/SEASAT combinations. The principal work was upon digitally and optically correlated imagery of south west England and Iceland. Interpretative scales ranged between 1:100 000 scale and 1:1 000 000 scale. The total imaged area examined was about 300 000 km^2.

17.2.2 United Kingdom study areas

It became evident during these studies that the United Kingdom geology posed certain interpretative difficulties. Over Scotland and much of England and Wales there has been extensive glaciation which has resulted in widespread distribution of glacial deposits and modification of landforms by ice action. These tend to subdue or blanket morphological expression of bedrock lithologies. In south west England comparable modifications have been produced by marine action. Agricultural patterns, woodland and urban areas also obscure lithologies. In summary it may be said that interpretation for lithostratigraphy was generally poor; but useful structural information could be identified.

(a) Northern Scotland and Eastern Yorkshire

The first imagery available for interpretation was from Orbit 762 which traversed northern Scotland and then followed the eastern coastlands of England. Imagery covering about 7000 km^2 between Cape Wrath and Montrose and 2000 km^2 of eastern Yorkshire was used for preliminary evaluation.

These studies showed that whilst SEASAT can be a useful mapping tool, because of its expression of topography, drainage and cultural features, it is of limited value for lithological distinctions for these parts of the United Kingdom. Where lithologies do give rise to distinctive landforms (e.g. the Old Red Sandstone and Jurassic sediments at Golspie, the granite syenite at Ben Loyal, the Corallian limestones of the North York moors) appropriate lithostratigraphic correlations can be made; but these are exceptions. An example of the morphological expression of lithologies is given in Fig. 17.3. In many cases lithologies with varied characteristics could not be differentiated. However, during the interpretation it was possible to identify many lineaments on the SAR imagery: significantly more lineaments than are recorded as faults on 1:63 360 scale geological maps.

(b) South Western England
For this area mosaics of both digitally and optically correlated imagery at 1:100 000 scale were used. The study area covered approximately 8000 km² and included parts of Cornwall, Devon and Somerset, and was bounded on the north coast by Port Isaac and Lynton and on the south coast by Fowey and Exmouth. Within this block are the high moorlands of the Dartmoor and Bodmin granites which rise above a variously dissected tableland mainly of Devonian and Carboniferous rocks.

Fig. 17.3 — Golspie Area, Sutherland, Scotland. In areas where the rocks are concealed by soil and vegetation the interpretation of lithology often depends upon the development of a characteristic morphology. In this scene the hills immediately to the west of the inlet are of Old Red Sandstone which are morphologically distinct from the granites inland.

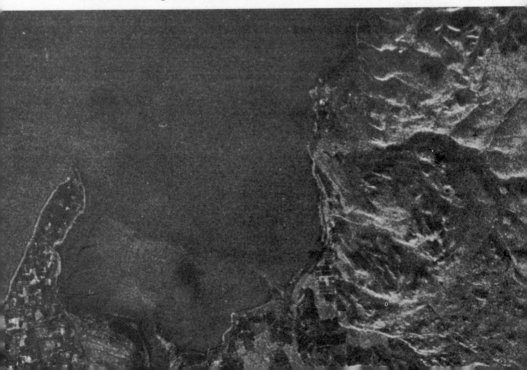

Some difficulty in distinguishing lithological units was also experienced in this area. Twenty-five radar mapping units could be identified on the basis of tone, texture and shape. A large proportion of the area could not however be differentiated into significant units.

Four of the twenty-five units have clear boundaries that match those on the geological maps. The nature of these units could be interpreted on the basis of morphology without reference to maps. For example, a small coastal zone around Barnstaple Bay has a mid-grey signature with many bright highlights and no clear drainage pattern. The unit is an area of irregular sand dunes that have a maximum height of approximately 15 metres. Other formations that can be recognised are alluvial deposits of the river floodplains and Palaeozoic metamorphics and sediments.

Four other units could be identified on the basis of morphology including areas of high open moorland. However, the boundaries of the interpreted units conform only loosely with boundaries on the geologic maps. This is because the radar signatures are modified by agriculture and it is not possible to identify the exact lithological boundary on the image.

Three other units show some partial coincidence with geological formations. The remaining fourteen radar mapping units and the undifferentiated areas had no significant correlation with mapped solid geology.

(c) North Pennines

A mosaic at 1:100 000 scale was constructed from optically correlated imagery of Revolution 719, covering an area of approximately 15 000 km^2 between the towns of Newcastle, Hexham, and Carlisle in the north, and Blackburn, Bradford, and York in the south.

The three major morphologic units of the area (the Alston and Askrigg Blocks, and part of the Lake District) are clearly visible on the mosaic. The Lower Palaeozoic slates and volcanics of the Lake District produce more rugged landforms than the Carboniferous formations of the Pennines. The faulted nature of the north, west, and south limits of the Alston Block is clear. Other linears mark the Teesdale faults and the Great Sulphur vein system. The Burtreeford monocline is depicted by a line of hills and in the south west the Lotherdale and Clitheroe anticlines can be identified. Lithologies cannot usually be differentiated, although at the eastern edge of the Howgill Fells the Coniston Grit forms a particularly distinctive unit with a coarse texture on the SAR image.

(d) The Scottish Tertiary volcanic district

A mosaic at 1:500 000 scale was produced of this area from optically correlated imagery. The region has comparatively well exposed geology with bold morphologic expressions. At this scale volcanic and sedimentary terrains were separable. Volcanic and intrusive centres could be differentiated on the basis of irregular mountain forms and, in some cases, circular outlines. Distinctions could not readily be drawn within these main categories. The extensive dyke swarms of this area could not be seen.

(e) SEASAT mosaic of the United Kingdom
The SEASAT mosaic of the United Kingdom ($270\,000\ km^2$) presents a different view of the structure of the islands, and it is in this respect that the SAR may prove to have made its most important geological contribution to the United Kingdom region.

In the Highlands of Scotland the Caledonian fracture trends and a conjugate set of faults are clearly illustrated. Southward many of the well-known faults of the Borderland, northern, and central England, and Wales are recognisable. Perhaps more interesting are the large numbers of lineaments of southern and central England, that are not clearly apparent on other forms of imagery. Numerous areas of broad tonal continuity can be detected. Particularly conspicuous examples include features extending from the Wash to the Severn and from Kent and Sussex to the Bristol Channel. These features would seem to reflect basement structure, but a full analysis has not been carried out.

17.2.3 Iceland
It was necessary to look outside the United Kingdom for a region of extensive rock exposure. Iceland is almost entirely covered by SAR imagery, and although there are widespread areas of drift cover, there are also large areas of exposed lithology. Comprehensive geological maps are also available.

A 1:500 000 scale mosaic was constructed from optically correlated imagery and systematically interpreted. The results were compared with the published geological maps of Iceland and are summarised in Table 17.1. This shows that the major lithostratigraphic units mapped by the Geological Survey of Iceland could be distinguished on the imagery. Some features mentioned in the legend of the published map (e.g. hot springs, fossil localities, and lignite) would not be recognised at this scale. Some others (including granite and gabbro) are ordinarily difficult to identify on any type of imagery unless characterised by particular textures or associations. Bedding can often be seen in the Plateau Basalts. In some cliff sections as many as 40 successive beds can be counted. Some of these beds are less than 20 metres in thickness (Fig. 17.4).

The main structural elements of the island — the young fracture zone of the Mid Atlantic Ridge and the flanking older volcanic lavas — are readily distinguished on the radar mosaic. During interpretation these zones could be identified without reference to other sources of information. Fracture lineaments are abundant and are particularly well depicted in the Mid Atlantic Ridge zone.

Table 17.1

Recognition of geologic units and morphologic features (optically correlated imagery) in Iceland

Unit/feature†	Expression in radar mosaic
1. Plateau Basalt	Well expressed
2. Old Grey Basalts	Poorly differentiated from (1)
3. Palagonite Formation	Readily distinguishable
4. Young Grey Basalts	Sometimes distinguishable
5. Gabbro	Not recognised
6. Granophyre and Granite	Not recognised
7. Rhyolite	Difficult to distinguish
8. Pliocene sediments	Not recognised
9. Basalt lavas	Well expressed
10. Acid lavas	Not distinguished
11. Coarse pyroclastics	Recognisable but not mapped separately
12. Pumiceous sheets	Not recognised
13. Glacial drift	Distinguishable
14. Alluvium	Readily distinguishable
15. Braided alluvium	Readily distinguishable
16. Aeolian sand	Occasionally distinguishable
17. Transgressed areas	Sometimes distinguishable
18. Post glacial volcanoes	Distinguishable if of sufficient size
19. Shield volcanoes	Readily distinguishable
20. Pseudo craters	Not distinguished
21. Hot springs	Not distinguished – too small
22. Hot spring deposits	Not distinguished
23. Terminal moraine	Well expressed
24. Young terminal moraine	Not separable from 23
25. Medial moraine	Well expressed
26. Direction of glacial striae	Features that may be of this category are recognisable but are of much larger scale than normally regarded as striae
27. Raised beach	Not identified
28. Tectonic fractures	Well expressed
29. Rock slides	Not separated
30. Fossil localities	Not applicable
31. Lignite	Not applicable

†Numbers refer to Supplementary Notes to the legend of the Geological Map of Iceland Published by: The Cultural Fund, Reykjavik.

Fig. 17.4 – Coastal section of Tertiary floodbasalts, Djupivogur, Iceland. The image shows a thick succession of near-horizontal basalt lava flows.

17.2.4 Other areas

Imagery covering the Navan and Avoca areas of the Republic of Ireland, glaciated terrain in Ireland, glaciated terrain in Greenland, and parts of the Vosges, Alps, and Sardinia were also examined. Imagery of the high Alps and mountainous parts of Sardinia contain very little interpretable information owing to the radar foreshortening and layover problems, but elsewhere the results supported the work carried out in the more detailed study areas.

17.2.5 Interactive computer processing

Interactive computer processing of SEASAT was carried out using Computer Compatible Tapes (CCTs) produced by the Royal Aircraft Establishment, Farnborough, of scenes from southwest England, the Irish Republic, and Iceland. This work was completed using the IDP 3000 Interactive Image Processor at the RAE. In the case of the southwest England images it was found that various contrast stretching techniques could be used to make the image subjectively more pleasing. However, these appear to have little effect on geologic interpretability. Images from two-look digitally correlated imagery were found to be better than one-look images because of the reduction in radar speckle No interpretational benefit resulted from the examination of enlargement of the imagery on the TV monitor screen. After enlargement the broad synoptic view was lost. Simple computer classification techniques were ineffective on these radar data.

The use of the IDP 3000 for scenes of Iceland did not significantly improve the interpretation derived from photographic prints of the optically correlated imagery.

17.2.6 SEASAT/LANDSAT comparison

SEASAT and LANDSAT images were compared by displaying similar areas from different images on the TV monitor screen of the IDP 3000 image processor at Farnborough. The screen was then photographed so that the two images could be compared in detail.

The relative information content in cartography and geology of the two data sets from the Plymouth area of southwest England is summarised in Table 17.2. The quality of expression of each feature was assessed subjectively. In most cases the false colour composite of LANDSAT multi-spectral scanner data contains more information than the SEASAT SAR. The SEASAT shows broad river channels particularly well as do the LANDSAT MSS near infra-red bands. The visible LANDSAT MSS bands tend to be less clear in this respect but contain some information on bathymetry.

SEASAT SAR generally gives a much better expression of morphology than LANDSAT owing to the highlight and shadow effects on the radar. LANDSAT generally gives better expression to lithology than SEASAT owing to the colour variations possible with multi-spectral data.

17.2.7 LANDSAT/SEASAT superimposition

LANDSAT scene 2126811355 of 13.7.1978 (Path 235 Row 015) covering part of Iceland was used for digital superimposition with SEASAT. For this purpose the Royal Aircraft Establishment, Farnborough, resampled the SEASAT data to fit the LANDSAT. Reorientation was achieved on the basis of orbit and attitude data. The scenes were not warped to fit each other exactly, and some residual skew remained between them. Nevertheless reasonably precise matching was obtained for the flatlands of the test areas.

The superimposition highlighted the extent of layover distortion; depending upon relief the horizontal displacement of hilltops and ridges can amount to several kilometres. The examples illustrated are the tongue and terminal moraine

Table 17.2

Expression of cartographic and geologic features on digitally correlated SEASAT imagery and LANDSAT MSS imagery of the Plymouth area, SW England

	Principal drainage channels	Minor drainage channels	Urban areas	Communication routes	Morphology	Lithology	Structure
SEASAT							
1 Look Imagery	A	C-D	B	D	B	C	B
2 Look Imagery	A	B-D	A-B	D	A-B	B-C	A-B
LANDSAT							
False Colour Composites	A	B-D	B	D	C-D	C-D	C
Ratio 4/6	A	C-D	D	C	D	B	C-D
Ratio 4/5	C	D	C	D	D	C	C-D
Ratio 4/7	B	D	B	C	D	B-C	C
Ratio 5/6	B	D	B	C-D	D	B-C	C
Ratio 5/7	B	D	B	C-D	D	B-C	C
Band 4	B	C-D	C-D	D	D	B-D	C-D
Band 5	A-B	C-D	D	D	D	B-D	C-D
Band 6	A	B	B-C	D	D	B-C	C
Band 7	A	B	B	C-D	D	B-C	C

Key

A = Good
B = Moderate
C = Poor
D = Unexpressed

of the Skeidararjokull, an alluvial area and adjacent volcanic terrain. SEASAT shows the rough morainic material clearly and also provides some ice stratification information. Apart from these features the substitution of SEASAT for one of the LANDSAT bands brings no significant benefit in this particular scene. The results of some image combinations are shown in Colour Figs. 17.5 and 17.6.

17.2.8 Conclusions of the geological studies
The following conclusions can be drawn from the various studies undertaken during the geologic evalution of SEASAT Synthetic Aperture Radar imagery.
 (i) Optically correlated SEASAT imagery can be successfully interpreted to show geological structures and morphology. In addition many linear features can be identified on the imagery which are not discernible on other forms of imagery or on existing maps.
 (ii) Lithological identification is only possible in a few special cases in the types of terrain investigated, although more units could be interpreted in the exposed rock areas of Iceland.
 (iii) Digitally correlated imagery, particularly multi-look images, offers advantages for interpretation particularly by improving the definition of boundaries between units and reducing tonal variations with units.
 (iv) Interactive image processing, apart from allowing LANDSAT/SEASAT combinations, did not greatly improve the interpretability of SEASAT SAR imagery for geological purposes.
 (v) SEASAT SAR did highlight some features but LANDSAT MSS imagery seems to contain more information than the equivalent radar image. This comparison between LANDSAT and SEASAT does, however, assume that LANDSAT imagery is readily available. Where there is presistent cloud cover, as in Europe, the all-weather potential of radar offsets many of its disadvantages. Also SEASAT L-band radar was designed to monitor oceans and is not necessarily the best configuration for observations over land.

17.3 LAND USE STUDIES
17.3.1 Outline of the study
The L-band SAR imagery used in the land use studies was acquired by SEASAT in August and September 1978. In total nine separate study areas were examined in order to determine the extent to which detailed land use information can be mapped from SEASAT imagery (Table 17.3). The locations of UK test sites are given in Fig. 17.7. These studies were completed using both optically and digitally correlated SAR imagery. The digitally correlated imagery was processed by the Royal Aircraft Establishment, Farnborough, who made available both precision photographic products and Computer Compatible Tape (CCT) data.

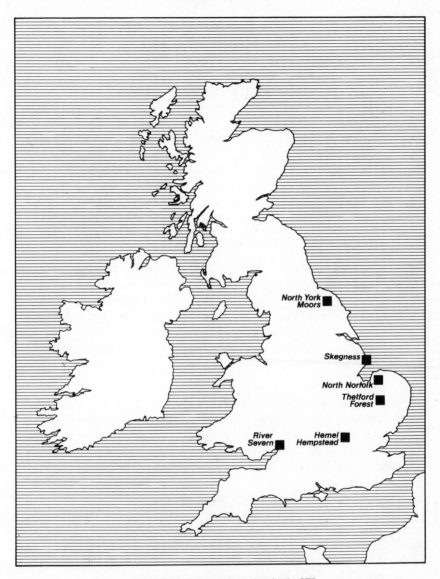

Fig. 17.7 – Land use test sites in the UK.

A considerable part of these analyses was undertaken using SAR imagery covering nothern East Anglia, UK, because the area was considered to be a model test site for establishing a SAR-based land use classification. To complete the study, the classification was tested in a wide variety of areas with significantly different land use patterns. Where possible the land use interpretations from SAR imagery were checked in the field.

Table 17.3

Details of land use study areas

Study area		SEASAT orbits	Image acquisition date	Type of imagery
Nothern East Anglia	UK	762	19/8/78	Optical & Digital
		785	20/8/78	Optical
Thetford	UK	762	19/8/78	Optical
N. York moors	UK	762	19/8/78	Optical
River Severn estuary	UK	834	24/8/78	Digital
		963	2/9/78	Digital
		1307	26/9/78	Optical
Skegness	UK	762	19/8/78	Optical
Hemel Hempstead	UK	719	14/8/78	Optical
Brittany coast	France	785	20/8/78	Optical
		791	21/8/78	Optical
Toulouse	France	590	7/8/78	Optical & Digital
		791	21/8/78	Optical & Digital
Turin	Italy	762	19/8/78	Optical & Digital

When this study was conceived in early 1978 it was proposed that several test sites, each characteristic of a particular land use system, would be established for continuous monitoring throughout the duration of the SEASAT mission. The early failure of the satellite meant that a monitoring programme could not be carried out and in its final form the project contained the following elements:

(a) Construction of a SAR-based land use classification

The first optically correlated imagery to become available for examination covered part of northern East Anglia (Orbit 762 acquired 19th August 1978). This coverage was very close to the area that has been originally chosen as a primary SAR test site. The relatively flat terrain, large fields and simple land use patterns are ideal for the construction of a basic SAR land use classification.

The optically correlated imagery was photographically enlarged to 1:50 000 scale. A preliminary interpretation of this imagery was made using image tone and texture differences as the basis for delineating units. These units were then checked against existing maps and a preliminary classification was established.

By the time that the imagery had been made available and interpreted almost a year had elapsed since the original acquisition of the imagery. It was therefore not possible to obtain detailed information about ground conditions at the time of SEASAT overflight. Broad crop information was available from local farmers, but the effect of variables such as soil moisture content and wind speed couuld not be determined, and it was not possible to establish fundamental links between ground conditions and tone/texture units on the radar imagery.

Using the SAR interpretation, supported by information from existing maps and fieldwork, it was possible to devise a broad classification for mapping land use in northern East Anglia. This classification was further tested by interpreting imagery covering Thetford in Suffolk.

(b) Visual comparison of optically and digitally correlated imagery
Digitally correlated imagery was processed by the RAE and made available as 1:50 000 scale precision photographic imagery in both single and multi-look formats. These images were visually compared with optically correlated imagery to determine the extent to which digital correlation improves the quality of information available for interpretation on SEASAT SAR imagery.

Improvement in image quality was assessed in terms of increases in the variety of tone/texture units that could be consistently recognised, enhanced definition of boundaries, and ease of interpretation. The detailed comparison was carried out on imagery covering northern East Anglia, adjacent to the areas of preliminary interpretation.

A rapid visual assessment of East Anglian imagery indicated, however, that for some of the other areas investigated, particularly over mainland Europe, the digitally correlated imagery would be significantly superior to optically correlated data. In such cases little detailed interpretation was carried out on the optically correlated imagery. As the preparation of digitally correlated imagery takes some weeks the locations of the images required for analysis had to be fixed early in the Project. The RAE produced single and multi-look digitally correlated SAR imagery, in photographic format, for the Brittany, Turin and River Severn estuary land use study areas.

In most cases the potential for mapping land use seemed greater with digitally correlated than with optically correlated imagery and so a quantitative assessment of the information content of SEASAT SAR imagery was made by comparing digitally correlated imagery with existing maps.

(c) Digital analysis of SEASAT imagery
In some study areas, particularly nothern East Anglia and the River Severn estuary, the IDP 3000 interactive image processor at the Royal Aircraft Establishment, Farnborough was used to analyse the SEASAT imagery. The CCTs of digitally correlated imagery were processed and supplied by the RAE.

IDP 3000 analysis was carried out in areas of East Anglia that had previously been investigated using optically correlated imagery in order to assess what additional interpretable information could be extracted using a computer-based image processor. Amongst the techniques employed were contrast stretch enhancements, density slicing, and edge enhancement. In addtion both single and multi-look digitally correlated imagery were compared.

The results of each enhancement process were recorded by photographing the monitor screen of the IDP 3000. Some of these transparencies were later used as an aid to visual interpretation of the imagery.

(d) Evaluation of the SAR-based land use classification under various land use systems

In order to evaluate fully the SEASAT SAR data, the classification that had been derived from SAR imagery covering a simple land use system had to be tested in areas where the topography, land apportionment system and farming practices present greater problems for mapping from radar. Therefore a number of other sites were examined using both optically and digitally correlated imagery that had been photographically enlarged to 1:50 000 scale.

The additional sites were located in the following areas:

(i) North York moors

This region of moderate relief was chosen as a study area in order to test the applicability of the SAR-based classification in places less ideal than northern East Anglia.

The interpretation of this imagery (Orbit 762) was checked in the field during the summer of 1979. As with the East Anglia imagery the length of time between image acquisition and field verification was too long to allow a detailed assessment of factors that influence radar reflectance to be made. Some simple relationships could, however, be identified.

(ii) Brittany coast

Part of the north Brittany coast was imaged by SEASAT from two different look directions (Orbits 785 and 791). As this area has a significantly different land use pattern from that of East Anglia it was considered appropriate to extend the evaluation of SEASAT by analysing some of this imagery. In Brittany a high proportion of the arable land is devoted to fodder crops. Wheat, barley, and oats are also grown. The topographic maps show that small blocks of woodland, used as windbreaks, are evenly scattered over the whole region.

(iii) Toulouse

SEASAT imagery of this area, which represents a different climatic regime to East Anglia was analysed (Orbits 590 and 791). The pattern of agriculture is closely related to climate; the most important crops are grain sorghum, maize and lucerne. Vineyards are also common.

(iv) Turin
A section of imagery from Orbit 762 covering a relatively flat region southwest of Turin, in northern Italy, was interpreted, and in this case some field checking of the results was carried out. Maize is the most important crop, vines are also grown and poplar plantations are common.

(v) Hemel Hempstead
In order to assess the possibility of mapping urban land use, a section of imagery from Orbit 719 covering Hemel Hempstead was analysed. An urban land use classification was worked out but its validity was not tested in other urban areas.

(e) Multi-temporal imagery
The principal advantage of using radar rather than other remote sensing instruments is that imagery can usually be acquired regardless of cloud and daylight conditions. Thus the possibility of obtaining regular repeated coverage is much higher than for optical sensors such as the LANDSAT multi-spectral scanner (MSS). For a complete evaluation study, therefore, it was necessary to make an assessment of any changes that could be detected in analysing sequential SEASAT SAR data of one area. The only area with repeated SEASAT coverage in the UK is the River Severn estuary.

SEASAT data from Orbits 834, 963 and 1307, that were acquired on 24 August, 2 September, and 26 September respectively, provided the imagery for this part of the study. It was found that in general terms the classification worked out for East Anglian imagery was applicable in the area surrounding the River Severn estuary. Examination of the optically correlated imagery from the three orbits revealed that in a number of cases the radar reflectance of mapping units changed markedly between different acquisition dates.

In order to investigate these areas of change more closely, CCTs of the digitally correlated imagery were analysed on the IDP 3000 at RAE Farnborough. The results were then checked in the field in an attempt to explain how the reflectance of some mapping units could change within a very few days.

It was expected that changes in crop height (including harvesting operations), row direction, soil moisture content, and wind speed would account for most of the observed differences. Without detailed ground measurements at the time of image acquisition it was only possible to speculate at relationships that might cause rapid changes of radar response.

(f) Comparison with other types of remotely sensed imagery
A brief comparison was made between SEASAT SAR imagery (from Orbit 762) LANDSAT MSS imagery, and X-band, real aperture, Side-Looking Airborne Radar imagery. The analysis was carried out using the IDP 3000 interactive image processor. The imagery used covered an area immediately surrounding Skegness

and the northern part of the Wash. A CCT of Band 7 LANDSAT data (Path 217 Row 023 16 November 1978) and a photographic print at 1.250 000 scale of X-band SLAR (acquired by Motorola Aerial Remote Sensing Inc. in June 1976) were used for this comparison.

As a result of the considerable time differences between the three images it was not appropriate to compare spectral responses of areas; the comparison was limited to differences in boundary information and the overall level of detail contained on each image.

17.3.2 Evaluation of the imagery

The weakness of this evaluation of SEASAT SAR imagery is that it is essentially an historic assessment. Fieldwork at, or very soon after, the time of image acquisition would have produced data that would have allowed an accurate classification to be worked out and tested in several regions with different land use patterns.

What has emerged is an indication of the patterns that a satellite L-band radar system can distinguish in areas of different land use types. These differences both in spatial and temporal terms can be identified and mapped. The results point to the possibility of using a satellite-borne SAR for land use studies.

From the study elements discussed above the following broad conclusions can be drawn:

(a) Image characteristics

The results and conclusions presented in this section on image characteristics are based upon preliminary visual interpretation of both optically and digitally correlated imagery and digital image analysis using the IDP 3000.

Visual interpretation of several sections of SEASAT SAR imagery, both digitally and optically correlated, by different interpreters showed that a maximum of six tone/texture radar mapping units could be consistently recognised. These six units formed the basis of the land use classification that is discussed below.

The number of units, which form a gradation between dark with no white speckles, through an increasing density of white speckles to evenly white tones, was not greater on digitally correlated imagery but the differences were more distinct. Separation of units at the preliminary mapping stage was, therefore, easier and more rapid using digitally correlated imagery. Contrast stretching using the IDP 3000 further enhanced differences between units.

When density slicing was attempted using the IDP 3000 it was found that the within unit tonal variability was so great that individual units could not be identified using this technique. Some interpreters did find that the consistent mapping of SAR units was easier when trying to identify combinations of different colours rather than combinations of different grey tones. However, the use of density slicing techniques was of only marginal benefit and the method was not used extensively.

Multi-look SEASAT radar imagery, i.e. imagery produced by summation of the returns from several segments of the synthetic radar antenna, has a less speckled appearance than single-look SAR imagery. This reduction in the tonal variation caused by radar speckle improves the appearance of the image and makes visual interpretation and separation into distinct tone/texture units more accurate. Wherever possible SAR imagery that had been correlated from three or more 'looks' was preferred for land use interpretation in the study. The application of a low-pass spatial filter, during the digital correlation process carried out by the RAE, improved image quality still further.

The final classification of SEASAT SAR imagery was completed using a photographic copy of a digitally correlated 3-look image that had been further enhanced using a low-pass spatial filter.

In all cases the SEASAT SAR images used for the evaluation of the land use data content were examined at approximately 1:50 000 scale. The scale was chosen as the most appropriate for mapping land use and, in general, boundaries could still be clearly identified despite the considerable photographic enlargement required to achieve this image scale. Unless considerable enhancement of LANDSAT has been carried out using a digital image processor, it would be impossible to use this form of satellite imagery at 1:50 000 scale. The higher resolution of SEASAT imagery and the clearer expression of boundaries allows interpretation at 1:50 000 scale even with optically correlated data that have been photographically enlarged.

For the purpose of accurately comparing SEASAT SAR interpretations with existing land use and topographic maps digitally correlated imagery is essential. This is because optically correlated imagery has considerable distortions in both the across-track and along-track directions.

Scale distortion was a particular problem when optically correlated imagery from different look-directions was compared (both in East Anglia and Brittany). Digital correlation used to produce imagery for the sequential cover of the River Severn estuary largely overcame this problem.

(b) Classification consistency

Table 17.4 shows the final classification worked out for northern East Anglia, and Fig. 17.8 shows an interpretation example. Some of the units listed can be unambiguously interpreted from the imagery; such units include mature woodland, the airfield, and certain bright reflectors such as isolated large buildings. It is often possible for the interpreter to identify the feature on the basis of its shape and its relationship with other features; for example, a linear series of bright spots on an image is most likely to be a series of pylons carrying electricity cables. In other examples the interpretation of radar mapping units must be supplemented by information from field studies and existing land use and topographic maps before a positive identification can be made.

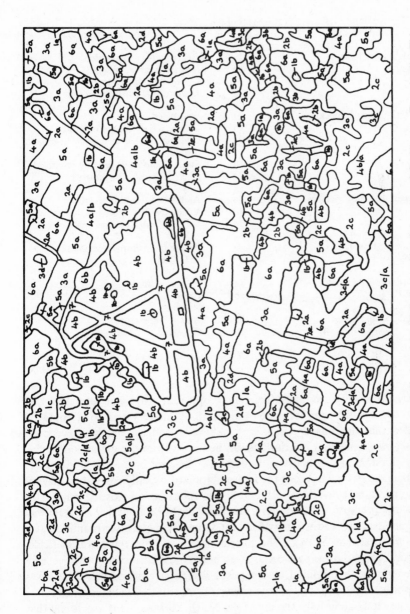

Fig. 17.8 — Land use interpretation example based on digitally correlated imagery of Sculthorpe, East Anglia. This example is shown at 1:50 000 scale and the interpretation key is given in Table 17.4 (next page) (O.S. Grid reference of northeast corner is TF 886 346).

Table 17.4
Land use classification for SEASAT interpretation Figure 17.8

Description of radar mapping unit	Symbol on Fig. 17.8	Land use description
Very bright even tone	1a	Recently harvested fields
	1b	Isolated large buildings
	1c	Closely spaced buildings
	1d	Unidentified strong reflectors
Bright with few dark speckles	2a	Woodland
	2b	Village/widely spaced buildings
	2c	Undifferentiated woodland/ heathland/ meadow
	2d	Some arable fields (usually potatoes)
	2e	Lines of trees/embankments
Mixed light and dark speckles	3a	Arable (root crops/vegetables)
	3b	Villages/widely spaced buildings
	3c	Meadow/rough pasture/heath
Dark with bright speckles	4a	Arable (root crops/vegetables)
	4b	Grassland
Dark with mid-grey speckles	5a	Arable (grain crops)
	5b	Grassland
Dark with faint speckles	6a	Arable (grain crops)
	6b	Grassland
Dark with faint speckles	7	Airfield runways

Only broad crop groupings (i.e. grain crops, root crops and vegetables, and grass) could be distinguished from the SEASAT imagery and so it could be concluded that this imagery is of little value for land use mapping purposes. However, two significant conclusions can be drawn from the evaluation of this imagery. First, a high proportion (around 60 per cent in the UK study areas) of the boundaries drawn during the interpretation of digitally correlated imagery were found to be coincident with boundaries appearing on existing maps. This level of interpretative detail suggests that radar might be a suitably accurate way of mapping current land use provided that some parameters, for example wavelength, are altered to improve the discrimination of crop types within fields. The second conclusion is that certain very bright areas correspond with recently harvested fields; this pointed to the possibility, later confirmed by the River Severn study, of monitoring land use changes from repeated radar coverage.

In the moderately hilly terrain of the North York moors it was found that the effect of slope was to modify considerably the tones on the imagery so that little, if any, land use information could be mapped on antenna-facing valley sides. Provided that the extent of radar shadowing is not too high, some information can still be mapped on away-facing slopes. In the flatter areas the land use of the York moors could be classified in a similar way to the East Anglian imagery.

The experience gained during the interpretation of SEASAT imagery from the various areas of the UK improved the detail of the interpretation of the Brittany and Toulouse imagery where no field verification was carried out. In general terms the classification of land use units in these areas was similar to that of East Anglia in that only broad crop groupings could be identified and additional information was required in order to be able to identify many of the units.

On the Brittany image the drainage patterns, towns, major villages, and large woodland blocks can be consistently identified, but very little detailed agricultural information can be mapped. The difference in detail between the East Anglia and Brittany images is related to land use patterns rather than significant variations in SAR image quality.

In Brittany the fields are small and only a few types of crop are grown. Much of the land devoted to agriculture appears on the image as a dark grey unit with some speckle (equivalent to Units 5a and 5b on Table 17.4). In only a few parts of the image can individual fields be identified.

During the interpretation of the Toulouse imagery it was not possible to differentiate between forestry plantations, poplar plantations, and mature woodland as all three have the same bright signature on the SEASAT SAR. Young poplar plantations that have incomplete canopy cover, were often confused with arable farming areas. Similarly vineyards have moderately speckled signature that can be confused with woodland plantations, rough pasture, meadow, and certain crop types. As with the East Anglian imagery, field boundary information is good, but on the Toulouse image the discrimination of crop types is generally poor.

In general the Turin imagery conforms to the pattern established in other study areas. However, there are some anomalies which could only be explained after ground verification. The radar interpretation showed the area of woodland to be much greater than is shown on the maps; when these areas were checked on the ground they were found to be poplar plantations that are not classified as woodland on the topographic maps. A relatively high proportion of fields with very bright signatures (equivalent to Unit 1a on Table 17.4) were found on the Turin image. These were found to correspond with fields of mature maize. These two features demonstrate the importance of checking interpretations on the ground before a final classification is worked out.

The interpretation of imagery covering the town of Hemel Hempstead is based upon the same six tone/texture radar mapping units as the rural land use classification. The close proximity of very bright tones from good radar reflectors is unlike any pattern found in agricultural areas. Large urban centres can, there-

fore, be distinguished from agricultural land. It is on the city margins that the
greatest risk of mis-classification occurs because the radar signatures of suburban
areas, which consist of a background of dark tones with some scattered bright
areas, are very similar to those of agricultural areas. Within the city little information on detailed urban structure can be identified; this is primarily because the
bright returns from buildings often obscure the low returns from roads and
gardens. The boundaries of larger enclaves within the built-up areas such as parks,
school playing fields, and cemeteries, can usually be accurately mapped.

(c) Change detection
In general the classification developed on the East Anglian imagery can be
applied to images of the River Severn estuary, although the mapped land use
pattern of the Severn area appears to be a little simpler because fewer types of
crops are grown.

Both the visual examination of the multi-temporal SAR coverage of the
River Severn area and analysis on the IDP 3000 have confirmed that the radar
signatures of certain features can change in a relatively short time. Image superimposition and density slicing using the IDP 3000 has been particularly successful
in demonstrating these changes.

Three particular types of features were found to exhibit tonal changes
between image acquisitions. Using density slicing techniques on the IDP 3000,
it was found that the distribution of bright point reflectors varies between
images. The look-direction and incidence angle of radar are important in determining whether a good reflector shows clearly on an image. For example one
group of disused aircraft hangers in the Severn area are very bright when their
longest sides are aligned parallel to the flight direction of the satellite. On orbits
where their orientation is between 45° and 90° to the flight direction the same
hangars appear much less bright on the imagery.

The radar signature of water in the estuary was found to change between
images. After consulting meteorological and tide data it could be confirmed
that this change is associated with surface roughness. At low water and in calm
conditions the whole estuary, both mudbanks and open water, has a uniformly
dark signature. At high water and in windy conditions the larger waves of the
deep water channels are marked by a very bright radar signature.

In rural areas the radar responses of several fields were found to have
changed significantly between image acquisitions: some fields changed from dark
to very bright signatures and others from mid-grey to dark. Soil moisture variations and the motion of crops and trees in different wind conditions can cause
radar responses to be altered. However, these explanations are likely to affect
several fields at the same time. On the River Severn images, several fields had
changed significantly, whilst the radar responses in neighbouring fields had not
changed at all. The reasons for change must, therefore, be associated with

specific agricultural treatments carried out in some fields between image acquisitions. Such treatment would include harvesting, ploughing, re-planting, and silage cutting.

It is not possible to isolate the exact causes of change without detailed fieldwork at the time of image acquisition, but the fieldwork carried out in this study did show that each of the treatments mentioned did account for radar reflectance changes in some fields.

Colour Figs. 17.9 and 17.10 show examples of imagery covering the River Severn estuary where individual fields have markedly changed radar signatures on different acquisition dates. In all about four per cent of the agricultural land area was found to have changed. This part of the study has pointed to the possibility of using a satellite-borne SAR system for monitoring agricultural areas, showing changes that have occurred over a very short time. However, an operational monitoring system would also require the SAR to discriminate crop types. SAR imagery of the type available from SEASAT cannot, at present, satisfy this requirement.

(d) Comparison with other remotely sensed imagery

As a result of the lack of same-date imagery and field information only a superficial comparison betwen SEASAT, LANDSAT, and SLAR could be made. Some of the major features that can be interpreted from remotely sensed data are listed in Table 17.5. Several interpreters were asked to rank the three types of imagery according to the suitablity of each for mapping a particular feature. These results are presented in summary form in Table 17.5.

Table 17.5

Ranking of SEASAT, LANDSAT (single MSS Band) and SLAR for
interpretation suitability, Skegness area

Feature	SEASAT	LANDSAT (Band 7)	SLAR (X-band)
Sea surface state	1	2	3
Coast	3	2	1
Urban areas	2	2	1
Roads/railways/canals	2	3	1
Field boundaries	2	3	1
Tonal contrasts	3	1	1
Rivers	Insufficient information on these images		

In most respects the higher resolution X-band SLAR appears to be superior to the satellite acquired images for land use mapping. However, the LANDSAT was only examined in one band, and additional information from colour composite images would alter some of the ratings. It is the higher resolution of SEASAT, compared with LANDSAT, that places it second in some of the rankings.

17.3.3. Conclusions of the land use studies

The following conclusions can be drawn from the various studies undertaken during the land use evaluation of SEASAT Synthetic Aperture Radar imagery:

(i) A total of six tone/texture units could be consistently recognised on SEASAT imagery. On the basis of shape, location and their relationship with other units a number of radar mapping units could be positively identified as land use units. Others required additional information from fieldwork and existing maps for positive identification.

(ii) Interactive image processing on the IDP 3000 did improve the accuracy of interpretation.

(iii) Multi-look digitally correlated imagery shows improved definition of boundaries between units and reduces within-unit tonal variations.

(iv) The examination of sequential SEASAT coverage showed some possibility for monitoring short-term land use change.

Acknowledgement

We gratefully acknowledge support from the National Remote Sensing Centre, Farnborough, in meeting the production costs of the colour plates included in the colour section from this chapter.

REFERENCES − Land use

Banyard, S. G., (1979). Radar Interpretation based on photo-truth keys. *ITC Journal,* **79** (2), 267-276.

Berger, D. H., (1970) Texture as a discriminant of crops on radar imagery. *IEEE Trans. Geo. Electron.* **GE-8** (4), 344-348.

Brisco, B., & Protz, R., (1980) Corn field identification accuracy using airborne radar imagery. *Canadian Journal of Remote Sensing,* **6** No. 1, 15-25.

Bush, T. F., & Ulaby, F. T., (1978) An evaluation of radar as a crop classifier. *Remote Sens. Environment* **7** 15-36.

Bush, T. F., & Ulaby, F. T., (1978) Crop inventories with radar. *Canadian Journal of Remote Sensing,* **4**, No. 4, No. 2, 81-87.

Coleman, A. I., (1963) Second Land Utilisation Survey of Britain: *Map Sheet 525, Fakenham* Isle of Thanet Geographical Association.

Daus, S. J., & Lauer, D. T., (1971) SLAR Imagery for evaluating wildland vegetation resources. *Proc. Am. Soc. Photog.* Fall Convention, 386-392.

Haralick, R. M., Caspall, F. C., & Simonett, D. S., (1970) Using radar imagery for crop discrimination: a statistical and conditional probability study. *Remote Sens. Environment,* **1**, 131-142.

Henderson, F. M., (1975) Radar for small-scale land-use mapping. *Photogrammetric Engineering and Remote Sensing,* **41** (3), 307-319.

Henderson, F. M., (1977) Land use interpretation with radar imagery. *Photog. Eng. and Remote Sensing,* **43** (1), 95-99.

Henderson, F. M., (1977) A comparison of cell and synoptic techniques for land use analysis with radar imagery. *Proc. Am. Soc. Photog. 43rd. Annual Meeting,* 15-26.

Henderson, F. M., (1979) Land-use analysis of radar imagery. *Photo. Eng. and Remote Sensing,* **45** (3), 295-307.

McDonough, M. & Deane, G. C., (1980) *Preliminary results of an evaluation of the land use and geologic data content of SEASAT imagery.* European Space Agency, Special Publ. 154.

Morain, S. A., (1976) Use of radar for vegetation analysis. *Remote Sensing of the Electomagnetic Spectrum,* Vol. 3, Ch. 4, 61-78.

Hunting Geology and Geophysics Limited, (1981) *The evaluation of the data content of overland SEASAT-A SAR imagery.*

Natural Environment Research Council, (1974) *Remote sensing evaluation flights 1971.* Eds. Curtis, L. F. & Mayor, A. E. S.. Publication Series C, No. 12.

Schmugge, T. J., (1978) Remote sensing of surface soil moisture. *J. of Appl. Meteor.* **17** 1549-1557.

Ulaby, F. T., (1975) Radar response to vegetation. *IEEE Trans.* **AP-23 36-45.**

Ulaby, F. T., & Bush, T. F., (1976) Monitoring wheat growth with radar. *Photogrammetric Eng. and Remote Sensing* **42** 557-568.

Ulaby, F. T., & Bare, J. E., (1979) Look direction modulation function of the Radar backscattering coefficient of agricultural fields. *Photog. Eng. and Remote Sensing,* **45** 1495-1506.

Geology

A. D. Beccasio, & J. H. Simons, (1965) Regional geologic interpretation from side-looking airborne radar (SLAR) Presented at the 31st Annual Meeting, American Society of Photogrammetry, Washington, 28 March to 3 April. *Photogrammetric Engineering,* **31** (3) 507.

B. J. Blanchard, (1980) Some confusion factors in radar image interpretation. *Radar geology: an assessment.* JPL Publication 80-61.

J. Bodechtel, K. Hiller, & U. Munzer, (1979) *Comparison of SEASAT and LANDSAT data of Iceland for qualitative geological application.* European Space Agency SP-154, 1979.

R. D. Brown, Jr., (1966) *Geologic evaluation of radar imagery: San Andreas fault zone from Stevens Creek, Santa Clara County, to Musael Rock, San Mateo County, California.* U.S.G.S. Technical Letter – NASA-45, August, 15 pp. (NTIS No. N70-38893).

M. L. Bryan, (1979) *Bibliography of geologic studies using imaging radar.* J.P.L. Publication 79-53. NASA/JPL, Pasadena, Ca.

I. Cindrich, C. Dwyer, A. Klooster, & J. Marks, (1980) Optical processing of SEASAT-A SAR data. *Proc. Working Group Meeting, ESRIN,* Frascati, Italy, 13-14th Dec. 1979 (ESA SP 1031, Sept. 1980).

M. Daily, C. Elachi, T. Farr, W. Stromberg, S. Williams, & G. Schaber, (1978) *Applications of multispectral radar and LANDSAT imagery to geologic mapping in Death Valley.* JPL Publication 78-19, Pasadena, Ca., 1978.

L. F. Dellwig, (1968) Pulses and minuses of radar in geological exploration. *Earth resources aircraft program status review,* Vol. 1, Geology, Geography and sensor studies, NASA, MSC, Houston, September pp. 14-1 — 14-25. (NTIS No. N71 16126).

C. Elachi, (1980) Spaceborne imaging radar: geologic and oceanographic applications. *Science,* **209** (4461).

A. M. Feder, (1960b) Interpreting natural terrain from radar displays *Photogrammetric Engineering,* **26,** (4), 618-630.

J. P. Ford, (1980) SEASAT orbital radar imagery for geologic mapping: Tennessee-Kentucky-Virginia. *Amer. Assoc. Petroleum Geologists Bull.,* (In press).

P. L. Jackson, (1980) Multichannel SAR in geologic interpretation: an appraisal. *Radar geology: an assessment.* JPL Publication 80-61.

A. J. Lewis, & W. P. Waite, (1972) Relative relief measurements from radar shadows: methods and evaluation. *Proc. 68th annual meeting of association of American Geographers,* Kansas City, MO., 23-26 April, **4,** 65-70.

H. C. MacDonald, & W. P. Waite, (1972) Remote sensing practicality: radar geology. *Proc. Technical Papers, Electro-Optical Systems Design Conference,* New York, 12-14 September, 68-78.

H. C. MacDonald, P. A. Brennan, & L. F. Dellwig, (1967) Geologic evaluation by radar of NASA sedimentary test site. *IEEE Transactions, Geoscience Electronics,* **GE. 5,** (3) December 72-78.

P. H. A. Martin-Kaye, & A. K. Williams, (1973) Radargeologic map of Eastern Nicaragua. *Memoriria de lat IX conferencia Inter-Guyanas, Boletin de Geologia,* Publicacion Especial No. 6, Caracas, Venezuela, pp. 600-605.

F. Sabins, R. Blom, & C. Elachi, (1980) SEASAT radar image of San Andreas Fault, California. *Amer. Assoc. Petroleum Geologists Bull.,* **64** 619-628.

D. S. Simonett, (1968) *Potential of radar remote sensing as tools in reconnaissance, geomorphic, vegetation and soils mapping.* U.S. Geological Survey Interagency Report — NASA-125, July, 19 pp. (NTIS No. N69 28154).

H. E. Stewart, R. Blom, M. Abrams, & M. Daily, (1980) Rock type discrimination and structural analysis with LANDSAT and SEASAT data. *Radr geology: an assessment.* JPL Publication 80-61.

R. S. Wing, W. K. Overbey, Jr., & L. F. Dellwig, (1970) Radar lineament analysis, Burning Springs Area, West Virginia — An aid in the defination of Appalachian plateau thrusts. *Bulletin of the Geological Society of America,* **81,** (11) November, 3437-3444.

Chapter 18

Visual interpretation of SAR images of two areas in the Netherlands

Th. A. de BOER
Centre for Agrobiological Research, Wageningen, Netherlands.

18.1 INTRODUCTION

The all-weather capability of synthetic aperture radar is of particular interest to applied ecology. Ecologists are interested not only in the pattern of land units and their individual characteristics but also in dynamic processes such as plant-biomass and ground-water table fluctuations. Other remote-sensing techniques may fail because of atmospheric conditions in different parts of the world.

As a user I am not familiar with radar techniques, and therefore we do radar research in the Netherlands in an interdisciplinary team with physicists, referred to as the ROVE team (Radar Observation of VEgetation).

In this team our Centre works together with members of the Physics Laboratory TNO at the Hague with the Delft University of Technology, the National Aerospace Laboratory in Amsterdam, and the Agricultural University at Wageningen. Previous radar research of vegetation, crops, and soils made it easier for us as users to interpret the SAR images. The images used for the interpretation were processed by DFVLR at Munich, German Federal Republic. Both SAR images were recorded on rev 1493 — flown on 9 October 1978, one day before the mission came to an untimely end.

The first area (I), see Fig. 18.1, is situated in the SAR frame with its centre indicated by N 052-56-14 and E 005-54-09. It concerns an almost flat transition area from sand to peat (height difference about 7 m at most). This area is mainly under grassland and its size is about 20 by 25 km.

A second area (II) is situated in the frame with its centre at N 052-23-60 and W 008-57-49. This area mainly consists of light marine clay and has an even surface; the major part is arable land.

The resolution of the images is 25 m and the scale 1:250 000. For the interpretation the images are viewed under a lens magnifying 2X, so they are eventually studied at a scale of 1:125 000.

Fig. 18.1 shows the location of the areas.

Fig. 18.1 – Situation of the test areas in the Netherlands

18.2 INTENTION

The main objective of the first area studied was to investigate if soil use could be detected from the image. In addition we were interested in the possible detection of ground-water level in the grassland areas.

In the second area, soil use detection was the principal aim, but we were also interested in whether or not arable crops grown that year could be detected.

However, we must keep in mind that on 9 October, the date on which the satellite images were made, all crops except beet and onions had already been harvested. Therefore only by indirect measurements of stubble height and soil tillage might some crop detection by SAR be possible.

18.3 RESULTS

18.3.1 Grassland Area

The grassland area concerned is situated in the Province of Friesland, surrounding the town of Wolvega (see Fig. 18.1, area I). For this size of area (distances up to 25 km) the image was geometrically correct.

The interpretation proceeded from a number of training sites in which various types of land use occurred and these are shown in Fig. 18.2. Besides densities of grey levels, the textural differences in the images were also used.

Ditches in the areas, which were filled with water, are comparatively narrow (2 m wide), and apparently affect the direction of linear textures. The sides of the ditches facing the satellite appear to affect the backscatter of the radar like a corner reflector, so that even at a given resolution of 25 m these small sides still affect the texture and, because of this, areas with different field patterns confined by ditches are clearly discernible.

One striking feature is that ditches parallel to the satellite's track frequently have distinctly higher backscatter than those running in other directions, with density values corresponding to those of towns and roads. In the first instance the 'antennae parallel ditch areas' can be confused with areas in which the grassland fields are separated by green belts. However, the line pattern here is more coarse, and distinct black lines occur beside the green belts (radar shadow). Here again the most distinct returns occur when the shelter belts run parallel to the satellite's track.

This phenomenon of the strength of the image being dependent on the direction of the elements with respect to the satellite's look angle was observed for a variety of features including not only ditches and green belts but also dikes, ditch-side vegetation (distinctly visible with the lakes in Friesland), river banks, trees along roads, etc.

Differences in backscatter were also observed from one field to another. Since this experiment was originally planned for the spring of 1979, no field observations on the grassland situation had been carried out on 9 October 1978. For this reason we cannot explain in detail the observed differences. Earlier studies with SLAR and ground-based radar have shown that there is a difference in radar backscatter between grasslands short-cut or just grazed, and fields with a grass height of about 15 cm. The same situation may have occurred in certain places at the beginning of October 1978. Differences in backscatter may also occur after applying slurry and, certainly in this area, slurry was regularly applied as we observed during field visits in October 1979.

Below we show the land use types finally identified. The numbers indicate where they occur in Fig. 18.2 interpreted from the SAR image (Fig. 18.3). The types of use have been arranged according to increasing density (decreasing backscatter), which is not proportionally distributed over the density classes.

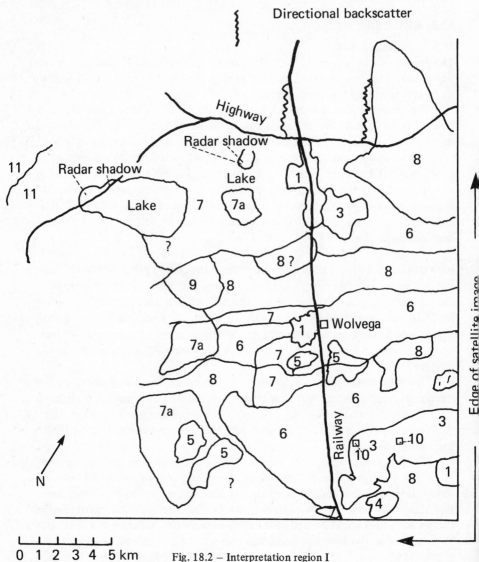

Fig. 18.2 – Interpretation region I

Fig. 18.3

Density
1. Towns, villages (concentration of houses), and sometimes
   grassland fields (no explanation)   1
2. Roads, usually distinguishable by adjacent dikes, houses, or trees   1
3. Woods dominated by coniferous trees   2
4. Arable land (only to be distinguished from 3 by texture and shape)   2
5. Deciduous woods (on wet soil)   3
6. Grassland areas with many green belts and thickets
   (texture formed by changing densities)   3 & 4
7. Open grassland areas with narrow ditches, well-drained   4
7a. Wet grassland with open water (other texture)   4
8. Open grassland areas with wide ditches, rather wet   5
9. Small open waters (lakes, canals) and moist fields   6
10. Arable land in wooded area (radar shadow?) and outside   7
11. Ditch-side vegetation (long radar shadows)   7

18.3.2 Arable crop area

This area is one of the Zuyderzee polders reclaimed after the Second World War, and is called 'Oostelijk Flevoland' (see Fig. 18.1, area II and Fig. 18.4).

The problems in interpreting this SAR-image are much greater, because the main area is used for arable cropping. Grassland consists of a permanent vegetation which looks almost identical to the conditions at the end of the growing season. Annual crops are grown on arable land, and by 9 October most of these crops have been harvested. The only crops present by then are sugar beet and onions, and since the latter are very seldom grown they can be neglected in the interpretation.

Yet on the well separated arable fields on the SAR image (Fig. 18.4) about 5 densities can be distinguished, similarly as on the SLAR images of the same area during summer, even without standing crops.

This phenomenon is caused by differences in the condition of the soil following the growing of various crops. When potatoes are grown the soil is turned up rather heavily after lifting and is still bare and rather rough. For seed potatoes, which have been lifted earlier, the soil is already more settled, more even, and partly overgrown with weeds. Where wheat has grown the field is covered with stubble and partly with green manure crops and weeds.

These differences in the roughness of the soil with or without vegetative cover cause differences in the backscatter which are shown on the image by differences in density.

At a scale of 1:250 000 in which the image is available, the arable fields cannot be indicated with symbols, so an interpreted map of area II cannot be presented.

Below we give a description of the density classes, with a similar legend to that presented for area I.

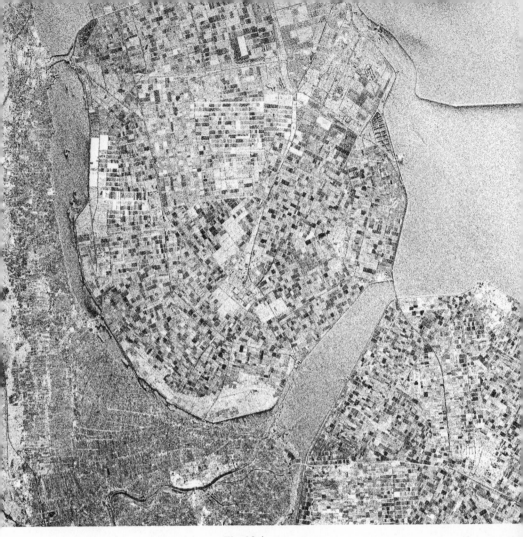

Fig. 18.4

		Density
1.	Airstrip and short-cut grasslands, houses, dikes, ditch sides	1 (almost white)
2.	Woods (deciduous)	2
3.	Orchards — 2 and 3 only distinguishable by differences in texture and shape	3
4.	Sugar beet (still present or harvested and leaves still present) and wheat (stubble or green manure)	4
5.	Newly sown grassland — 5 and 4 to be distinguished by adjacent areas and a slightly different texture	4
6.	Fully grown potatoes, bare soil rather rough (does not always apply, for it may have been another crop)	5
7.	Seed potatoes, lifted early, bare soil somewhat collapsed (does not always apply, for it may be wheat soil)	6

PART 4

Radar altimeter

Chapter 19

Altimeter measurements of ocean topography

D. E. CARTWRIGHT and G. A. ALCOCK

Institute of Oceanographic Science, Bidston, Birkenhead

19.1 INTRODUCTION

The amazing instrumental precision of the satellite-altimeter from the order of a metre in SKYLAB and GEOS-3 to about a decimetre in SEASAT, in measuring a distance equivalent to that between London and Toulouse, puts this instrument into a class of its own as a new tool for global oceanography. When one adds its ability to measure wave height and surface roughness, its unambiguity, and its modest data rate, it easily becomes the 'star' of the four microwave sensors aboard SEASAT, remarkable though the others are. The wealth of dynamic features extracted from the altimetry during the short active life of SEASAT by Cheney & Marsh and others at the Venice COSPAR meeting (Gower 1981) makes us look forward with impatience to the next generation of altimeter-bearing satellites promised for the late 1980s.

At the same time, the SEASAT experiment has also taught us that the altimeter cannot be used in isolation. Some form of radiometer is necessary, in order to sound the water vapour in the air column, hence to provide a non-negligible correction to the return time. The precision of the altimeter height is valueless unless supported by equal precision in the computed orbital elevation, which requires a detailed knowledge of the gravity field and of non-gravitational forces. Steady currents cannot be deduced from the sea surface topography without an accurate map of the gravitational geoid as reference surface, although variations in currents including tides may be deduced without such a map. All these matters require the collaboration of scientists with expertise in several branches of geophysical knowledge besides oceanography.

Perhaps the simplest lesson learnt has been the vital importance to oceanography of fixing the satellite's orbital period so that its Earth-track repeats itself every few days. Before SEASAT, scientists were divided between supporters of this type of orbit and those who preferred one whose Earth-track progresses in small increments of longitude to give ultimately a very dense spatial coverage. The latter is perhaps best for dense definition of the altimetric geoid, but the results achieved from the fixed 'Bermuda' orbit of 13 September – 10 October 1978 showed to all users the vital advantage of having a constant repeated profile against which to assess dynamic changes in the ocean surface. Future missions should of course preferably aim for a longer repeat period than the three days of the 'Bermuda' orbit, in order to achieve a finer spatial grid, but periods with a tidal 'alias' must be avoided; somewhere in the range 6–12 days would probably be best for time- and space-resolution of most ocean features.

19.2 WORK ON OCEAN ALTIMETRY AT IOS

Oceanographic interpretation of the SEASAT altimetry in the USA has mainly concentrated on the famous detection of a Gulf Stream eddy (Cheney & Marsh 1981), the Gulf Stream itself (Kao & Cheney 1982), and the global variability of dynamic height, (Cheney, Marsh & Grano 1981)[†] In all these studies, ocean tides were assumed to be removed from the data by means of a global numerical model. Current systems in the eastern north Atlantic are relatively slight, so our work at IOS and elsewhere among the SURGE group has concentrated on the special problems of removing tides from shelf-sea data and on direct evaluation of the tides from the altimetry over the ocean. (We except here the work on dynamic variability in various ocean areas by Menard (1981).)

The authors' work is divided into three distinct phases, each associated with a distinct sea area, as follows:

(1) Removal of tides and meteorological effects from altimetry over the North Sea, as a contribution to the geoidal studies at IAG Frankfurt and ITG Hannover.
(2) Study of residual errors and evaluation of semidiurnal tides from the altimetry of the northeast Atlantic (5–30°W).
(3) Expansion of (2) with new JPL data set, and tidal evaluation for the central north Atlantic (30–60°W)

In this chapter, we briefly review researches (1) and (2) with references to more detailed accounts published elsewhere, and finally present some recent results from (3).

19.3 THE NORTH SEA

From the time of formation of SURGE in 1977 the major objective of its altimetry team has been an intensive study of the data from the North Sea as a test of the validity of the altimetric equation. This study, which is now practically completed, made use of the accurate geoid available for the North Sea, the European satellite-tracking network, a good model for tides and weather-induced variations in sea

---

†See also other US studies in Gower (1981), and in EOS **62** (17).

level, and good coverage of coastal tide-gauges. The result (Brennecke & Lelgemann Chapter 24 of this book; multi-author paper in preparation), are very satisfactory, except for some uncertainty near the coast of Norway. Our part consisted in computing the dynamic deformation of the sea surface from a nearly geoidal surface, as one of the several elements in the altimetric equation.

Our calculations took account of:
(a) the vertical tidal motion of the solid Earth,
(b) the vertical marine tide relative to the solid Earth,
(c) surges caused by wind stress and atmospheric pressure,
(d) slow circulation and steric changes associated with movements of the Atlantic Ocean,
(e) a quasi-static slope which balances the mean meridional gradient of atmospheric pressure, and
(f) local steric differences caused by freshwater influx.

We ignored increments less than 0.05 m because several elements in the altimetric equation are only to decimetric precision. The Earth tide correction (a) was a straightforward calculation of the 'body tide' of the solid Earth, because loading and self-attraction from the M2 ocean tide in the North Sea has been shown to be less than 0.02 m (Baker 1980).

Tides (b) and surges (c) were estimated to at least decimetre precision from a numerical model devised by Flather (1976, 1981) for the forecasting of floods and now used routinely by the UK Meteorological Office for this purpose. The tides calculated by global numerical models, though adequate in the deep ocean, are of poor accuracy in shallows such as the North Sea because of the small horizontal scale of such motions. Flather's model represents the principal components of the tides over the shelf seas more correctly, as free waves generated by the tides at the shelf edge which have been accurately measured by IOS (Cartwright *et al.* 1980). Shallow water effects are self-generated by a full complement of nonlinear terms used in the dynamical equations. Some small tidal constituents not included in the model, notably those of diurnal period, were added from an array of constants from an adequate tidal map. The dynamical effects of wind and pressure are computed in the model from an input array supplied by the 10-level weather forecasting model run by the UK Meteorological Office. These were not severe during the period of operation of SEASAT.

Estimation of steric slopes (f) was shown by some applications of Bowden's formula (Bowden 1960) to typical density distributions in the North Sea to be negligible (less than 0.05 m) except in the extreme margins of the German Bight and near the entrance of the Baltic Sea and the Rhine river. We finally made a small correction to allow for (d) and (e), occasionally of order 0.10 m, by fitting a least-squares plane surface to the low-passed sea levels measured at nine coastal stations round the North Sea (including Shetland), using for each individual datum the mean sea level recorded during a simultaneous year, corrected for the mean atmospheric pressure during that year. The year chosen for this reference datum was the only recent 12-month period during which all the chosen tide-gauges operated without obvious fault, namely the period October 1972–September 1973. We considered that geodetic information relating this surface to a true equipotential level was insufficiently reliable to attempt further reduction, but the correction if known would probably amount to less than 0.1 m.

Fig. 19.1 illustrates a typical set of dynamic 'corrections', calculated as described above, for the altimeter surface topography on the descending track of orbit no. 1158, crossing the North Sea from southern Norway to East Anglia and the English Channel from near Portland to Brittany. Points on the abscissae represent the passage over England. Similar corrections were made to all valid passes over the North Sea, and the results used in the overall altimetry exercise are described by Brennecke & Lelgemann in Chapter 24.

19.4 PRELIMINARY WORK ON NORTHEAST ATLANTIC

Our earlier work on the northeast Atlantic Ocean was based on the first edition of the SEASAT altimetry released to SURGE members, based on preliminary assessment of the orbital elevations by JPL. Our results are adequately presented in Cartwright & Alcock (1981), and we give only a brief summary here.

We at first applied corrections for the ocean surface topography in terms of known physical influences, as for the North Sea but with different techniques. The body tide of the solid Earth was calculated as before, but the oceanic loading effects are not negligible here. The loading tide was approximated simply by −0.04 × the local ocean tide, which was known accurately. In this area, the tidal signals supplied from the models of Parke & Hendershott or Schwiderski are probably adequate, but we calculated a complete tidal spectrum from our own maps of the area deduced from a major programme of measurements (Cartwright *et al.* 1980). Finally, weather effects in deep water were represented by the inverse barometer effect, using the good coverage of this area by 6–hourly isobaric charts, and a possible small steric anomaly was corrected in terms of tide-gauge records at Cascais, Brest, Stornoway, and Reykjavik.

Our statistical analysis of the altimetric topography, corrected for the above effects, during the period of repeated tracks (Bermuda orbit), confirmed what has probably been found by all investigators. The altimetric profiles (Fig. 19.2) show amazingly repeatable detail at short wavelengths, much of it correlated with bathymetric features, but there are evident errors in the orbital corrections at very long wavelengths, causing successive estimates of the surface height at any given point to vary within a range of 1–2 metres. The long wavelength errors may however be largely removed by pre-whitening operations, of which the simplest is working with the differences between surface elevation at two given points, L kilometres apart. The standard deviation of such pre-whitened measures is of order 0.1–0.2 m, which is certainly in the region at which one

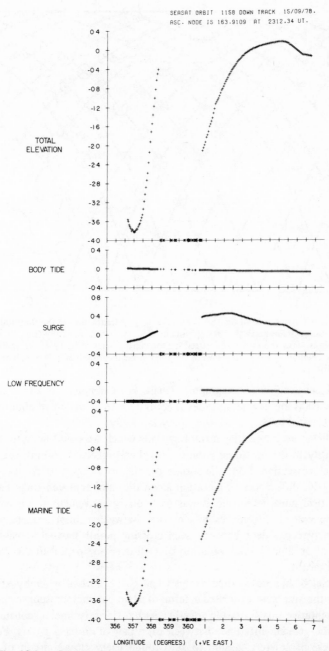

SEASAT ORBIT 1158 DOWN TRACK 15/09/78.
ASC. NODE IS 163.9109 AT 2312.34 UT.

TOTAL ELEVATION

BODY TIDE

SURGE

LOW FREQUENCY

MARINE TIDE

LONGITUDE (DEGREES) (+VE EAST)

Fig. 19.1 – Computed increments in metres to altimetric sea surface recorded during SEASAT orbit no. 1158 over the North Sea and the English Channel, 15 September 1978. The bottom scale is in degrees of east longitude.

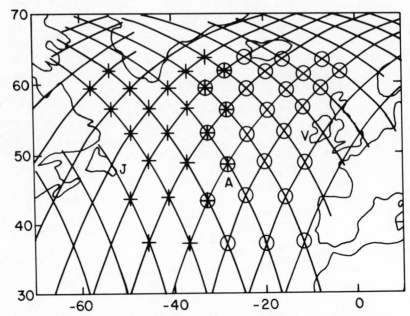

Fig. 19.2 – Earth-tracks of SEASAT over the north Atlantic during the 'Bermuda' orbit period. Circles mark the cross-points used for the northeast array (Fig. 19.3), and + signs mark the cross-points used for the central array (Fig. 19.4). 'A' marks the point where tidal constants were supplied from measurements. V = Valentia, J = St John's.

hopes to detect genuine oceanographic signals. For example, a dynamic change of 0.1 m in 1000 km at mid-latitudes is equivalent to a current of about 0.012 m/s which is useful to be able to measure synoptically.

There being no interesting current systems which we could hope to resolve unambiguously in the northeast Atlantic, we then proceeded to omit the semi-diurnal tidal correction from the altimetry, and to attempt to recover it by analysis. Again, differences of elevation along the same repeated track had the lowest practical noise level, and it was particularly convenient to analyse the tidal component of differences in elevation between orbital crossing-points. Plots of uncorrected data between such crossing points showed a reasonable semblance of an aliased tidal variation of well over one period in the 25 days of repeated record.

A special technique of tidal analysis had to be devised whereby all tidal elements in the area were expressed in terms of a two-parameter response operator on a representative tidal variation together with a third parameter (constant) to allow for the geoidal difference between each pair of crossing points. Forcing each of these parameters to sum to zero round every closed circuit of inter-crossing-point segments was a necessary and useful constraint on the overall solution.

Applying this procedure to the (then) total available data set (after omission of a few obvious anomalies), covering a network of 18 segments, of which 12 were independent and the remaining 6 were determined by the zero-constraint, required solution of a 36 × 36 matrix. The resulting tidal-difference parameters were then converted to vectors of elevation at the crossing-points themselves by addition of a single tidal vector for the known tidal elevation at one of the crossing points, derived from one of our bottom-pressure records. The array of M2 amplitudes and phases, so derived, agreed surprisingly well with the known tidal map of the area, considering the small sample of data used and its noise level. An expanded result is shown in Fig. 19.3. This greatly encouraged the prospect of being able to use longer sections of future altimetry data to derive the tidal parameter for regions of the oceans where they are still very uncertain.

19.5 RECENT TIDAL ANALYSIS FOR THE NORTHEAST ATLANTIC

With the arrival in 1981 of a global set of altimetry from SEASAT, newly edited by JPL, we decided to extend the method of tidal analysis outlined above to larger areas. Again, we were perforce restricted to the 'Bermuda' period of repeated ground tracks. Our procedure was the same, except that, for the atmospheric loading correction, in place of the barometric pressure array derived from UK sources, we used the FNWC data set supplied on the altimeter tape. Also, we omitted the steric correction based on tide-gauge data, because this would not in general be available in wider ocean areas. On the whole, there were fewer items of dubious data requiring elimination.

After applying the same method to precisely the same network of crossing-points described earlier, and obtaining similar though not identical results, we expanded the network to include cross-points at about 32.8°, 7.7°, and 3.5° W longitudes and at 63.8° and 36.9°N latitudes. This entailed 39 x-point differences, of which 25 were independent, and 14 determined by constraints, leading to inversion of a 75 × 75 matrix of independent solution elements. The associated x-points are marked by circles in Fig. 19.2, and the matrix of solutions for M2 is displayed in Fig. 19.3.

As before, the 26 amplitudes and phases follow, by and large, the pattern known from the underlying tidal map, with amplitudes increasing ·eastwards towards the European shelf and phases progressing steadily northwards and westwards round the major amphidrome off the western side of the map. The reasonably good additional values at the x-point in the SE corner are interesting because the vector difference from the nearest x-point at 43.3°N has no constraint, owing to the absence of data from the x-point over northwest Spain preventing a closed loop. The evaluation of this difference is therefore independent of the rest of the array, and its satisfactory value is encouraging. Another area of interest is in the northeast corner, where the amplitudes and phases correctly represent the rapid changes round the secondary amphidrome east of Iceland.

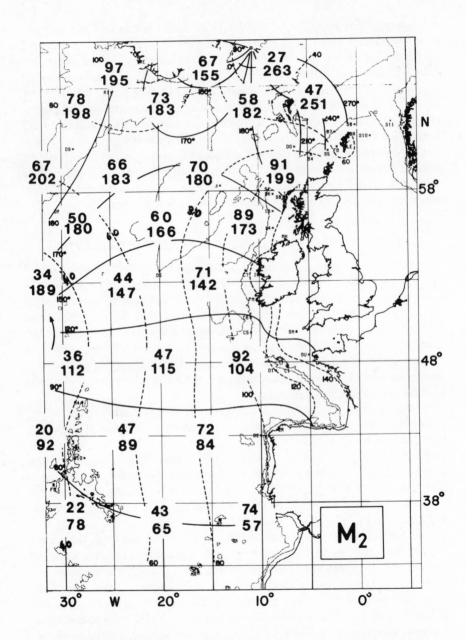

Fig. 19.3 – Large pairs of numbers denote amplitude in centimetres (upper) and phase lag in degrees (lower) of M2 tidal component at x-points of 'Bermuda' orbit, deduced from altimetry by method described in text. The full-line contours are the true phase lags, the broken line contours amplitudes, derived from previous pressure measurements.

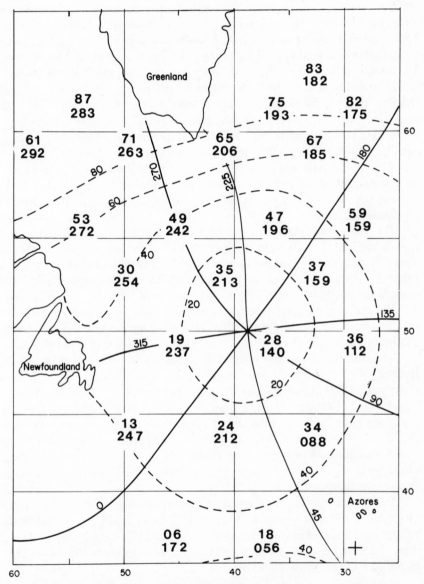

Fig. 19.4 − As Fig. 19.3 for central north Atlantic, except that contours are from Schwiderski (1980).

19.6 EXTENSION TO CENTRAL NORTH ATLANTIC

Our most recent exercise in tidal analysis applies the procedure of section 19.4 to a more westerly sector of the north Atlantic, whose x-points are indicated by + in Fig. 19.2. Six x-points are here used in common with our northeast array, partly in order to use the same tidal point (marked A) as origin for converting differences to elevations, and partly for a comparison test. The nature of the zero-constraints applied in our method results in a different solution for a given x-point difference vector according to the matrix array from which it was derived. To attempt solution for one huge array covering the whole north Atlantic would probably stretch the conditioning of the matrix of normal equations, although it would be worth considering when future altimeter data sets with longer time span and lower noise level become available. Besides this, the single vector used to parameterise the tides, necessitated by the present short span of data, is best referred to another tidal regime in the western Atlantic.

Whereas our tidal vectors in the northeastern array were referred to the tide at Valentia, southwest Ireland, for our array in the central north Atlantic we referred the vectors to the tide at St John's, Newfoundland. Apart from that the procedure was identical.

The results are displayed in Fig. 19.4, superimposed on a tidal map transcribed from Schwiderski (1980). They are significantly less good than in our notheastern array. There is some qualitative agreement with the rapid change of phase as one passes the main amphidrome to the west, and a correct increase in amplitude towards the north, but there are areas with errors of order 90 degrees in phase and 50 per cent in amplitude, especially in the southern sector. It is possible that the geocentric tide as seen by the altimeter differs significantly from the Earth-relative tide plotted in the tidal map, especially in the position of the amphidrome, but this is only likely to make vector differences of a few centimetres amplitude. However, the main cause of deterioration is almost certainly the high noise level in the area of Gulf Stream meanders south of Newfoundland, clearly identified in the altimetry of Cheney, Marsh & Grano (1981), and by Menard (1981), accompanied by generally lower tidal signals than in our notheastern zone. Another possible source of error is the steep gradient of the geoid west of 30°W, which would produce apparent errors in surface elevation related to slight lateral variations in the nominally fixed orbit. The latter source could be removed by further calculation.

Tidally uncorrelated noise would be less harmful to tidal analysis as the time-span of the data increases. However, Fig. 19.4 usefully sets a limit to the amount of meaningful tidal data which can be extracted from the 25-day 'Bermuda' orbit data from SEASAT. We plan similar exercises restricted to the quieter areas identified by Cheney, Marsh, & Grano (1981).

REFERENCES

Baker, T. F. (1980) Tidal gravity in Britain: tidal loading and the spatial distribution of the marine tide. *Geophys. J.R. astr. Soc.* **62** 249–267.

Bowden, K. F. (1960) The effect of water density on the mean slope of the sea surface. *Bull Géodesique* **55** 93–96.

Cartwright, D. E. & G. W. Alcock (1981) On the precision of sea surface elevations and slopes from SEASAT altimetry of the northeast Atlantic Ocean. *In Oceanography from Space,* Plenum Press ed. J. F. R. Gower, 885–895.

Cartwright, D. E, A. C. Edden, R. Spencer, & J. M. Vassie (1981) The tides of the northeast Atlantic Ocean. *Phil. Trans. R. Soc. London* A **298** 87–139.

Cheney, R. E. & J. G. Marsh (1981) SEASAT altimeter observations of dynamic ocean currents in the Gulf Stream region. *J. Geophys. Res.* **86** (Cl) 473–483.

Cheney, R. E., J. G. Marsh, & V. Grano, (1981) Global mesoscale variability from SEASAT collinear altimeter data. *EOS Trans. AGU* **62** (17) 298.

Flather, R . A. (1976) A tidal model of the northwest European continental shelf. *Mem. Soc. Roy. Sci. Liege* **6 ser, 10** 141–164.

Flather, R. A. (1981) Storm-surge prediction using numerical models. *In* 'Monsoon Dynamics', ed. M. J. Lighthill & R. Pearch. Cambridge Univ. Press, 659–687.

Gower, J. F. R. (ed) (1981) *Oceanography from space.* Plenum Press, New York, 987 pp.

Kao, T. W. & R. E. Cheney, (1982) The Gulf Stream front: a comparison between SEASAT altimeter observations and theory. *J. Geophys. Res.* **87** (Cl) 539–545.

Menard, Y. (1981) Study of variability of dynamic ocean topography west of the mid-Atlantic ridge, 30–55°N. *Annal. Geophys.* **37** (1) 99–106.

Schwiderski, E. G. (1980) Ocean tides, part II; A hydro-dynamical interpolation model. *Marine Geodesy* **3** 219–255.

Chapter 20

A survey of some recent scientific results from the SEASAT altimeter

JOHN R. APEL

Applied Physics Laboratory, The Johns Hopkins University, Laurel, Maryland, USA

20.1 INTRODUCTION

The radar altimeter on SEASAT was a remarkable instrument in many ways, yielding a variety of information on geophysical phenomena such as sea surface topography, wave heights, and wind speeds. This chapter discusses selected regional and global fields derived from that instrument which serve to illustrate its utility in marine science.

The altimeter makes basic measurements of three types:

(a) The time delay, T, between the transmission and return reception of the 3 ns compressed radar pulse, along with precision orbit determinations, can be used to construct the topography of the sea surface as a function of the distance along the satellite sub-orbital path; from a sequence of these measurements, information can be deduced on the shape of the marine geoid, underlying gravimetric features, surface current speeds, tides, and other distortions and distensions of the ocean surface.

(b) The broadening of the leading edge of the returned pulse can be used to derive significant wave height, $H_{1/3}$, and the pulse shape can be used to glean information on higher-order moments of the probability distribution function for sea surface heights, also as a function of along-path distance.

(c) The amplitude of the returned pulse determines the normalized radar cross-section per unit area of the sea surface, σ°, from which one can calculate surface wind speed along track, u_s, under certain assumptions.

For all these basic types of measurements, much ancilliary information as well as highly sophisticated mathematical algorithms are needed in order to arrive at the geophysical quantities of interest; the concomitant precisions vary greatly.

The SEASAT Data Utilization Project at the Jet Propulsion Laboratory (NASA 1981a) gives the following r.m.s. precisions for various quantities.

(a) Satellite-to-surface height precision: ±5–8 cm for $H_{1/3} < 5$ m; ±10 cm for $H_{1/3} < 10$ m; ~±15 cm for $H_{1/3} < 15$ m.

(b) Significant wave height $H_{1/3}$ (defined as 4 times the standard deviation of a Gaussian sea): the lesser of ±0.1$H_{1/3}$ or 0.3 m, for $H_{1/3} < 8$ m.

(c) Radar backscatter cross-section, $\sigma°$: ±1 dB for $2 < u_s < 25$ ms^{-1};

(d) Radial component of orbit determination ±70 cm globally, with ±50 cm ultimately determinable.

(e) Corrections to ancilliary data: height bias, 11 cm; electromagnetic mean sea level bias, $-0.05H_{1/3}$; error in correction for propagation through the ionosphere, ±3 cm; error in correction for propagation through the wet troposphere, ±3 cm.

From the three months of SEASAT data have come a variety of very interesting geophysical results, some of which are grouped below according to topography, currents, and surface variability.

20.2 SEA SURFACE TOPOGRAPHY

Marsh & Martin (1982) have constructed a global mean surface topography from the corrected altimeter heights and precision satellite orbits for an 18-day period between 28 July and 15 August 1978. This surface is illustrated in Fig. 20.1, which shows the topography relative to an ellipsoid of revolution equivalent to a mean radius of the earth of 6,378.137 km and a flattening of 1/298.257; contours are at 2-meter intervals. The radial orbit precision is of order ±70 cm (Lerch *et al.*, 1982). This topography resolves small-scale geoidal features such as deep ocean trenches, island arcs and even mid-ocean ridges (signatures of 2–20 m) as well as the larger geodial undulations obtained from satellite orbit perturbations and spherical harmonic analyses. While this topography includes oceanographic 'noise' due to currents, tides, and the like, the r.m.s. contribution of these quantities is thought to be of order 20–40 cm. Thus, this surface can be taken as an estimate of a marine geoid having an internal precision of order 1 m. Longer data sets and removal of the oceanographic signals in the future are likely to improve this type of analysis considerably. It is worth noting that the orbit 'crossover' method, which minimises the r.m.s. altitude differences at points where two orbits cross each other, was not used in arriving at the topography of Fig. 20.1.

A more detailed view of sea surface topography showing evidence of an oceanographic signal on top of the geoid is given by Fig. 20.2 (Cheney & Marsh, personal communication). This illustrates the northwestern Atlantic off the U.S. east coast and is contoured in 1 m intervals at 0.25° in latitude and longitude. While most of the topographic variation is geoidal, there is a generalised anomaly in the region of the mean position of the Gulf Stream with a height change of order one metre, which is the characteristic set-up of a western boundary current. This surface has been computed from radar altimetry data using the method of minimisation of altitude differences as measured where ascending and descending orbits cross over each other. The r.m.s. internal precision of this surface is of order 20 cm, much of which is due to oceanic variability. Using the same method in a dynamically quiet ocean area, the northeastern Pacific, Marsh *et al.* (1982) have derived a surface topography having an internal r.m.s. crossover difference

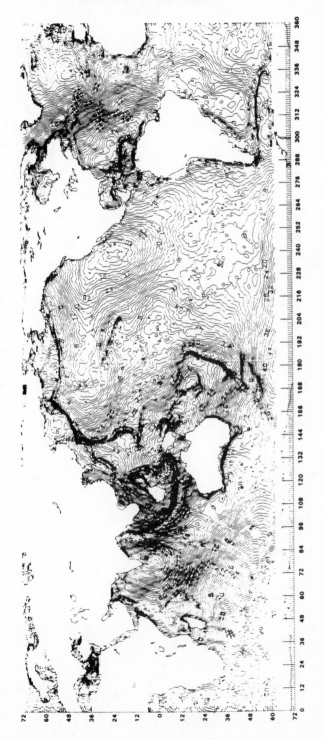

Fig. 20.1 — Mean sea surface topography based on 18 days of SEASAT altimeter data and orbits obtained from an improved gravity model. Contours are in 2 m intervals. Deep ocean trenches, island arcs and mid-ocean ridges are visible as small-scale geoid anomalies.

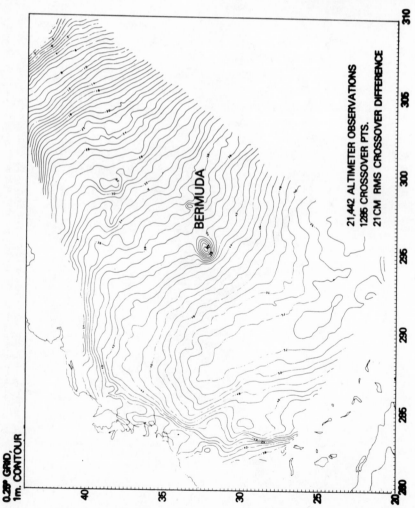

Fig. 20.2 – Mean sea surface topography for the Western North Atlantic based on SEASAT altimeter data and the 'orbit-crossing-minimisation' procedure. Contours are in 1 m intervals. Visible are the geoid anomalies due to the continental shelf edge and the Bermuda seamount, and the surface set-up at the northwest wall of the Gulf Stream.

of 12 cm, which should be divided by $\sqrt{2}$ to account for the fact of two passes per crossover, thus obtaining an internal precision for this regional sea suface topography of well below 10 cm. At this level of accuracy, time-varying oceanic features with signatures above 10–15 cm should be readily observable in a single altimeter pass as departures from such a reference surface. Indeed, it appears that the intrinsic accuracies of radar altitude and satellite orbit determination will ultimately allow the goal of obtaining sea topograph to a few cm to be obtained; this goal is the object of the future NASA TOPEX Program (NASA 1981b).

20.3 OBSERVATIONS OF CURRENTS

Beyond observations of topography, one hopes to measure currents via altimetry, or more precisely, those components of surface geostrophic current velocities at right angles to the satellite sub-orbital path; these include both baroclinic (vertically varying) and barotropic (vertically constant) currents. The general feasibility of this from SEASAT has been demonstrated by several teams of investigators, but limitations of intrinsic accuracy and lack of an independent marine geoid have prevented either a broader or a deeper analysis.

Cheney & Marsh (1981a) published observations of time-varying topographic features in the Gulf Stream region, derived from eight passes taken during the time SEASAT was in an orbit that repeated itself nearly exactly every three days. These clearly showed what was interpreted as meandering of the western boundary of the Stream and motion of mesoscale eddies. A subsequent theoretical analysis of similar data by Kao & Cheney (1982), using a quasi-geostrophic model on an f-plane, shows that six surface profiles, representing a broad range of conditions and taken across the northwest wall of the Stream, may be reduced to a single similarity solution. The vertical coordinate is scaled by the maximum

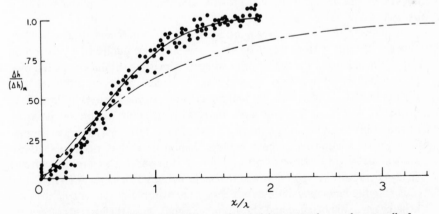

Fig. 20.3 – Profile of normalised sea surface height across the northwest wall of the Gulf Stream as a function of cross-stream distance. Solid curve is Kao's theory, and dots indicate SEASAT data.

elevation of the Stream, $\Delta h_m \approx 100$ cm, which in turn depends on cross-stream inflow, density contrast, and thermocline depth. Very good agreement is obtained between the Kao–Cheney model and the altimeter profiles, indicating the likelihood of reasonably correct physics (Fig. 20.3).

Byrne & Pullen (1981) have clearly demonstrated the coincidence between the locations of the edge of the Stream and anticyclone rings to its north on one hand, and altimeter height signatures of these features on the other; these authors compare thermal infrared imagery from the NOAA–5 environmental satellite with near-simultaneous SEASAT altimeter data, for several passes taken during the repeat orbit phase. Fig. 20.4 is a line drawing interpretation of one of their analyses, which shows the Stream edge, the positions of three eddies to its north, the suborbital tracks on 6 and 8 October 1978, and surface geostrophic speeds normal to those tracks, $u_\perp$, the last indicated by arrows. These speeds were obtained by subtracting the corrected altimeter heights from a fine-scale marine geoid which was constructed by NASA for this area out of marine gravity and orbit perturbation data. The geostrophic current relation was used,

$$u_\perp = (g/f)\tan\alpha \sin\theta,$$

where $\tan\alpha$ is the slope of the sea surface relative to the geoid, θ is the angle between sub-orbital track and surface current streamlines, g is the acceleration of gravity, and f is the Coriolis parameter. The currents so obtained range up to 200 cm s^{-1}, and are altogether reasonable; however, the use of the fine-scale geoid mentioned above leads to errors in these speeds which could easily be as large as ±50%. This points up one of the major unresolved problems in altimetric determinations of ocean currents: arriving at a precision marine geoid, either by independent means or as an element of a solution to a larger problem. This is discussed below.

A more detailed comparison of SEASAT measurements over a western boundary current with in-water observations has been published by Bernstein *et al.* (1982). During the repeat-orbit interval, three sub-orbital tracks fell over the Kuroshio Extension Current, and eight complete passes of altimeter data were obtained for each track. By looking at time-varying changes in the surface topographic data and by comparing these with dynamic heights obtained by air-expendable bathythermographs (AXBT's) corrected through historical temperature-density relationships, these authors demonstrate that the two methods yield surface height variabilities associated with time-varying geostrophic currents that agree to within ±10 cm r.m.s. It should be noted that the AXBT method yields only the baroclinic current, but that the barotropic current also contributes to the geostrophic balance and the altimeter signal. Nevertheless, this agreement is considered good, and the experiment is the most definitive test of current measurement via altimetry that has been made to date.

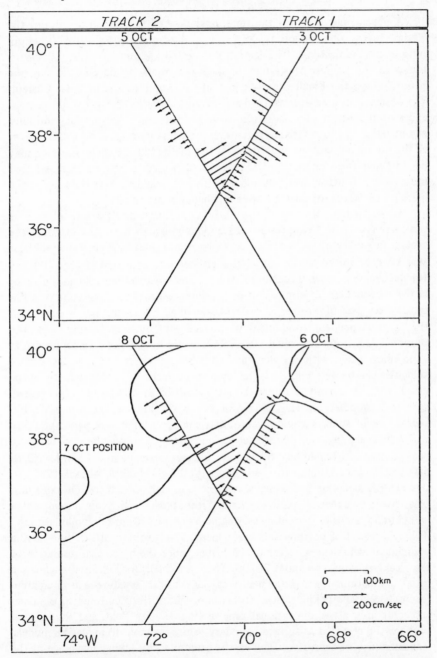

Fig. 20.4 – Surface current speeds for the Gulf Stream from altimetry. Upper: speeds for 3 and 5 October 1978; Lower: speeds for 6 and 8 October 1978, with positions of the northwest wall of the Gulf Stream and detached rings indicated by curved lines.

A general indication of the time variability of the sea surface during the 27-day repeat-orbit period of SEASAT is illustrated in Fig. 20.5 (Chelton, private communication). The largest standard deviation (of order 20 cm) is found in the regions of western boundary currents — the Gulf Stream, the Kuroshio and the Brazil Currents, and in the Antarctic Circumpolar Current. The observation period is less than a characteristic period for mesoscale motions and so represents a very short record length for fully resolving such motions. Nevertheless, the high-variance regions on the figure correspond to active regions of the ocean and indicate how the statistics of the time-dependent motion may be obtained from altimetry. GEOS-3 has delivered a 1.5-year data set that provides much better statistics and which shows standard deviations about the mean of 35–50 cm for the Gulf Stream (Cheney & Marsh 1981b).

To summarise the results on current measurements via altimetry: it is considered that SEASAT results are good enough to have demonstrated the general feasibility of the method and to have more clearly framed the problems that must be solved before considering altimetry to be a tool of general utility for determining ocean surface circulation. The most severe problem remaining is the independent determination of a reference geoid to a precision of a few centimetres over all horizontal scales greater than 20–30 km; this is essential if the time-independent component of surface velocity is to be measured. The proposed U.S. program called GRAVSAT would be a major step in arriving at a precision geoid with wavelengths >200 km. However, time-varying surface velocities can be determined via the repeat-orbit technique independently of the geoid, such as was done during the last month of SEASAT's life. This general scheme is proposed for TOPEX (NASA 1981b); in this case, the proposed orbit-repeat time of 10 days would preclude determinations of shorter-period motions, if not their aliasing over into longer period signals. However, the most energetic oceanic motions (excluding tides) have characteristic periods of order 20 to 500 days and wavelengths of order 20 to 1000 km (Munk & Wunsch 1982), so the TOPEX sampling procedure would serve to delineate these. Overall r.m.s. orbit errors of under 10 cm in the vertical seem possible with this system, which would imply a steady current-measuring precision of ± 50 cm s^{-1} at mid-latitudes. This is a useful, if not precise measurement. The precision of the time-varying component would be a factor of 10 higher, since the horizontal resolution of the altimetry would be about 20 km. There are difficulties in disentangling the geoid undulations from those signals associated with steady ocean circulation (Wunsch & Gaposchkin 1980). Futhermore, the surface currents so obtained represent only a two-dimensional time-varying velocity field, and thus should probably be regarded as surface boundary values for input to numerical models. The inference of subsurface flow can be attempted by taking advantage of the fact that most of the variance in oceanic flow is generally found in the barotropic and the first baroclinic mode; alternately and probably preferentially, the techniques of acoustic tomography can be brought to bear in a combined

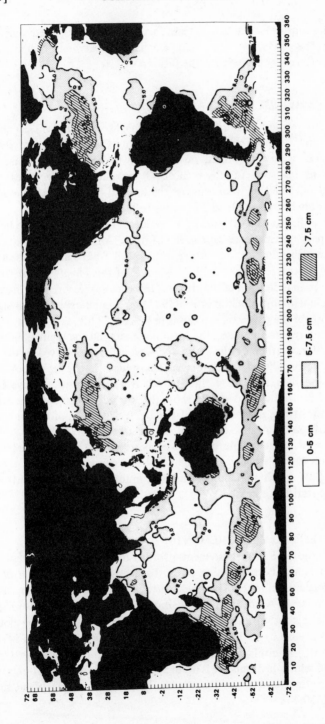

Fig. 20.5 – R.M.S. Surface variability in sea surface heights as derived during the 27-day repeat period SEASAT. The high-variance areas are found in the meander regions of western boundary currents and in the Antarctic Circumpolar Current. This record length is short compared with typical meander periods.

satellite and subsurface acoustic system, each of whose components complement the other. The broad outlines of this are barely discernible (Munk & Wunsch 1982) but offer great promise for the future.

20.4 WAVE HEIGHT OBSERVATIONS

The GEOS–3 mission established the wave height measurement capabilities of altimetry, and before the end of that satellite's useful life, the short-pulse measurements were being used operationally by the U.S. National Weather Service in its sea state forecasts. SEASAT extended the theoretical range of measurable wave heights to encompass the interval 0.5 to 20 m, although verification has only been possible to 8 m owing to limited ground truth.

Fedor & Brown (1982) present an evaluation of both the wave height and the wind speed measurement capabilities of the SEASAT altimeter and a description of the algorithm for computing wind speed from radar backscatter. The measurement accuracies cited in the beginning of this chapter are derived from this work. As an indication of the utility of this type of measurement in eastablishing wave climatology, Chelton *et al.* (1981) have composited global significant wave heights for the entire three months of SEASAT's life, as shown in Fig. 20.6a. This time interval encompassed the Austral winter, and the figure concomitantly reflects winter weather by the very large wave heights observed in the 'roaring 40's' and 'fierce 50's' of the Southern Hemisphere. In the region southwest of Australia, the data indicate a three-month-average significant wave height in excess of 5.3 m. They also clearly show the low wind regime in the equatorial region of the Asian archipelago, and the brisk wind area of the Indian Ocean monsoon off Saudi Arabia. Since the measurement of wave height is relatively straightforward, considerable confidence exists as to the correctness of these data. It is clear that a one-year altimeter mission can yield monthly global wave climatology of much value to mariners; a multi-year mission would serve to refine these statistics considerably.

20.5 WIND SPEED MEASUREMENTS

The third basic type of measurement afforded by the altimeter is that of scalar wind speed along the subsatellite track. Earlier work by Brown *et al.* (1981) and by Mognard & Lago (1979) had established the GEOS-3 measurement capability for scalar wind, and Fedor & Brown (1982) extended this to SEASAT. Using similar algorithms, Chelton *et al.* (1981) compiled an average global wind speed map analogous to the wave height data of Fig. 20.6a; this is shown in Fig. 20.6b which displays isotachs of wind speed. The asymmetry between summer and winter hemispheres is again apparent, with the Antarctic circumpolar region exhibiting some regions where the three-month-average wind speed was in excess of 10.4 m s^{-1}.

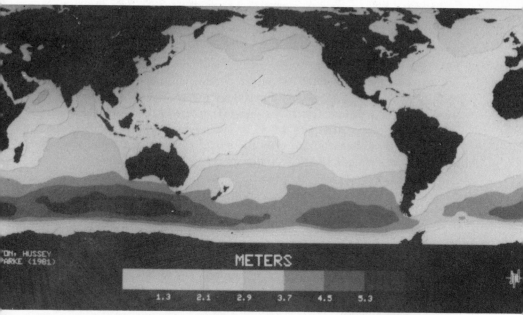

Fig. 20.6a – SEASAT altimeter wave heights for 7 July – 10 October 1978.

Fig. 20.6b – SEASAT altimeter scalar winds for the same conditions as for Fig. 20.6a.

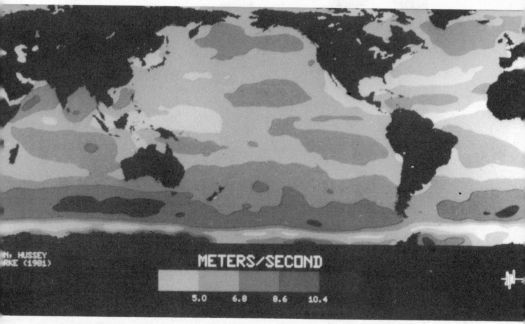

These workers have also calculated a global wind speed-wave height cross-correlation field, which exhibits the obviously high but not perfect correlations apparent between Figs. 20.6a and 20.6b (Chelton, private communication). Very good correlation is not expected, because the altimeter measures the significant wave height due to the combination of wind waves and swell, and the latter will not be correlated with local but rather distant winds at earlier times. Once again, the value of these data to climatology is high, but in addition, surface wind speed is a meteorological variable of importance in weather and wave forecasting. However, the lack of wind direction reduces the altimeter wind measurements considerably but does not render them useless. Radar scatterometry such as practiced on SEASAT attempts to correct this deficiency.

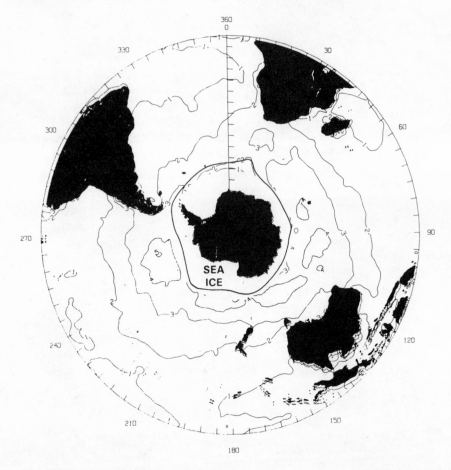

Fig. 20.7 – Minimum swell heights derived from (a) altimeter measurements of significant wave height due to wind waves and swell together, and (b) calculations of wind wave heights from altimeter wind observations.

Mognard *et al.* (1982) have exploited the simultaneous wind speed-wave height observational capability on SEASAT to estimate lower bounds to the swell height and thereby to observe the propagation of swell fields away from their generation region. Reasoning that the wind waves observed at a point cannot readily exceed that of a fully developed wind wave field due to the local wind (which can be calculated from the wind speed alone), they attribute any excess observed by the altimeter to swell; since the wind waves may not be fully developed, this establishes a lower bound to the swell.

Fig. 20.7 shows minimum swell heights in metres for the last several weeks of the SEASAT mission, as compiled from all useful passes in the southern hemisphere during that interval. Using this technique, these investigators have been able to track the propagation of coherent swell fields out of the wind/wave generation region in the Antarctic Circumpolar Current, northward to the equator. The smoothed group velocities agree with what one would expect for long-wave swell. The ability to distinguish swell from wind waves is a novel and useful feature of the altimeter. It should also be noted that the altimeter can delineate the position of the ice edge with considerable precision, because of the convergence of orbits in this region.

20.6 CONCLUSIONS

It is clear that the altimeter can produce precision measurements of a wide variety of geophysical features, including an ocean geoid, western boundary current speeds and positions, wind speeds, wind wave and swell heights, and polar sea ice edge positions. Not discussed here are the measurements of Greenland ice cap profiles (Zwally *et al.* 1982), tidal amplitudes and phases (Cartwright & Alcock 1981, Parke 1980) and evidence for cyclostrophic flow of ocean currents driven by hurricane winds (Fedor 1981). In addition, there are possibilities for mapping land features whose steepnesses are not so great as to preclude tracking by the altimeter. The value of these measurements to geophysics is generally quite high.

Because of the relatively short duration of the SEASAT data set, analyses of extended, time-varying phenomena have not been possible (with the exception of waves, perhaps). However, several new altimetric satellites are in the offing, including spacecraft to be launched by ESA, France, the USSR, Japan, and the USA.

Extended data sets and a wealth of geophysical understanding should flow from these missions, and participating earth scientists are eagerly awaiting their appearance.

REFERENCES

Bernstein, R. L., Born, G. H., & Whritner, R. H. (1982) SEASAT altimeter determination of ocean current variability, *J. Geophys. Res.* 87 3261.

Brown, G. S., Stanley, H. R., & Roy, N. A. (1981) The wind speed measurement capability of spaceborne radar altimetry, *IEEE J. Oceanic Engr.* **OE-6** 59.

Byrne, H. M. and Pullen, P. E. (1981) Western boundary current variability derived from SEASAT altimetry data, in *Oceanography from Space,* 877, J. F. R. Gower, ed., Plenum Press.

Carwright, D. E. and Alcock, G. A. (1981) On the precision of sea surface elevations and slopes from SEASAT altimetry over the N. E. Atlantic Ocean, in *Oceanography from space,* 885, Plenum Press.

Chelton, D. B., Hussey, K. J. & Parks, M. E. (1981) Global satellite measurements of water vapour, wind speed and wave height, *Nature* **294** 529.

Cheney, R. E. & Marsh, J. G. (1981) personal communication.

Cheney, R. E. & Marsh, J. G. (1981a) SEASAT altimeter observations of dynamic ocean currents in the Gulf Stream region, *J. Geophys. Res.* **86** 473.

Cheney, R. E. & Marsh, J. G. (1981b) Oceanic eddy variability measured by GEOS-3 altimeter crossover differences, *EOS Trans. Amer. Geophys. Union* **62** 743.

Fedor, L. S. (1981) personal communication.

Fedor, L. S. & Brown, G. S. (1982) Waveheight and windspeed measurements from the SEASAT altimeter observations and theory, *J. Geophys. Res.* **87** 539.

Lerch, F. J., Marsh, J. G., Klosko, S. M., & Williamson, R. G. (1982) Gravity model improvement for SEASAT, *J. Geophys. Res.* **87** 3281.

Marsh, J. G., Cheney, R. E., Martin, T. V., & McCarthy, J. J. (1982) Computation of a precise mean sea surface in the eastern North Pacific using SEASAT altimetry, *EOS, Trans Amer. Geophys. Union* **63** 178.

Marsh, J. G. & Martin, T. V. (1982) The SEASAT altimeter mean sea surface model, *J. Geophys. Res.* **87** 3269.

Mognard, N. M., Campbell, W. J., Cheney, R. E. & Marsh, J. G. (1982) Southern Ocean mean monthly waves and surface winds for winter 1978 by SEASAT radar altimeter, (in press).

Mognard, N. M. & Lago, B. (1979) The computation of wind speed and wave heights from GEOS-3 data, *J. Geophys. Res.* **84** 3979.

Munk, W. & Wunsch, C. (1982) Observing the oceans in the 1980s, (in press).

NASA (1981a), SEASAT Data Utilisation Project Final Report, NASA/JPL, (30 Sept. 1981).

NASA (1981b), *Satellite altimeter measurements of the ocean; Report of the TOPEX Science Working Group,* NASA/JPL (1 March 1981).

Parke, M. E., (1980) Detection of tides from the Patagonia Shelf by the SEASAT satellite radar altimeter: an initial comparison, *Deep Sea Res.* **27A** 297.

Wunsch, C., and Gaposchkin, E. M., (1980) On using satellite altimetry to determine the general circulation of the ocean with applications to geoid improvements, *Revs. Geophys. and Space Phys.* **18** 725.

Zwally, H. J., Bindscadler, R. A., Brenner, A. C., Martin, T. V. & Thomas, R. H.,
     Surface elevation contours of Greenland and Antarctic ice sheets, *J. Geophys.
     Res.* (accepted for publication, 1982).

Chapter 21

SEASAT tracking over Europe

P. WILSON
Institut für Angewandte Geodasie, Frankfurt am Main, F. R. Germany
L. AARDOOM
Delft University of Technology, Department of Geodesy, Delft, The Netherlands

21.1 INTRODUCTION

One of the most promising space geodetic techniques is satellite-borne radar altimetry, the only operational space-to-earth technique. When applied over extended areas of the oceans the technique addresses important scientific issues concerning both geodesy and oceanography. To determine the mean sea surface in a global context is certainly an objective of contemporary geodesy, also because it approximates the geoid, thus being a significant input to the determination of the Earth's gravity field. The geopotential difference between mean sea level and the geoid is an invaluable input to studies of ocean dynamics. The evaluation of the data may be performed in an absolute (global) or in a relative (regional) framework.

A prerequisite for the interpretation of altimetry data in terms of the shape of the global sea surface is the determination of the satellite's orbit, precise to a level compatible to that of the altimetry. Since, mainly for power reasons, altimetry satellites have been and will likely be launched into relatively low orbits, the vertical control of these requires extensive and painstaking force field modelling and precise tracking.

For SEASAT it was considered that ground-based 162/324 MHz Doppler (integrated) range-rate and laser range measurements would be available for tracking. The tracking was designed to serve two purposes:

- precise orbit determination for the global SEASAT investigations;
- precise orbit determination specifically in support of the regional investigations proposed by SURGE and, as regards the altimetry, mainly concerning the North Sea.

The North Sea altimetry experiment (SURGE 1977) aimed at:

- a comparison of a terrestrially derived astro-gravimetric geoid with the altimetry-derived sea surface;
- the calibration of the altimeter range bias;
- an independent solution for the altimetry-derived sea surface.

In this connection these, and the global objectives, called for three distinct tracking campaigns:

- SEATOC, to determine the spatial positions of all European SEASAT-1 tracking stations in a unique common reference frame;
- NORSDOC, to relate the astro-gravimetric geoid to that same reference frame;
- SEATRACK, to track SEASAT-1 and contribute to the determination of its orbit.

The execution of these campaigns is reported in sections 21.4, 21.5, and 21.6 respectively. The results of the campaigns are summarised in section 21.7.

21.2 EUROPEAN TRACKING FACILITIES

21.2.1 Doppler

NNSS Doppler positioning was selected as the means of reference frame definition and geoid positioning (SEATOC) and NORSDOC). Doppler tracking of SEASAT was to provide the all-day, all-weather facility to determine the satellite's orbit.

Distinction should be made between three classes of instrumentation:

- instruments receiving at the NNSS 150/400 MHz frequency pair to be exclusively used in the SEATOC and NORSDOC campaigns: CMA-722B, CMA-751, JMR-1, Magnovox-GEO-II;
- instruments receiving at the SEASAT 162/324 MHz tracking frequencies, exclusively useful during the SEATRACK campaign: CMA-725;
- receivers operating at both frequency pairs and thus operable during all three campaigns: TRANET-doppler, Magnavox GEOCEIVER.

The technical specifications of these instruments are well known. Tables 21.1, 21.2, and 21.3 indicate to which stations the instruments were permanently or temporarily deployed. Fig. 21.1 depicts the locations of these stations.

21.2.2 Laser ranging

Though weather-restricted, laser ranging was to provide the necessarily intermittent means of precise tracking to determine the SEASAT orbit to the best achievable standard of accuracy.

Table 21.4 lists the laser ranging stations which took part in the SEATRACK campaign, together with some of their relevant technical characteristics applicable at the time of the compaign. The locations of the laser stations appear in Fig. 21.1.

21.3 ORGANISATION OF TRACKING ACTIVITIES

The organisation of both tracking components (Doppler and laser ranging) drew on the structure and experience established during previous campaigns.

As concerns the Doppler positioning and tracking, these related to the European Doppler Observation Campaigns EDOC-1 (Nouel 1975) and EDOC-2 (Boucher *et al* 1981) and a campaign to interconnect European laser ranging stations: EROSDOC (Schluter *et al* 1979).

The laser ranging component of SEATRACK could be conveniently organised in the framework of the European Range Observations of Satellites (EROS) cooperation which adopted SEATRACK as its second campaign: EROS-2 (Aardoom 1981). Adverse weather and difficulties in communicating orbital manoeuvres performed during the unexpectedly short lifetime of the satellite were to lead to a limited laser data yield. The authors were in charge of the organisation and management of the campaigns.

Table 21.1

Instrumentation and observing teams for SEATOC

Station	Instrumentation	Observing Teams
Barton Stacey	TRANET, Dir. Mil. Surveys – UK	Dir. Mil. Surveys – UK
Brussels	TRANET, ORB, – Brussels	ORB. – Brussels
Dionysos	CMA-722B, TU – Graz	NTU – Athens
		TU – Graz
		IfAG – Frankfurt
Florence	TRANET, IROE – Florence	IROE – Florence
Geldingaholt	CMA-751, Univ. – Hannover	IfAG – Frankfurt
		Univ – Hannover
		Icelandic R. C.
Grasse	CMA-722B, IfAG – Frankfurt	CERGA – Grasse
		IfAG – Frankfurt
Kootwijk	CMA-751, TU – Delft	TU – Delft
La Laguna	JMR-1, IGN – Madrid	IGN – Madrid
Metsahovi	CMA-751, NGO – Oslo	Geod. Laitos – Helsinki
	CMA-722B, IfAG – Frankfurt	NGO – Oslo
San Fernando	JMR-1, IGN – Madrid	IGN – Madrid
Toulouse	JMR-1, IGN – Madrid	GRGS – Toulouse
		IfAG – Frankfurt
Trondheim	CMA-751, NGO – Oslo	IKU – Trondheim
	MAGNAVOX GEO-II, IKU-Trondheim	NGO – Oslo
Wettzell	CMA-722B, IfAG – Frankfurt	IFAG – Frankfurt

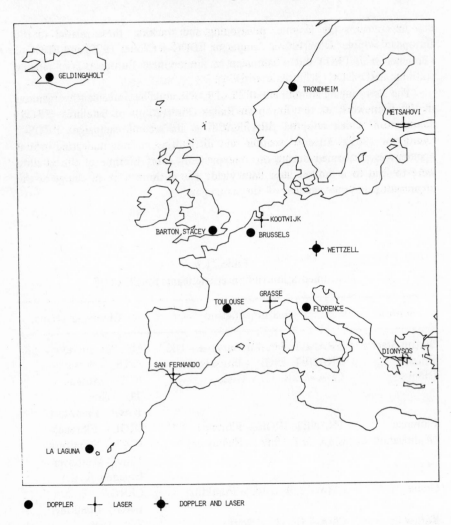

Fig. 21.1

21.4 SEATOC

The SEASAT Tracking-network Observation Campaign (SEATOC) took place
in the period 24 June to 4 July, 1978.

Its objective was to determine the precise positions of the thirteen 162/324
MHz Doppler and laser ranging stations, preparing to track SEASAT, in a common
spatial frame of reference. This operation was deemed necessary in order to meet
the requirements of the SURGE North Sea altimetry experiment, and desirable
in view of SURGE's contribution to the entire SEASAT mission. The method

chosen for this purpose was NNSS Doppler positioning, applying simultaneous occupation of all projected SEATRACK stations by 150/400 MHz Doppler receivers. This required the temporary deployment of such equipment to those SEATRACK stations which were not normally equipped with such instruments. Fig. 21.1 shows the projected SEATRACK network, featuring both 162/324 MHz Doppler and laser ranging facilities. Table 21.1 lists the assignment of NNSS positioning instruments and observing responsibilities to the thirteen tracking locations.

Campaign management resided at the Institut fur Angewandte Geodasie (IfAG) in Frankfurt am Main, FRG, where the observation data were collected and a database established.

The results of the campaign are given in section 21.7.

21.5 NORSDOC

NORSDOC took place in the 17–27 April, 1979 timeframe, after the untimely demise of SEASAT's active functions. Its purpose was to relate the vertical position of the astro-gravimetrically derived North Sea geoid to the reference frame defined by means of SEATOC. The method adopted for this comprised two steps (Lelgemann *et al* 1980):

(a) the connection of mean sea level marks at selected tide gauges in the vertical sense to nearby 'transformation points';

(b) determination of these transformation points in the vertical sense in the reference frame defined by the combination of results obtained for four stations occupied during both SEATOC and NORSDOC (see section 21.7).

Step (a) was realised by spirit levelling if the required height information was not already available. For step (b) NNSS Doppler positioning was again chosen in a network consisting of both the eleven transformation points and the four stations common to the SEATOC network.

NORSDOC and SEATOC were used to implement step (b). NORSDOC required the temporary deployment of 150/400 MHz Doppler receivers to the eleven transformation points depicted in Fig. 21.2, together with the four stations common to the SEATOC network. The transformation points were given the designations used to identify the tide gauges. The deployment of instrumentation and the responsibility for local operations has been documented in Table 21.2.

The method adopted draws on the following assumptions:

(1) the geoid coincides with the surface of mean sea level;

(2) local mean sea level is indeed indicated by the mean sea level mark of the selected tide gauges.

The validity of both assumptions can be approached optimally by proper selection of those coastal tide gauges which feature:

- a minimum of effects which cause local additional deviations of mean sea level from the geoid;
- a maximum length of reliable recording of sea level.

The tide gauges and their associated transformation points were selected on the North Sea periphery on the basis of these criteria, taking into account logistic considerations. Campaign management again resided at the IfAG, where the observational data were collected and a database established. Results of the campaign are presented in section 21.7.

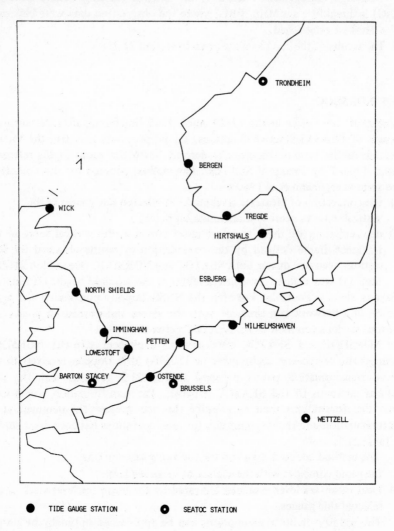

Fig. 21.2

Table 21.2

Instrumentation and observing teams for NORSDOC

Station	Instrumentation	Observing Teams
Bergen	CMA-751, NGO – Oslo	NGO – Oslo
Esbjerg	JMR-1, Geod. Inst. – Charlottenlund	Geod. Inst. – Charlottenlund
Hirtshals	JMR-1, Geod. Inst. – Charlottenlund	Geod. Inst. – Charlottenlund
Immingham	JMR-1, Oxford Univ.	Ordnance Survey
Lowestoft	JMR-1, Decca Surveys – UK	Ordnance Survey
North Shields	JMR-1, Univ. Nottingham	Ordnance Survey
Ostende	CMA-722B, TU – Graz	IGN – Brussels IfAG – Frankfurt
Petten	CMA-751, TU – Delft	TU – Delft
Tregde	CMA-751, NGO – Oslo	NGO – Oslo
Wick	JMR-1, IGN – St Mandé	Ordnance Survey
Wilhelmshaven	CMA-751, Univ. Hannover	Univ. Hannover
Barton Stacey	TRANET, Dir. Mil. Surveys – UK	Dir. Mil. Surveys – UK
Brussels	TRANET, ORB – Brussels	ORB – Brussels
Trondheim	MAGNAVOX GEO-II, IKU-Trondheim	IKU – Trondheim
Wettzell	CMA-722B, IfAG – Frankfurt	IfAG – Frankfurt

21.6 SEATRACK

SEASAT tracking commenced on 26 June 1978 immediately after the satellite was launched and continued until the power failure on board the spacecraft on 10 October 1978, when the satellite lost all its active functions. Laser ranging was continued until 10 November 1978.

The purpose of the SEATRACK campaign was to:
- assist in precise determination of the satellite's orbit, both globally and over the SURGE altimeter experiment area in particular;
- contribute to the calibration of altimeter biases.

The campaign provided two distinct elements:
- Doppler tracking at the 162/324 MHz nominal frequencies;
- laser ranging.

The overall network is shown in Fig. 21.1. Table 21.3 gives information on tracking equipment and responsibilities for its operation.

Table 21.3
Instrumentation and observing teams for SEATRACK

Station	Instrumentation	Observing Teams
Barton Stacey	TRANET, Dir. Mil. Surveys — UK	Dir. Mil. Surveys — UK
Brussels	TRANET, ORB — Brussels	ORB — Brussels
Dionysos	LASER	NTU — Athens
Florence	TRANET, IROE — Florence	IROE — Florence
Geldingaholt	CMA-725/722B, Univ. New Brunswick	IfAG — Frankfurt Icelandic R.C.
Grasse	LASER	CERGA — Grasse
Kootwijk	LASER	TU — Delft
La Laguna	MAGNAVOX GEOCEIVER, DMA — USA	IGN — Madrid
Metsahovi	LASER	Geod. Laitos Helsinki
San Fernando	LASER	IOM — San Fernando GRGS — Toulouse
Toulouse	CMA-725, GRGS — Toulouse	GRGS — Toulouse
Trondheim	CMA-725/722B/RB-Clock, TU — GRAZ/IfAG — Frankfurt	IKU — Trondheim
Wettzell	LASER/CMA-725	IfAG — Frankfurt

The tracking covered all phases of SEASAT's active life during which altimeter data were taken. The changes in orbit which set up these phases, led to considerable difficulties in laser tracking over Europe owing to the delay which occurred in communicating details of the changes to the tracking stations. This was a primary reason contributing to the relatively low output of laser tracking from the European stations (see section 21.6.2).

21.6.1 Doppler tracking

Doppler tracking of SEASAT-1 commenced on 26 June 1978, and was continued until the on-board power failure ended the active functions of the spacecraft. By that time the satellite had been tracked over hundreds of passes across the area covered by the European Doppler network.

IfAG once more provided management of the campaign and established a data base to collect observations for circulation to other users of SEASAT data upon request, as well as for use in the North Sea experiment.

Results of computations made in support of the North Sea altimeter experiment have been reported by Lelgemann *et al*. (in press), Lelgemann & Brennecke (1982) & Rummel *et al*. (1982).

21.6.2 Laser ranging

The European laser ranging contribution to the tracking of SEASAT was organised as the second campaign, EROS-2, in the framework of the European Range Observations of Satellites cooperation (Aardoom 1981). The first pass of laser data was taken on 11 July, the last on 10 November 1978.

After the power cut of 10 October the ranging became increasingly difficult partly because of the degrading quality of predictions, partly because of loss of vertical stabilisation of the spacecraft. Unfavourable weather and orbital manoeuvres led to a disappointing data yield, despite devoted efforts of the laser ranging stations and their observer teams. Although five of the six stations reported observations, only data taken by Kootwijk, San Fernando, and Wettzell could finally resist rejection against quality criteria. The resulting total number of qualified single station passes became 39 (Kootwijk:25, San Fernando:7, Wettzell: 7); however, roughly one third were observed after 10 October.

The EROS Prediction Centre at the Kootwijk station in the Netherlands assumed special responsibilities as regards data exchange with the US SEASAT Project Altimetry Team, to which, through the National Space Science Data Centre (NSSDC) at NASA's Goddard Space Flight Center, the data were made available directly by the European stations. In return, the US laser data were submitted to Kootwijk, from where it was disseminated to European investigators upon request.

The European laser data, though small in numbers, were used by several investigators, Wakker *et al* (1982), for example.

Table 21.4

SEATRACK laser ranging systems/main characteristics

	Dionysos	Grasse	Kootwijk	Metsahovi	S. Fernando	Wettzell
Wavelength (nm)	694	694	694	694	694	530
Max. energy (J)	0.75	4	3	1	1	0.25
Pulse width (ns)	5	2	4	25	12	0.2
Repetition rate (ppm)	60	15	15	4	60	240
Min. beam div. (arcsec)	80	5	60	200	60	5
Receiver aperture (cm)	45	100	50	63	36	60
Single shot r.m.s (cm)	20	20	20	100	75	5
Daylight capability	yes	yes	yes	no	no	yes

21.7 RESULTS OF THE CAMPAIGNS

21.7.1 SEATOC/NORSDOC station positioning

Tracking station coordinates from the three campaigns were computed at the IfAG using sequentially the programmes GEODOP (Kouba & Boal 1975) and ORBDOP (Hauck, personal communication). Using the four stations Barton Stacey, Brussels, Trondheim, and Wettzell, which were common to both SEATOC and NORSDOC, the coordinates of both station sets could be combined into a common system using both sets of normal equations plus the covariance matrices derived for the individual solutions for the four stations to obtain a Helmert blocked solution. The procedure adopted was to compute best estimates of the station coordinates, frequency offsets, etc. with GEODOP, which were then used as input parameters to the short arc solution derived using ORBDOP. The initial state vectors were taken from the Precise Ephemeris. Based on standard deviations

$$\sigma\,(r) = 1.5 \text{ m} \qquad \sigma\,(\dot{r}) = 1.5 \text{ mm/s}$$
$$\sigma\,(G/R_0) = 1.5 \text{ mm/s} \qquad \sigma\,(R_0.u) = 1.5 \text{ m}$$
$$\sigma\,(R_0.i) = 1.5 \text{ m} \qquad \sigma\,(R_0.\Omega) = 1.5 \text{ m}$$

of the short arc orbital (Hill) elements (Lelgemann *et al.* 1980) spatial Cartesian coordinates of the SEATOC and NORSDOC stations are given in Table. 21.5. The corresponding ED-50 coordinates of the NORSDOC stations are summarised in Table 21.6. The authors are indebted to the investigators, who kindly gave their permission for the presentation of these results prior to publication (Amberg *et al.* 1982).

Table 21.5

Coordinates of SEATOC and NORSDOC tracking stations in the unified coordinate system. Results are also given for the SEATRACK stations derived from the surveyed corrections to the NNSS Doppler positions and quoted in the unified coordinate system. The positions are given for NNSS (150/400 MHz) Doppler antenna phase centres (D1), together with either 162/324 MHz phase centre (D2) and/or laser ranging system (L). M_X, M_Y and M_Z are estimated standard deviations in metres.

Station	Obs. types	X (metres)	Y (meters)	Z (metres)	M_X	M_Y	M_Z
Dionysos	D1	4 595 234.24	2 039 437.19	3 912 616.47	0.24	0.33	0.25
		− 5.50	+ 10.73	− 4.01			
	L	4 595 228.74	2 039 447.92	3 912 612.46			
Florence	D1	4 522 410.65	897 985.17	4 392 484.79	0.19	0.26	0.20
Geldingaholt	D1	2 625 126.10	−969 353.12	5 712 459.02	0.22	0.36	0.21
		+ 0.32	− 0.12	+ 0.71			
	D2	2 625 126.42	−969 353.24	5 712 459.73			

Table 21.5 continued

Station	Obs. types	X (metres)	Y (meters)	Z (metres)	M_X	M_Y	M_Z
Grasse	D1	4 581 917.16	556 563.40	4 389 074.61	0.20	0.28	0.20
		−220.46	− 421.88	+ 282.69			
	L	4 581 696.70	556 141.52	4 389 357.30			
Kootwijk	D1	3 899 223.60	396 752.80	5 015 071.98	0.17	0.22	0.16
		+ 4.53	−25.21	+ 0.90			
	L	3 899 228.13	396 727.59	5 015 072.88			
La Laguna	D1	5 384 985.39	−1 576 501.84	3 023 837.68	0.33	0.46	0.33
Metsähovi	D1	2 892 591.65	1 311 788.60	5 512 616.52	0.21	0.30	0.19
		+10.52	+ 8.65	− 5.68			
	L	2 892 602.17	1 311 797.25	5 512 610.84			
San Fernando	D1	5 105 592.62	−555 245.64	3 769 671.60	0.22	0.33	0.23
Toulouse	D1	4 627 844.52	119 847.44	4 372 990.78	0.19	0.27	0.19
		+ 2.50	− 5.40	− 3.30			
	D2	4 627 847.02	119 842.04	4 372 987.48			
Barton St.	D1	4 004 967.00	−096 584.81	4 946 538.73	0.19	0.24	0.19
	D2						
Brussels	D1	4 027 837.50	306 999.63	4 919 535.14	0.17	0.22	0.17
	D2						
Trondheim	D1	2 814 806.16	516 794.81	5 681 074.35	0.16	0.22	0.15
		+ 0.32	+ 0.06	+ 0.64			
	D2	2 814 806.48	516 794.87	5 681 074.99			
Wettzell	D1	4 075 541.33	931 805.98	4 801 607.58	0.15	0.19	0.15
		+15.22	− 5.87	− 10.61			
	D2	4 075 556.55	931 800.11	4 801 596.97			
		− 5.01	--40.38	+ 9.35			
	L	4 075 536.32	931 765.60	4 801 616.93			
Bergen	D1	3 144 969.76	291 950.12	5 522 698.58	0.20	0.28	0.19
Esbjerg	D1	3 581 032.17	537 022.25	5 233 049.26	0.23	0.32	0.21
Hirtshals	D1	3 374 659.49	592 767.65	5 361 700.17	0.23	0.33	0.21
Immingham	D1	3 786 316.88	−022 026.81	5 115 528.95	0.22	0.31	0.21
Lowestoft	D1	3 896 697.16	116 698.02	5 031 236.82	0.20	0.27	0.19
N. Shields	D1	3 666 200.82	−098 153.19	5 200 862.82	0.26	0.39	0.24
Ostende	D1	3 996 832.46	204 213.39	4 949 757.39	0.24	0.33	0.23
Petten	D1	3 854 340.99	314 036.90	5 055 156.55	0.20	0.26	0.20
Tregde	D1	3 358 084.40	445 350.01	5 386 153.25	0.21	0.28	0.19
Wick	D1	3 330 465.82	−188 684.36	5 418 168.12	0.23	0.33	0.21
Wilhelmshav.	D1	3 762 459.46	538 475.63	5 104 873.50	0.20	0.26	0.19

Table 21.6
Coordinates of the NORSDOC tracking stations transformed to ED-50 and
supplemented by the levelled height on the quoted reference datum.

Station	Latitude	Longitude	Height (metres)
Bergen	60° 21′ 01″349	5° 18′ 19″679	48.09 NN 1957
Esbjerg	55° 29′ 55″484	8° 31′ 48″777	30.93 DNN/GM
Hirtshals	57° 35′ 41″555	9° 57′ 50″755	11.33 DNN/GM
Immingham	53° 40′ 36″438	0° 19′ 53″874	35.86 ODN
Lowestoft	52° 25′ 00″328	1° 43′ 01″136	47.18 ODN
North Shields	54° 59′ 28″786	1° 31′ 54″512	52.22 ODN
Ostende	51° 13′ 55″808	2° 55′ 35″264	31.88 TAW
Petten	52° 46′ 15″394	4° 39′ 34″162	16.88 NAP
Tregde	58° 00′ 25″009	7° 33′ 22″206	5.54 NGO
Wick	58° 33′ 10″267	3° 14′ 26″152	55.25 ODN
Wilhelmshaven	53° 30′ 58″967	8° 08′ 45″558	

21.7.2 SEATRACK short arc solution
The short arc solution computed for SEATRACK using ORBDOP was based on
tracking from the stations Brussels, Geldingaholt, Toulouse, Trondheim, and
Wettzell, because only these stations had 162/324 MHz Doppler receivers.
Corrections were applied to reduce the Doppler data from the antenna phase
centre (A) to the mass-centre (C) of the satellite. The corrections thus applied in
the sense 'A–C' were:

  along-track: 0.22 m
  across-track: 5.08 m
  radial: –3.50 m.

  Short arc ephemerides were computed for all 24 passes tracked between
15 September and 9 October 1978 when the satellite was in the exact repeat
'Bermuda' orbit. Station coordinates were taken from the unified SEATOC/
NORSDOC solution (see 21.7). The short arc solution derived satellite position
coordinates in terms of corrections to the orbital elements, together with an
empirical, dimensionless, correction to the tropospheric refraction, a correction
to the nominal value of the receiver delay, and a correction to the nominal
frequency offset. A comparison with altimeter data yields results in the form of
a model expressed in terms of corrections to the Precise Ephemeris coordinates
and velocities for each pass.

  The results of the short arc computations will be presented by Brennecke
et al. (1982).

21.8 CONCLUSION

Although the operational phase of the SEASAT-1 mission was much shorter than anticipated, the observing teams and the tracking management acquired additional experience in cooperative satellite geodetic observing activities, this time jointly using two different techniques; (i) Doppler, both with (NNSS, 150/400 MHz) and without (SEASAT-1, 162/324 MHz) Broadcast Ephemeris; and (ii) laser ranging.

It was demonstrated that cooperative arrangements made independently for Doppler positioning and laser ranging, on previous occasions, could be conveniently used, coordinated, and expanded. It was also demonstrated again that European satellite geodetic observation facilities are indeed in a position to join cooperatively in global programmes.

The yield of European laser range data was severely limited primarily by the short active lifetime of SEASAT, the adverse weather, and orbital manoevres during a substantial part of that time. Consequently only limited use could be made of the data over the area of the North Sea; see, however, Wakker *et al.* (1982).

The participation in the work of SURGE encouraged various European satellite geodetic investigator groups involved in laser ranging to join in further cooperative programmes with NASA: the LAGEOS (NASA, 1978, ELSI 1978) and Crustal Dynamics (NASA 1980, WEGENER 1981) projects, the latter involving also Very Long Baseline Interferometry (VLBI) to extra-galactic radio sources.

The participation in SURGE brought European (satellite) geodesists in closer touch with the technique of satellite radar altimetry and with its multidisciplinary users, in particular the oceanographers. This led the way to possibly continued involvement in future altimetric satellite missions such as ERS-1 and TOPEX, which will again call on precise global and regional tracking support.

ACKNOWLEDGEMENTS

The work accomplished would have been impossible without the cooperation of the following organisations and institutes:

Centre d'Etudes et de Recherches Géodynamiques et Astronomiques, Grasse, France

Decca Surveys, Leatherhead, Surrey, United Kingdom

Defense Mapping Agency Hydrographic and Topographic Centre, Washington, D.C., USA

Director of Military Surveys of the United Kingdom, Feltham, Middlesex, United Kingdom

Geodaetisk Institut, Charlottenlund, Denmark

Geodeettineh Laitos, Helsinki, Finland

Groupe de Recherches de Géodésie Spatiale, Toulouse, France
Icelandic Research Council, Reykjavik, Iceland
Institut Géographique National, Brussels, Belgium
Institut Géographique National, St Mandé, France
Instituto Geografico Nacionale, Madrid, Spain
Instituto y Observatorio de Marina, San Fernando, Cadiz, Spain
Istituto di Ricerca sulle Onde Elettromagnetiche, Florence, Italy
Institutt for Kontinentalsokkelundersøkelser, Trondheim, Norway
Landmälingar Island, Reykjavik, Iceland
National Technical University, Athens, Greece
Norges Geografiske Oppmåling, Oslo, Norway
Observatoire Royal de Belgique, Brussels, Belgium
Ordnance Survey, Southampton, United Kingdom
Oxford University, Department of Surveying and Geodesy, United Kingdom
Technische Universität Graz, Austria
Universität Hannover, Federal Republic of Germany
University of New Brunswick, Fredericton, N. B., Canada
University of Nottingham, Department of Civil Engineering, United Kingdom

REFERENCES

Aardoom, L., (1981) European range observations to satellites (EROS) *CSTG Bull.* No. 3, 23-26.

Amberg, L., Hauck, K., Herzberger, K., & Schlüter, W., Final results of SEATOC and NORSDOC, *Deutsche Geod. Komm.*, in preparation.

Boucher, C., Paquet, P., & Wilson, P., (1981) Final report on the observations and computations carried out in the Second European Doppler Observation Campaign (EDOC-2) for position determination at 37 satellite tracking stations, *Deutsche Geod. Komm.* **B255**.

Brennecke, J., Hauck, H., & Lelgemann, D., (1982) SEASAT short arc computations over the North Sea region, *Deutsche Geod. Komm.*, in preparation.

EARSeL/SURGE, Final proposal for an oceanographic programme based on the use of data from SEASAT-A, submitted to NASA and NOAA via ESA, July 1977.

European Lageos and Starlette Investigator (ELSI) Group, proposal submitted to NASA through ESA, December 1978.

Hauck, H., ORBDOP programme description, personal communication.

Kouba, J., Boal, J. D., (1975) *Program GEODOP,* Geod. Survey of Canada Ottawa.

Lelgemann, D., Brennecke, J., Hauck, H., Herzberger, K., & Wilson, P., (1980) Geodätische Aspekte des SEASAT-1 Nordsee-Experimentes, *Zeitschr. f. Vermessungswesen* **105** 18-31.

Lelgemann, D., Brennecke, J., Torge, W., & Wenzel, H. G., Validation of SEASAT-1 altimetry using ground truth in the North Sea region, *J. Geoph. Res.*, in press.

Lelgemann, D., & Brennecke, J., Altimetric geoid in the North Sea, *SURGE Discussion Meeting*, London, England, 14-16 April 1982.

NASA (1978), Announcement of Opportunity, Laser Geodynamic Satellite (Lageos), A. O. No. OSTA 78-2.

NASA (1980), Announcement of Opportunity, Crustal Dynamics and Earthquake Research, A. O. No. OSTA 80-2.

Nouel, F., Campagne 'EDOC', Groupe de Recherches de Géodésie Spatiale Toulouse, report, November 1975.

Rummel, R., Strang Van Hees, G., & Versluijs, H., Gravity field investigations in the North Sea using SEASAT-1 altimetry data, SURGE *Discussion Meeting*, London, England, 14-16 April 1982.

Schlüter, W., Wilson, P., & Seeger, H., Final results of the EROS Doppler Observation Campaign, *17th General Assembly IUGG*, Canberra, Australia, 2-15 December 1979.

Wakker, K. F., Ambrosius., & Van der Ploeg, T., SEASAT orbit determination from laser range observations, *SURGE Discussion Meeting*, London England, 14-16 April 1982.

Working-group of European Geo-scientists for the Esatablishment of Networks for Earthquake Research (WEGENER), proposal submitted to NASA through ESA, March 1981.

Chapter 22

SEASAT orbit determination from laser range observations

K. F. WAKKER, B. A. C. AMBROSIUS, & T. VAN DER PLOEG
Delft University of Technology, Department of Aerospace Engineering, Kluyverweg 1,
2629 HS Delft, The Netherlands

22.1 INTRODUCTION

The SEASAT ocean monitoring satellite was the first American satellite designed specially for oceanographic observations. It represented a proof-of-concept mission whose objectives. included demonstration of techniques for global monitoring of oceanographic phenomena and features, provision of oceanographic data for both application users and scientific users, and determination of key features of an operational ocean-dynamics monitoring system. A description of the satellite and its mission is given by Cutting & Pounder (1978). During its operational life SEASAT transmitted an unequalled massive amount of information on surface winds and temperatures, currents, wave heights, ice conditions, ocean topography and coastal storm activities. These data were collected by the satellite's Synthetic Aperture Radar (SAR), Microwave Scatterometer System (SASS), Scanning Multichannel Microwave Radiometer (SMMR), Radar Altimeter (ALT), and the Visual and Infrared Radiometer (VIRR).

The satellite was launched on 26 June 1978, into a near-circular orbit at an altitude of about 790 km and with an inclination of 108°. It circles the earth 14 times daily. The satellite consists of a modified Agena-D rocket stage to which a sensor module is attached. It weighs about 2200 kg, measures about 13 m long, and is equipped with a 28 m² SAR antenna and two 7.4 m² solar panels. Two distinct orbits were used for SEASAT (Cutting et al. 1978). The launch orbit nearly repeated its ground track every 17 days. It had a 17-day closure at 1.67° sub-satellite track spacing providing rapid global sampling. A second orbit was achieved through a sequence of manoeuvres, executed by hydrazine thrusters near the satellite's centre of mass, which were initiated on 18 August and completed on 10 September 1978. This orbit almost exactly repeated every 3 days with one descending leg passing directly over the Bermuda laser site. This characteristic permitted near-zenith laser ranging from Bermuda for the purpose of calibrating the radar altimeter. A tide gauge, accurately surveyed relative to the tracking station, provided a precise tie to the sea surface in the vicinity of this island. On 9 October 1978, after completing 1503 revolutions, a catastrophic failure occurred in the satellite power subsystem, and contact with the satellite was lost. Efforts were initiated to remedy the problem, with commands being sent to the satellite to shut down all non-essential functions, but no further signals were ever received.

One of the main experiments performed with SEASAT was the Altimeter/ Precision Orbit Determination Experiment (ALT/POD). The altimeter part of this experiment had two objectives: to measure very precisely the altitude of the satellite above the ocean surface and to measure the significant wave height of the ocean surface at the sub-satellite point. The altimeter generated a 13.6 GHz chirp signal at 2 kW peak power and measured altitude with a precision of about 10 cm. In order to most effectively use the altimetry data to model the marine geoid and dynamic topography to decimetre accuracy the radial component of the spacecraft's position must be known to a comparable accuracy. The goal of the POD was, applying the most-advanced orbit computation techniques, to attain the highest precision and accuracy of the SEASAT ephemeris. Over selected ocean areas the original goal even was to have orbit estimates accurate to 10 cm root-mean-square (rms) (Cutting & Pounder 1978). The POD experiment utlized ground-based radio-frequency and laser tracking and onboard altimeter data. The satellite included an S-band transponder, a doppler beacon and a laser retro-reflector array. The Unified S-band (USB) range-rate and laser range data obtained by NASA, the Smithsonian Astrophysical Obersvatory (SAO) and several cooperating European stations are archived at the American National Space Science Data Centre. The tracking precision of the NASA lasers was about 10 cm, while the SAO lasers tracked with a precision of 50 cm to 1 m. The precision of the European lasers varied from 10 cm to 1 m. The USB data have a precision of 0.1 mm/s and, while the accuracy is degraded by atmospheric effects, they have the advantage of providing all-weather and expanded geographic coverage. The American doppler data, which are theoretically equivalent to range difference measurements with a precision of 10 cm at 30 s averaging intervals, were collected by the Department of Defense (DOD) and archived at the Naval Surface Weapons Center (NSWC) (Tapley & Born 1980). Because of the extensive global coverage of the DOD tracking network, a dense doppler tracking data set on the order of 200 passes per day was obtained, which allows a sequence of independent two-revolution orbit solutions (Tapley & Born 1980). While the inherent accuracy of the laser and USB data is higher, the sparsity of coverage requires longer arc lengths to generate the satellite ephemeris. This fact imposes more-stringent requirements on the accuracy of the dynamic models.

Also in Europe a SEASAT laser and doppler tracking campaign (SEATRACK) has been organized (Wilson & Aardoom 1982) from 26 June to 10 October 1978, to permit a North Sea altimetry project proposed by the SEASAT Users Research Group of Europe (SURGE). The positions of the European tracking stations involved have been determined in the SEATOC observation campaign (Wilson & Aardoom 1982) from 24 June to 4 July 1978 by processing doppler observations of the U.S. Navy Navigation Satellite System (NNSS). The data

acquired in these projects are processed and archived by the Institute fur Angewandte Geodasie (IfAG) at Frankfurt. In the Netherlands, the Working Group
for Satellite Geodesy (WSG) at the Kootwijk satellite observatory was actively
involved in the laser ranging to SEASAT. For many years there exists a close
cooperation between this group and the Section Orbital Mechanics (SOM) of
Delft University's Department of Aerospace Engineering. This cooperation aims
primarily at geodetic and geophysical studies which need accurate orbit determinations from laser ranging to the satellites LAGEOS, STARLETTE, GOES-3
and SEASAT (Wakker & Ambrosius (1981), (1982)). Apart from the laser range
observations also some SEASAT doppler data acquired at 5 European sites have
been processed by SOM to estimate the position of these stations relative to a
global network of laser stations. These post-mission precise orbit computations
for SEASAT have proven to be a good exercise to acquire the experience required
to perform similar computations for the forthcoming European ERS-1 satellite.
This satellite will be launched by the European Space Agency (ESA) in 1987
into a sun-synchronous orbit at an altitude of about 800 km, and will also carry
a radar altimeter and laser retroreflectors (Anon. 1981).

22.2 ARC SELECTION

Orbit determinations are conventionally divided into short-arc and long-arc
solutions. Typically, a short arc covers a period of a few minutes or part of a revolution. Long-arc solutions utilize tracking data from a worldwide network of tracking
stations. Short-arc solutions require intense tracking coverage by several stations
in the same geographical regions for intervals that are fractions of an orbital
period. The short-arc analyses are less influenced by small dynamic errors than
the analyses of long arcs, but they have the disadvantage of being rather dependent on the data distribution within the arc. The rather sparse SEASAT laser
tracking data necessitated the use of orbital arc lengths of several days, which in
turn required the accurate modelling of all perturbing forces acting on the satellite.
It is the limited accuracy of the current perturbation models, and in particular
of the earth's gravity and atmospheric drag models, that prohibited a global
orbital accuracy down to the 10 cm level.

   Accurate long-arc orbit determinations generally require the observations to
be evenly distributed in time, and taken at a large number of stations with a
good geographical distribution. In reality, however, one has hardly any choice
and one has to work with the available data. For this study the data were taken
from the Kootwijk databank, which is managed and regularly updated by WSG
and includes American and European observations. These data indicated that the
laser tracking coverage over the larger part of SEASAT's operational life was far
from perfect, which seriously limited the selection of arcs suitable for highly-
accurate orbit analyses. After careful examination of the available laser data and
avoiding the periods of orbital manoeuves, 7 arcs have been selected. For each

arc some hand-editing has been performed, deleting passes or parts of passes that contain too much obvious wild data points, and reducing the number of data points for some laser stations to prevent stations with a high laser pulse repetition rate to dominate the solution. In this way the few available passes over Wettzell (Fed. Rep. Germany) and Helwan (Egypt) were completely edited out. Though in one arc some good passes over Grasse (France) were present, these data were deleted because the global set of laser station coordinates used for the majority of the computations reported in this chapter did not contain this station.

Table 22.1
Arcs selected for laser data analysis

Arc id.	Start (yymmdd)	Stop (yymmdd)	Length (day)	Stations	Passes	Observations
1	780719	780722	3.0	7	25	1161
2	780802	780805	3.0	7	20	1047
3	780808	780813	5.8	8	45	2255
4	780829	780901	3.0	5	24	2282
5	780912	780917	5.4	10	35	1623
6	780919	780922	3.7	7	44	2324
6A	780919	780925	6.0	9	60	3482
7	780925	781001	6.1	9	47	2811
Total					256	14661

A summary of the laser data arcs is presented in Table 22.1. For each arc the following quantities are listed: the arc identification number, the start- and stop-time of the arc, the arc length, the number of stations contributing to the observations, the total number of satellite passes over the ground stations and the number of observations used in the orbit solutions. Arcs 1 to 3 refer to the launch orbit, arc 4 to an interim orbit and arcs 5 to 7 to the Bermuda orbit. Arc 6A is an extended arc 6 and is used especially to investigate a gravity field mis-modelling phenomenon (section 22.7). Fig. 22.1 shows the distribution of the arcs in time along with two quantities representing the solar and geomagnetic activities. These are the solar radio flux density at 10.7 cm wavelength, F, and the planetary geomagnetic index, A_p. According to this figure, only arc 7 contains a period of strongly enhanced geomagnetic activity. This arc was deliberately selected to enable an analysis of atmospheric drag modelling errors. The other arcs cover a period of relatively quiet solar and geomagnetic activity and can therefore be used to investigate the accuracy of the applied gravity models. The distribution of the passes over the ground stations is listed in Table 22. 2. The amount of laser tracking data averaged at 8 passes per day, with variations from about 6 to 12 passes per day. For comparison, the USB system provided an average of 18 passes per day (Tapley & Born 1980), while the

Table 22.2
Laser data pass summary

Arc id.	S. Diego 7062	Greenbelt 7063	Bermuda 7067	Gr. Turk 7068	Patrick 7069	S. Fernando 7804	Kootwijk 7833	Arequipa 9907	Mt. Hopkins 9921	Natal 9929	Orroral 9943
1	3	—	3	—	2	—	5	8	—	3	1
2	4	—	2	—	3	2	—	6	—	1	2
3	10	—	6	—	5	3	3	12	—	1	5
4	7	2	7	3	5	—	—	—	—	—	—
5	2	1	7	5	4	—	2	4	2	4	4
6	9	4	7	3	7	—	—	7	7	—	—
6A	12	4	9	4	7	—	1	11	11	—	1
7	9	3	5	6	7	—	1	7	7	—	2
Total	47	10	39	18	33	5	12	48	20	9	15

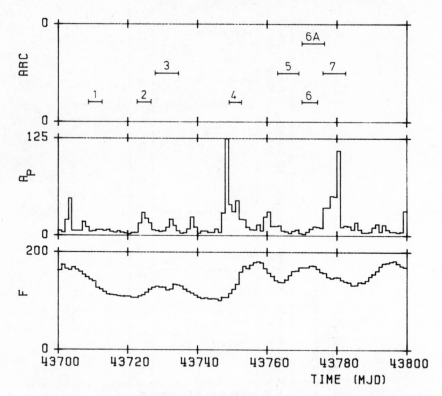

Fig. 22.1 – The distribution of the laser data arcs along with the solar radio flux, *F*, and the planetary geomagnetic index, A_p.

European doppler data presented in section 22.8 averaged 20 passes per day. As an example, Fig. 22.2 shows the sub-satellite tracks during the periods of laser tracking in arc 3, clearly illustrating the relatively poor distribution of the tracking stations.

22.3 COMPUTATION MODEL

All computations have been performed with the NASA Geodyn computer program (Martin 1978), which has been implemented on the Delft University Amdahl 470 V/7B computer.

Part of the data from Orroral Valley (Australia) had to be corrected for a range-dependent error caused by instrumentation problems at that station (Latimer 1979). To all laser observations preprocessing corrections, such as tropospheric refraction and transit time delay, were applied. In addition, all range data were corrected for the range offset between the laser retroreflectors and the satellite's center of mass. For the NASA stations an *a- priori* observation standard deviation

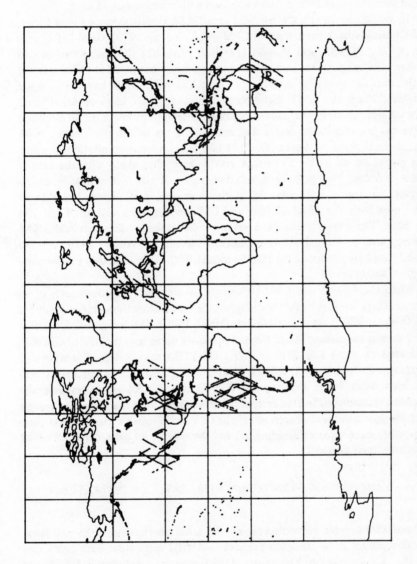

Fig. 22.2 – SEASAT ground tracks during the laser ranging periods of arc 3.

of 15 cm was assumed; the Kootwijk observations were assigned an accuracy of 25 cm and the SAO stations a value of 70 cm. For the numerical integration of the equations of motion and of the variational equations, a fixed-stepsize 11th-order Cowell predictor-corrector method was used with a stepsize of 75 s.

To model the earth's gravity field four NASA Goddard Space Flight Center (GSFC) models have been applied. GEM-9 (Lerch *et al.* 1979), GEM-10B (Lerch *et al.* 1978), PGS-S3, and PGS-S4 (Lerch *et al.* 1981). The latter two models represent the most recent in a series of progressively improved so-called tailored gravity models, specially derived for SEASAT. All models have been made available through the GSFC Geodynamics Branch. The GEM-9 model is based solely on optical, laser, and radio-frequency observations taken on 30 satellites. The model is complete to degree and order 20 in the spherical harmonics, with selected coefficients to degree 30. GEM-10B is a combination solution, containing a global set of $5°$ by $5°$ surface gravity anomalies along with the data in GEM-9 and over 700 globally distributed passes of GEOS-3 altimetry, and is complete to degree and order 36 in the harmonic coefficients. The PGS-S3 model uses both the SEASAT laser and USB tracking data with the GEM-10B data base. The combination of a set of 9600 globally distributed SEASAT altimetry data covering four 3-day arcs from 28 July to 9 August 1978, with the PGS-S3 data base produced the PGS-S4 model. Both tailored models are complete to degree and order 36.

When the GEM-9 and GEM-10B models were used, the coordinates of all stations except for Kootwijk were assigned the values from the so-called Modified New Orleans (MNO) set of coordinates (Marsh 1980). This is a global dynamical GSFC station coordinate solution based primarily upon laser tracking of GEOS-3, augmented by some LAGEOS and STARLETTE tracking data. For Kootwijk coordinates derived from LAGEOS observations (Wakker & Ambrosius 1981) have been used. When applying the PGS-S3 and PGS-S4 gravity models, the associated station coordinates derived simultaneously with the respective gravity field solution were used (Lerch *et al.* 1981). For the earth's gravitational parameter, GM, mean equatorial radius, a_e, and the velocity of light, c, the following values have been assumed:

$$GM = 398600.64 \text{ km}^3/\text{s}^2, a_e = 6378.138 \text{ km}, c = 299792.5 \text{ km/s}.$$

In addition, orbit perturbations due to atmospheric drag, solar and lunar attraction, direct solar radiation pressure and solid earth tides were accounted for. The Jet Propulsion Laboratory (JPL) planetary ephemeris DE-96 was adopted along with the Bureau International de l'Heure (BIH) polar motion and UT1 data. The luni-solar solid earth tides were modelled through a second-degree spherical harmonic, characterized by the Love number $k_2 = 0.29$ and a phase lag $\phi_2 = 2.5°$. The contribution from ocean tides was not considered in the computations. In the measurement model the geometric tracking station displacement due to luni-solar tidal effects was taken into account, and modelled by the Love and Shida numbers $h_2 = 0.60$ and $l_2 = 0.075$, respectively. Atmospheric drag perturbations were computed using the Jacchia 1971 reference atmosphere (Jacchia 1971).

For the computation of the accelerations due to the non-gravitational or surface forces, i.e. atmospheric drag and solar radiation pressure, the satellite's mass and an appropriate cross-sectional area have to be known. For the mass a value of 2213.6 kg was adopted. Owing to the complex geometry of SEASAT its cross-sectional area is highly dependent on the viewing direction. To model the surface forces accurately, these large variations have to be taken into account. For drag the cross-sectional area was therefore interpolated from a GSFC table, specifying the area as a function of the relevant viewing angles. For solar radiation a similar table of normalized accelerations was used. In the tables and figures presented in this chapter the application of these models is referred to by 'tables'. For comparison, some orbit computations have been performed in which the cross-sectional area for drag and solar radiation pressure was assigned a constant mean value of 25.31 m^2. This model is indicated by 'no tables'.

The 'solve-for' parameters in the computation process were the satellite's state vector at epoch, a solar reflection scaling factor (solar reflectivity), and one or more drag scaling factors (drag coefficient). The epoch for each arc was selected closely ahead of the first measurement of that arc. Three models were applied for the adjustment of atmospheric drag perturbations: a constant drag coefficient, a linearly-varying coefficient and multiple daily-constant coefficients. These are referred to in this chapter by respectively: 'constant C_D', 'linear C_D' and 'daily C_D'. Many different computer runs were made, each run being different in terms of perturbation model applied and observations processed. In the following discussion only the main results will be presented.

22.4 QUALITY OF AN ORBIT SOLUTION

Because the 'real-world' orbit is not known, it is very difficult to specify how accurate the computed orbit approximates the real orbit. In fact, the only criterion that is available is to confront the computed orbit with observations, and in particular with observations which are not used in the orbit solution. The latter approach is usually not attractive when determining orbits from laser range observations only, because of the very sparse data set. In addition, the observations generally refer to small parts of the orbit and were recorded in regions of the world where the gravity model errors probably are minimal. Consequently, the rms of fit of the tracking data in the orbit solutions is not a conclusive indicator of global ephemeris accuracy.

For SEASAT a very strong test on the global radial orbital accuracy is possible by using the altimeter data. These data have an accuracy generally better than

10 cm and the onboard data recording system permitted continuous coverage of the oceans around every orbital revolution. The altimeter data can be applied for this purpose in different ways (Goad *et al*. 1980, Marsh & Williamson 1980, Lerch *et al*. 1980), but the most powerful technique is based on processing altimeter data at ground track intersections. At the crossing points of ascending and descending tracks, the constant portion of the ocean surface height above the reference ellipsoid is the same on both tracks. The sea surface height differences at the intersection points, called crossover differences, will reflect unmodelled changes in time-dependent ocean topography, radial ephemeris errors, and time tag biases or height bias changes in the altimeter data. When many thousands of globally distributed crossover points are processed ocean tides are modelled and instrument errors are properly accounted for, the main contributor to the crossover differences turns out to be the radial ephemeris error (Marsh & Williamson 1980). But even this test of orbital accuracy is not absolute since some dynamic model errors may produce correlated orbit errors at the crossover points.

As the processing of altimeter data was not within the scope of this study, the quality of the orbit solution will, in this chapter, necessarily be judged primarily on the basis of the behaviour of the laser range residuals. These are defined as the actual measurements minus the range values computed from the orbit determined within the parameter estimation process. These residuals are a measure of how well the computed orbit fits the actual measurements. If the majority of the residuals plotted per pass as a function of time do not lie within a narrow band about zero, having a width in the order of the measurement's accuracy, it is a clear indication that the modelling was not optimal. On the other hand, if the residuals are nicely scattered about zero this does not necessarily mean that the solution is correct. It only proves that some solution has been obtained that fits the observations; but the recovered values of the individual parameters need not to be correct.

Using arcs of more than a few orbital revolutions, one may expect that the history of the residuals per pass will show some signature, reflecting the effects of model errors. As an example of the residual histories, Fig. 22.3 is included. These plots show residuals from arc 1 for six successive passes within an 8-hour period over the Kootwijk, Natal (Brazil), Arequipa (Peru), and Patrick AFB (Florida) laser stations. The results were obtained with the PGS-S4 gravity model, applying the geometry tables and solving for a linearly-varying C_D. In each plot the satellite and station number, the date of the pass, and the origin of the time scale for that pass are indicated. The plots clearly reveal a signature in the range residuals. To extract the orbit error information out of these residuals, the iterative parameter estimation process can, after convergence has been reached, be extended with a final iteration in which all physical parameters are held fixed and only apparent range and timing biases are determined for each pass over a station. These biases are intended to represent the actual radial and along-track errors of the orbit in the spatial region covered by that ground

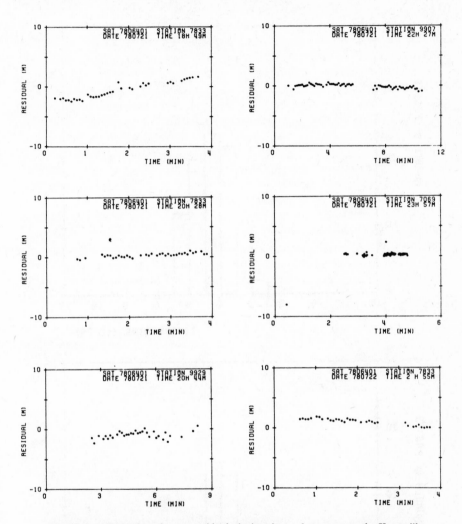

Fig. 22.3 — Examples of range residuals during six arc 1 passes over the Kootwijk (7833), Natal (9929), Arequipa (9907), and Patrick AFB (7069) laser stations within a period of 8 hours.

station, assuming no tracking system errors. As an example, Fig. 22.4 shows a histogram of the range and timing biases for all passes in arc 3 and arc 6A. Again, the PGS-S4 gravity model and the geometry tables have been applied, while a series of multiple daily-constant C_D's was solved for. If these biases are taken as a measure of the global accuracy of the orbital fit, it may be concluded that the orbit has been computed with an rms error of about 1 m in radial direction and about 0.27 ms along-track. This timing error corresponds for the orbital altitude of SEASAT to an along-track position error of 2.0 m.

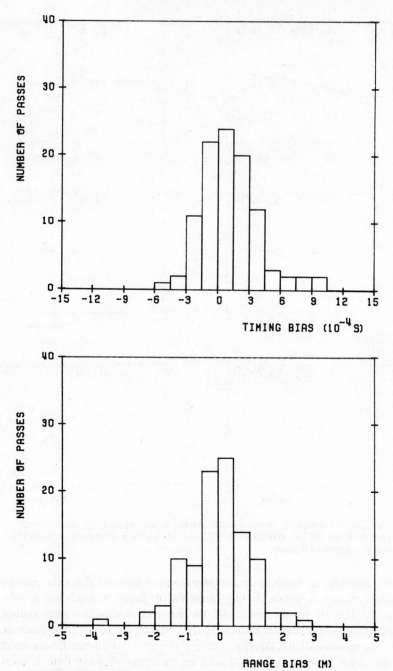

Fig. 22.4 – Histogram of the apparent range and timing biases for all passes in arcs 3 and 6A (PGS-S4, daily C_D, tables).

For a further indication of the strength of the orbit solution, also the values recovered for additional model parameters, such as the solar reflectivity and the drag coefficient may serve. By comparing these values with values that follow from physical theory or that were obtained in other independent studies at least some impression of the quality of the solution may be obtained.

22.5 DETAILED ANALYSIS OF THE ARC 6 RESULTS

Initially, only the 3.7-day arc 6 was analysed in detail to get an impression of the orbital effects of applying different models for the gravity field and atmospheric drag. This particular arc was selected because of its relatively large number of laser passes (an average of 12 passes per day) and because it fits into a period of relatively quiet solar and geomagnetic conditions (Fig. 22.1). During such periods the atmospheric drag modelling errors are a minimum, which makes the arc suitable for studies on the accuracy of gravity models.

Table 22.3
The results for arc 6 when applying different gravity models (const. C_D, tables).
For the biases mean/standard deviation values are listed

Gravity model	Datapoints	Residual rms (m)	Range bias (m)	Timing bias (ms)	C_D	C_R
GEM-9	2294	2.7	0.54/2.13	0.10/0.91	2.36	1.36
GEM-10B	2221	2.2	0.35/1.85	−0.04/0.72	2.35	1.40
PGS-S3	2213	1.3	0.00/0.93	0.18/0.60	2.18	1.36
PGS-S4	2293	1.8	0.10/1.17	0.06/0.44	2.17	1.31

Table 22.3 presents a summary of the number of datapoints remaining in the solution after automatic data editing, the residual rms's of fit, the mean and standard deviation values of the apparent range and timing biases and the solar reflection and drag coefficients recovered when applying different gravity models. In all cases a single C_D was solved for, and the geometry tables were used. The table clearly shows the superiority of the tailored PGS-S3 and PGS-S4 models over the general-purpose GEM-9 and GEM-10B models. The increase of the rms of fit for PGS-S4 with respect to PGS-S3 is partly due to the smaller number of datapoints rejected in the editing process, but was also found by Lerch et al. (1981) in some comparisons of rms fits between PGS-S4 and PGS-S3. This, in fact, is an illustration of what has already been mentioned, namely that due to the poor geographical distribution of laser tracking data, the rms of fit is not a good indicator of the global orbital accuracy. By confronting the tailored models with the global altimeter data it has been demonstrated (Lerch et al. 1981) that the overall rms crossover difference is reduced significantly from 1.7 m for PGS-S3 to 1.1 m for PGS-S4. It is believed that with this model the rms of the radial orbit errors alone is about 0.8 m. This proves that PGS-S4

approximates the real orbit globally much better than PGS-S3. Therefore, in the majority of the orbit computations presented in this chapter the PGS-S4 model has been used.

Fig. 22.5 shows a comparison of the history of the apparent range biases per pass for GEM-10B and PGS-S4. It is clear that the scatter in the range biases is smaller with PGS-S4, indicating a better radial orbit modelling. Comparing the history of the timing biases for GEM-10B and PGS-S4 reveals (Fig. 22.6) that the scatter has reduced for the tailored models, but that now a peculiar sinusoidal trend shows up. In order to verify if this phenomenon is only characteristic for that particular arc, arc 3 was processed using the same dynamic model. It turned out that also for this arc a similar oscillatory trend in the timing biases was present. Since both arcs are in a period of relatively quite geomagnetic conditions, it is unlikely that atmospheric drag modelling errors alone caused these long wavelength features. It is more plausible that this behaviour of the apparent timing biases is at least partly a result of mismodelled gravity coefficients. This subject will be discussed further in section 22.7.

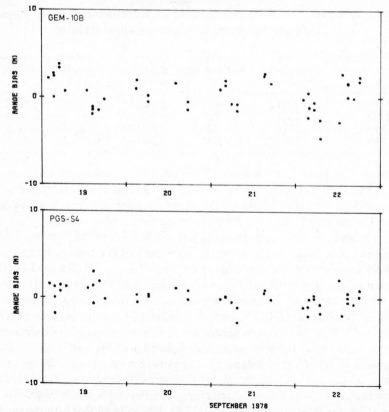

Fig. 22.5 – The apparent range biases for arc 6, computed with GEM-10B (top) and PGS-S4 (bottom) (const. C_D, tables).

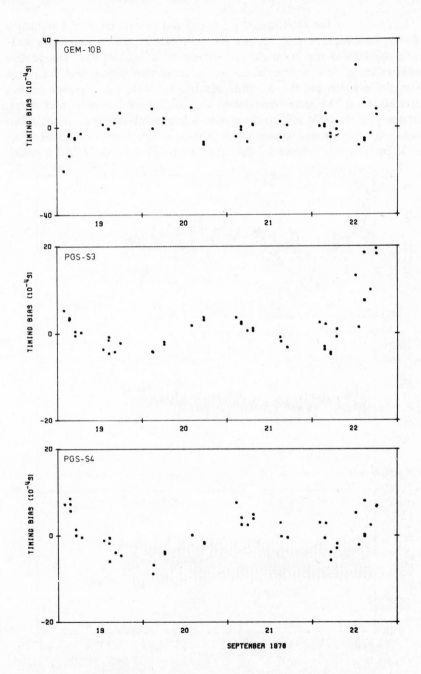

Fig. 22.6 – The apparent timing biases for arc 6, computed with GEM-10B (top), PGS-S3 (centre) and PGS-S4 (bottom) (const. C_D, tables).

To investigate the orbit sensitivity to the gravity field model the technique of orbit differences may be applied. For arc 6 first an orbit solution was made using the PGS-S3 gravity model and solving for a constant C_D. The satellite ephemerides at time intervals equal to the integration stepsize were extracted from the solution for the complete arc length. Then, the computation was repeated, using the same observations and solving for the same parameters, but applying the GEM-10B gravity model. Subsequently, the state-vector differences were transformed to position differences in radial, cross-track and along-track components, relative to the orbit computed with the PGS-S3 model.

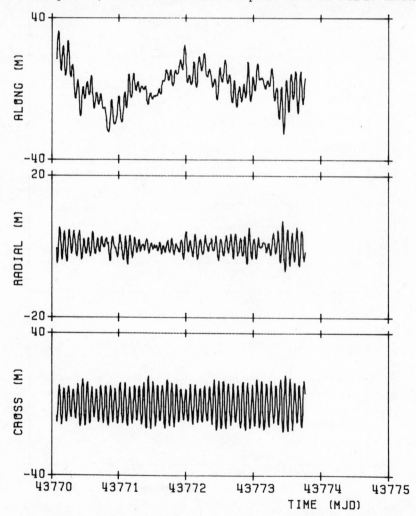

Fig. 22.7 — Arc 6 cross-track, radial and along-track position differences for GEM-10B versus PGS-S3 (const. C_D, tables).

Fig. 22.7 shows the results. A summary of results from comparisons between other gravity field models is listed in Table 22.4; in all computations the geometry tables were applied and a single C_D was solved for. These comparisons show that position differences of many metres may exist, in particular in the along-track direction. It is remarkable that even when comparing the two tailored models, along-track position differences of up to 16 m occur. So, assuming that the PGS-S4 model most closely approximates the real gravity field, the comparison suggests that with PGS-S3 the global radial and cross-track orbit errors reach more than 2 m and the along-track error even more than 10 m for this arc.

Table 22.4

Gravity model effects on the orbit solution for arc 6 (const. C_D, tables).
Listed are rms/maximum values of position differences in metres.

Comparison	radial	Orbit differences cross-track	along-track
GEM-10B vs. GEM-9	2.0/ 5.4	2.0/ 6.2	6.7/26.4
GEM-9 vs. PGS-S3	2.5/11.0	8.0/15.3	10.5/33.0
GEM-10B vs. PGS-S3	1.9/ 5.6	8.4/16.0	8.9/30.2
PGS-S4 vs. PGS-S3	1.1/ 2.8	1.4/ 3.8	4.1/16.2

Owing to the relatively low altitude and high area-to-mass ratio, the orbit of SEASAT is appreciably influenced by atmospheric drag. If the satellite's aerodynamic characteristics and the atmospheric conditions could be modelled perfectly, this would not hamper the precision of the orbit determination. However, the precise interaction mechanisms of the atmospheric particles with the satellite's skin are at present not fully understood, and the current atmospheric models are known to be far from perfect. The computed drag will show both systematic and secular or periodic errors. The first type of error does not affect the orbital accuracy if a drag scaling factor, C_D, is also solved for. The non-systematic errors in the computed drag perturbations can not be absorbed by solving for a single C_D. In particular, the occurrence of solar and geomagnetic disturbances may cause serious problems. A possible approach to handle the short-period density fluctuations resulting from these disturbances is to adjust both C_D and its time-derivative (linear model) or to apply the model of multiple adjustable daily-constant C_D's. Application of the latter technique has proved to permit the extension of SEASAT orbital arcs as long as 17 days with only a minor degradation in accuracy (Marsh & Williamson 1980). The availability of such long dynamically consistent orbits is advantageous for the interpretation of satellite altimeter data. The adjustment of multiple C_D's has, however, a disadvantage when one is interested in the geophysical explanation of the observed orbit behaviour. The multiple C_D's will absorb any slowly-varying along-track orbit error, such as the periodic along-track errors due to mismodelling of geopotential coefficients. For example, the peculiar behaviour of the timing bias history shown in Fig. 22.6 can in this way effectively be removed.

Table 22.5

Drag coefficient modelling effects on the orbit solution for arc 6 (PGS-S3, tables). For the biases mean/standard deviation values are listed. The C_D values for the linear and daily C_D models represent either the initial C_D and its time-derivative or the extremes of the range of multiple C_D's.

Model	Residual rms (m)	Range bias (m)	Timing bias (ms)	C_D	C_R
const. C_D	1.3	0.00/0.93	0.18/0.60	2.18	1.36
linear C_D	1.4	0.03/0.97	0.14/0.53	2.08,0.04	1.33
daily C_D	0.8	0.04/0.86	0.04/0.26	2.15-3.04	1.59

Table 22.5 shows the effects of the different C_D models on the orbit solution for arc 6. In all solutions the PGS-S3 model and the geometry tables have been applied. Even for periods of quiet solar and geomagnetic activity the adjustment of multiple C_D's obviously removes a large part of the geometric and dynamic model errors. This manifests itself in a significantly smaller rms of fit and reduced apparent timing biases. Table 22.6, finally, presents a summary of the orbit sensitivity to the application of the geometry tables and the different C_D modelling techniques. These results suggest that the modelling of the variable cross-sectional area of SEASAT considerably affects the orbit solution. In addition, the application of the linear, and in particular the multiple C_D models yield orbits that significantly deviate from the orbit computed with the single C_D model. This may be interpreted as follows: the application of a single adjustable C_D and a constant average cross-sectional area will, for this arc, lead to orbit errors of some decimetres in radial direction and several metres along-track. Experience with this arc, and a number of additional computations not described in this chapter, have led to our current practice of always solving for C_D and its time-derivative for arcs of up to 3 days length. For longer arcs and arcs covering a period of strongly enhanced solar or geomagnetic activity multiple C_D's are adjusted.

To investigate the overall model consistency a final test was performed in which arc 6 was divided into two 2.6-day arcs having a 1.6-day overlap. Each of these arcs was processed with the PGS-S3 gravity model and the geometry tables, and C_R and a single C_D were solved for. Over the common 1.6-day period the two orbits were subsequently differenced. It was found that the radial and along-track rms differences were 0.8 m and 2.3 m, respectively. These differences, which are due to dynamic model errors and the different distribution of laser data, may be considered as a lower boundary of the real orbital accuracy for the applied perturbation model.

Table 22.6

Drag modelling effects on the orbit solution for arc 6. Listed are
rms/maximum values of position differences in metres.

Model	Comparison	Orbit differences		
		radial	cross-track	along-track
GEM-9; const. C_D	no tables vs. tables	0.4/1.1	0.3/0.8	3.1/ 7.0
GEM-9; tables	daily C_D vs. const. C_D	0.3/0.8	0.1/0.1	2.9/ 7.1
PGS-S3; const. C_D	no tables vs. tables	0.3/0.6	0.4/1.1	3.2/11.8
PGS-S3; tables	linear C_D vs. const. C_D	0.1/0.3	0.1/0.2	1.0/ 4.2
PGS-S3; tables	daily C_D vs. const. C_D	0.8/2.0	0.1/0.2	5.6/26.2

22.6 RESULTS FOR THE OTHER LASER DATA ARCS

Also for the other arcs a number of solutions have been computed using different force models. In this section only the results for the best models will be presented. That means that for all arcs, except for arc 6, the PGS-S4 model was used and that the geometry tables were applied. For arcs with a length of up to 3 days the linear C_D model was used, while for longer arcs multiple daily-constant C_D's were adjusted.

Table 22.7 presents a summary of the main results. It may be concluded that for arcs shorter than 4 days the laser range residual rms's are below 1 m. This means that the radial orbital accuracy in the neighbourhood of the ground stations is also better than 1 m. The 5 to 6-day arcs have range residual rms's of 1 to 1.2 m. The apparent range bias standard deviations for these longer arcs is also less than 1.2 m, except for arc 5 where a value of 1.6 m is reached. The timing bias standard deviations are less than 0.46 ms, which corresponds to an along-track position error of 3.4 m.

Table 22.7 also shows some indication of a systematic variation of C_R with time. When these arcs were computed with other gravity models the predominant character of the variation of C_R was found to be unaffected. Thus this variation from arc to arc seems to represent an actual phenomenon. Since the force modelling does not account for albedo radiation and for changes of the satellite's reflectance as a function of its orientation relative to the sun and the earth, nor for a potential aerodynamic lift associated with the large SAR and solar panels, it is conjectured that the observed variations of C_R are due to these unmodelled phenomena. There are two additional arguments for this hypothesis. First, it can be seen in Table 22.7 that the lengthening of the 3.7-day arc 6 to a 6-day

Table 22.7

Summary of the best orbit solutions for each arc (tables). For the biases mean/standard deviation values are listed. The C_D values represent either the initial C_D and its time-derivative or the extremes of the range of multiple C_D's.

Arc id.	Length (day)	Model	Residual rms (m)	Range bias (m)	Timing bias (ms)	C_D	C_R
1	3.0	PGS-S4; linear C_D	0.65	–	–	4.47,0.006	0.60
2	3.0	PGS-S4; linear C_D	0.96	–	–	3.40,0.242	2.15
3	5.8	PGS-S4; daily C_D	0.91	0.01/0.82	0.02/0.24	2.75-4.44	1.64
4	3.0	PGS-S4; linear C_D	0.50	–	–	2.38,−0.224	1.51
5	5.4	PGS-S4; daily C_D	1.23	0.11/1.57	−0.16/0.46	2.11-3.77	1.40
6	3.7	PGS-S3; daily C_D	0.79	0.04/0.86	0.04/0.26	2.15-3.04	1.59
6A	6.0	PGS-S4; daily C_D	1.23	0.00/1.15	0.21/0.32	2.02-3.10	1.58
7	6.1	PGS-S4; daily C_D	1.13	−0.21/0.83	0.02/0.42	2.06-5.07	1.70

arc 6A does not change the value of the solved-for reflectivity signficantly. This makes it unlikely that the value of C_R is much affected by unmodelled along-track perturbations. A second argument is that a very similar variation of C_R has also been found by Schutz & Tapley (1980), who present the results of SEASAT orbit computations using different software but also neglecting albedo radiation, reflectance variations and atmospheric lift.

The solutions for C_D show that, although always a physically acceptable value of 2-5 was obtained, a significant variation of C_D during an arc may occur. It is recalled that this variation is due to the overall force model errors. For arc 7 these will be in particular errors in the atmospheric drag modelling, as this

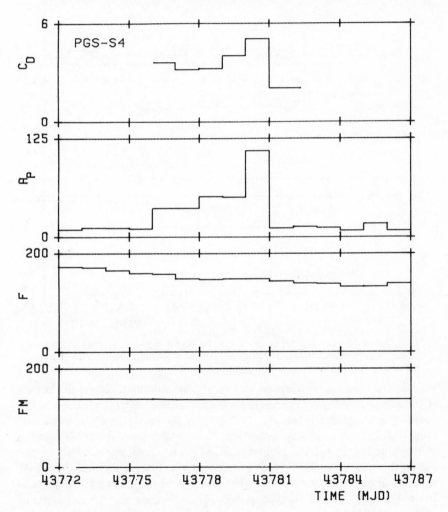

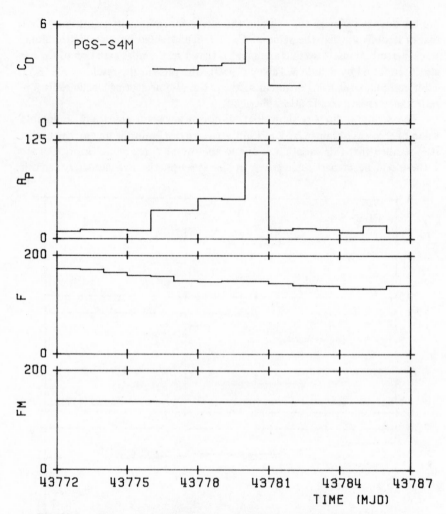

Fig. 22.8 – The actual solar radio flux, F, its 90-day mean value, FM, and the planetary geomagnetic index, A_p, along with the recovered arc 7 multiple C_D's for PGS-S4 (p. 373) and PGS-S4M (this page) (tables).

arc contains a geomagnetic disturbance. It this assumption is correct the history of the adjusted multiple C_D's has to show a correlation with the variation of the planetary geomagnetic index, A_p. In Fig. 22.8 the recovered C_D's are plotted along with the values of solar radio flux, F, its 90-day mean value, FM, and the geomagnetic index, A_p. The plot on p. 373 refers to a solution with PGS-S4, the one on this page to a solution with a somewhat modified gravity model PGS-S4M which is discussed in section 22.7. The plots reveal the expected strong correlation between C_D and A_p and provide a justification for the technique of adjusting multiple C_D's. To demonstrate the improvement of the orbital accuracy by using the multiple C_D model, the computations for arc 7 were repeated, now solving only for a single C_D. It was found that the rms of fit increased from 1.1 m for the multiple C_D model to 25.1 m for the single C_D adjustment. The two orbits, which were derived from exactly the same laser ranging data, were subsequently differenced. The position differences in radial, cross-track and along-track directions, relative to the orbit computed with the multiple C_D model, are plotted in Fig. 22.9. From this figure it may be concluded that if the

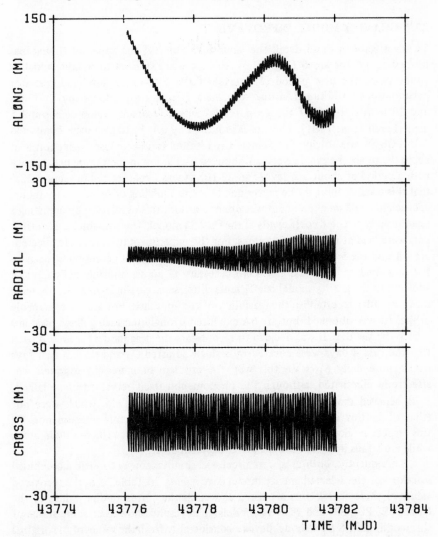

Fig. 22.9 – Arc 7 cross-track, radial, and along-track position differences for a constant versus a multiple daily-constant C_D model (PGS-S4, tables).

classical approach of solving for a single C_D had been adopted, the errors in modelling the effects of geomagnetic disturbances on the atmospheric density would have led to along-track position errors of more than 120 m. For the radial and cross-track components, errors of up to 10 m and 12 m, respectively, would have resulted. In case of orbit predictions, when the actual (future) solar and geomagnetic activity are not known, the orbit errors can, of course, even become much larger.

22.7 GRAVITY MODEL IMPROVEMENT

To investigate in more detail the sinusoidal trend of the apparent timing bias history found for arc 6 (Fig. 22.6), this arc was extended to 6 days, which is about twice the beat period (3.1 days) of the 14th-order dominant resonant perturbations. Additional strong resonance effects are produced by the 28th and 29th-order terms of the geopotential with a resonant period of about 1.5 days (Lerch *et al.* 1981). This arc is indicated by 6A. Fitting an orbit computed with PGS-S4 and solving for a single C_D resulted in a large laser residual rms of about 10 m and revealed a very-pronounced oscillation in the timing bias history with a period of about 3.2 days (Fig. 22.10). It was hypothesized in section 22.5 that this oscillation might be partly due to a mismodelling of some gravity model coefficients, and an experiment was done to absorb this oscillation by adjusting a few properly-selected coefficients of the PGS-S4 model. The non-adjusted coefficients were held at their PGS-S4 values. For this experiment two arcs were selected: arc 6A and the 5.8-days arc 3. During both arcs the level of geomagnetic activity is relatively low, so it did not seem necessary to adjust multiple daily C_D's in addition to the gravity model coefficients. First, separate single-arc solutions were made in order to establish the feasibility of the approach. For arc 3 a reasonable orbital fit was obtained, but arc 6A exhibited strongly increasing range residuals during the last day. It was decided to reconfigure the drag model for arc 6A such that the first 4 days were covered by a single adjusted C_D and the last two days by separate daily C_D's. In this way the problem of increasing residuals was effectively eliminated, although the phenomenon itself remained unexplained. It is believed that errors in other terms of the gravity field, which were not adjusted in this experiment, are partly responsible for this phenomenon. In this respect it should be recalled that arc 6A corresponds to the Bermuda orbit, while arc 3 falls into the time span of the launch orbit.

Subsequently, both arcs were processed simultaneously to yield a combined solution for the selected gravity model coefficients. In Table 22.8, the recovered values of these coefficients are listed as well as the corresponding values of the GEM-10B, PGS-S3, and PGS-S4 models. The results show that some terms of this modified PGS-S4M model deviate considerably from the values of the original PGS-S4 model, in particular the C(15,14), C(29,29), S(29,29), C(30,28), C(32,30), and S(32,30) terms. Interesting to note is that for a number of these terms the

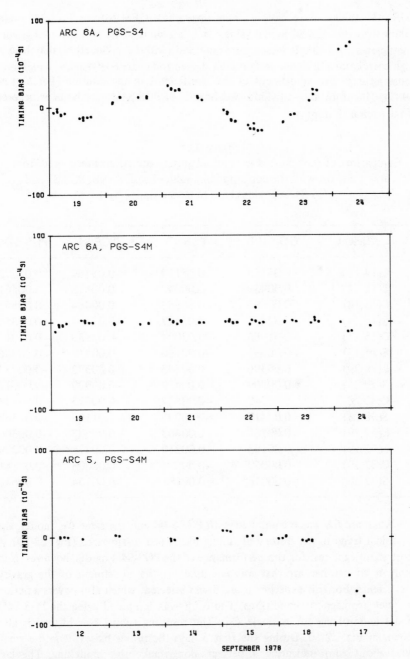

Fig. 22.10 – The apparent timing biases for arc 6A, computed with PGS-S4 (top; tables, const. C_D) and PGS-S4M (centre; tables, first 4 days const. C_D, last 2 days daily C_D), and for arc 5 computed with PGS-S4M (bottom; tables, const. C_D).

PGS-S4M values more closely approximate the GEM-10B or PGS-S3 model values than the PGS-S4 model values. It is known that the estimation of gravity coefficients by a simple least-squares method (without collocation) may lead to high correlations between certain high degree and order coefficients, which may cause an excessive adjustment of the coefficients in the solution (Lerch *et al.* 1981). Therefore, the individual coefficients of PGS-S4M may be in error, even if the orbital fit improves.

Table 22.8

Comparison of the adjusted normalized gravity coefficients recovered from arcs 3 and 6A with the corresponding values in GEM-10B, PGS-S3 and PGS-S4. All values have to be multiplied by 10^{-6}

Coefficient	GEM-10B	PGS-S3	PGS-S4	PGS-S4M
C(14,14)	−0.05193	−0.05153	−0.05369	−0.05672
S(14,14)	−0.00459	−0.00448	−0.00858	−0.00561
C(15,14)	0.00395	0.00363	0.00464	0.00855
S(15,14)	−0.02402	−0.02437	−0.02494	−0.02354
C(15,15)	−0.01966	−0.02004	−0.02533	−0.02917
S(15,15)	−0.00639	−0.00560	−0.00912	−0.00486
C(29,29)	0.00396	0.00543	0.00372	−0.00933
S(29,29)	−0.00809	−0.00472	−0.00920	0.00617
C(30,28)	−0.01742	−0.00837	−0.00773	−0.02136
S(30,28)	−0.03412	−0.00945	−0.01109	−0.01567
C(32,29)	0.00740	−0.00663	−0.00533	−0.00300
S(32,29)	0.00163	0.00210	0.00378	0.00209
C(32,30)	−0.00059	−0.00274	0.00008	−0.02141
S(32,30)	−0.00016	−0.00091	0.00134	0.00656

When arc 6A was recomputed with PGS-S4M and the same C_D model, the sinusoidal trend in the history of timing biases had disappeared (Fig. 22.10). A more significant test for the performance of the PGS-S4M model, however, is to apply it to another arc that was not used for the adjustment of the gravity coefficients. For this experiment arc 5 was selected, which also covers a period of quiet geomagnetic conditions. The orbit was computed using the PGS-S4M model and solving for a single C_D. The resulting timing bias history is also shown in Fig. 22.10. During the first 5 days the timing biases fit into a small band about zero indicating a perfect along-track orbit modelling. The last three passes on this plot clearly fall outside this band. Two other passes on 16 and 17 September, which are not indicated in the plot, were completely edited out during the iterations. This sudden departure from the band of timing biases was also observed for arc 6A if a constant C_D was solved for the whole arc. These phenomena are probably due to some remaining gravity field mis-modelling. Another indication that PGS-S4M models the gravity field pertur-bations slightly better than PGS-S4 came from an additional test. In this experi-ment the orbit for arc 7 was recomputed with the PGS-S4M model, and multiple daily C_D's were solved for. As shown in Fig. 22.8, the adjusted C_D's correlate slightly better with the A_p values than the C_D's recovered with the original PGS-S4 model.

22.8 POSITIONING OF EUROPEAN DOPPLER STATIONS

As already mentioned in section 22.1, SEASAT carried a dual-frequency doppler beacon, radiating at 162 MHz and 324 MHz. The two frequencies are used to enable a correction for first-order ionospheric effects on the signal. In Europe, a number of doppler stations have tracked SEASAT (Wilson & Aardoom 1982). The data acquired at 5 stations: Brussels (Belgium), Geldingaholt (Iceland), Toulouse (France), Trondheim (Norway), and Wettzell (Fed. Rep. Germany) have been made available to us by Lelgemann (1980), and were used in this study. The measurements represented raw data, not corrected for errors nor screened for evi-dent outliers, and consisted of so-called doppler counts, being the number of beat cycles between an analog combination of received frequencies and a frequency generated by a ground oscillator. From these observations a set of doppler data was selected that covered the time span of each laser data arc, except arc 5.

A summary of the doppler data passes is presented in Table 22.9, which lists, apart from the arc identification number and the arc length, the number of passes over each individual doppler station and the total number of observations for that arc. Comparing the number of doppler passes per day with the number of laser passes per day, one of the main disadvantages of satellite laser ranging immediately emerges: the data pass frequency for the all-weather doppler stations is much higher. Neglecting arc 1 where only one station is observing, the average number of passes per day is, even for the 5-station network, about 21. As an example, Fig. 22.11 shows the satellite tracks during the periods of doppler observations in arc 7. The high data pass frequency is, in general, very advantageous for orbit computations because it implies a good orbital coverage. In this case of doppler tracking from only five closely spaced European stations, however, it may not be expected that the doppler data add much information to increase the accuracy of the orbits computed from laser observations, especi-ally because the doppler measurements have a lower accuracy. This was confirmed by some numerical experiments, which indicated that by adding the doppler data the adjusted epoch state-vector position components changed only from 1 to 10 cm, while the standard deviations, being about 3 cm, were hardly affected. On the other hand, the doppler data may be used to determine the position of

Table 22.9
Doppler data pass summary

Arc id.	Length (day)	Brussels 021	Trondheim 506	Geldingaholt 510	Wettzell 643	Toulouse 7145	Observations
1	3.0	16	–	–	–	–	488
2	3.0	23	–	25	–	–	1322
3	5.8	38	–	45	–	–	2147
4	3.0	19	23	26	2	–	1689
6A	6.0	37	44	41	–	36	3776
7	6.1	34	44	45	5	37	3718
Total		167	111	182	7	73	13140

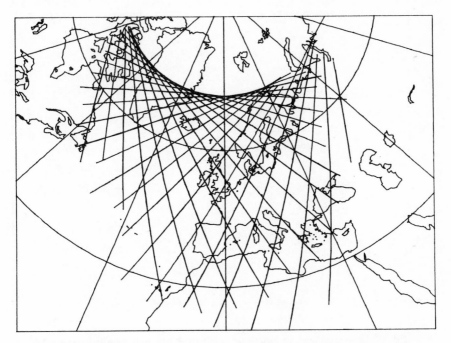

Fig. 22.11 – SEASAT ground tracks during the doppler tracking periods of arc 7.

the doppler stations relative to a reference frame defined by a global network
of laser stations. For the analysis described in this section the PGS-S4 laser station
coordinates (Lerch *et al*. 1981) were adopted.

To process the doppler counts with Geodyn the measurements were first
converted to average range-rate values; these range-rates are translated internally
by Geodyn into range differences. To all observations, preprocessing corrections,
such as tropospheric refraction and transit time delay, were applied and all data
were corrected for the offset between the doppler antenna and the satellite's
centre of mass. In all computations the laser data and doppler data were processed
simultaneously, applying the dynamic models described in section 22.3.

A number of different solutions has been obtained for the individual data
arcs. As an example, Fig. 22.12 shows the range-rate residuals for 4 passes over
4 ground stations within a period of 20 minutes during arc 6A. These results
were obtained from a single-arc solution for the doppler station coordinates
using PGS-S4 and solving for C_R, multiple C_D's and for each doppler pass for a
range-rate bias. The reason for the adjustment of range-rate biases is to absorb
the frequency drift of the doppler beacon and the ground receiver frequency
error. In the data reduction process the laser data were assigned their usual
accuracies but an infinitely large standard deviation was taken for the doppler
data. In this way it is dictated that the orbit is effectively determined from the

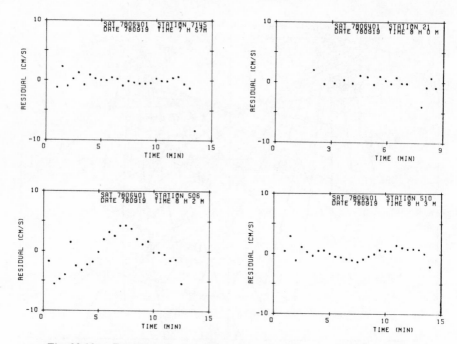

Fig. 22.12 – Examples of range-rate residuals during four simultaneous arc 6A passes over the Brussels (021), Geldingaholt (510), Toulouse (7145), and Trondheim (506) doppler stations.

global laser data only. When the doppler residuals are compared with the laser residuals as shown e.g. in Fig. 22.3, it is found that the signature in the band of residuals is sometimes much more pronounced. This is probably due to instrumental deficiencies and the higher sensisitivity of radio-frequency transmissions to tropospheric and ionospheric conditions, which can not be modelled perfectly.

Estimates of the doppler station coordinates were obtained from six single-arc and one 6-arc solution. Only the results of the 6-arc solution, in which the observations acquired during 221 laser passes and 540 doppler passes were processed simultaneously, will be presented here. In these computations the doppler data were assigned a 1 cm/s measurement standard deviation and for each arc linear or multiple C_D's, C_R, and pass-dependent range-rate biases were solved for. The coordinates of the doppler stations were considered adjustable parameters common to all arcs. The station coordinates solution is presented in Table 22.10. The formal coordinate standard deviations, which indicate only the coordinate uncertainties due to measurement noise, were found to be about 12 cm for Brussels, Trondheim, and Geldingaholt, 23 cm for Toulouse, and about 1.7 m for Wettzell. This large uncertainty is a direct consequence of the small number of passes processed for that station. It is emphasized that owing to dynamic model errors the actual uncertainty of the station coordinates may be considerably larger than the formal standard deviations. Brussels was the only station that had tracked SEASAT in all six arcs. When comparing for this station the coordinates from the six single-arc solutions with the coordinates from the combined 6-arc solution, it was found that the position in each single-arc solution deviated less than 2.5 m. This is an indication that the Brussels coordinates presented in Table 22.10 probably have an accuracy of 1 m, which is only slightly worse than the accuracy of the PGS-S4 laser station coordinates. A better test for the derived coordinates is to compare them with other independent solutions, such as those obtained in the European SEATOC doppler campaign (Amberg *et al.* (in press)). Because these coordinates refer to another coordinate system, which may be translated and rotated relative to the PGS-S4 system, instead of the station coordinates only the straight-line interstation distance (baseline) were compared. Although there may also be a scale difference between the two coordinate systems, this comparison of baselines will give at least some impression of the accuracy of the coordinates. It was found that the SEATOC baseline values for Brussels-Trondheim and Brussels-Geldingaholt differ less than 0.9 m from our solutions. The Brussels-Toulouse baseline differs 2.0 m and the baseline Brussels-Wettzell 9.1 m. This last large discrepancy, of course is due to the very small number of Wettzell passes in our solution.

Table 22.10

Solution for the doppler station coordinates (PGS-S4)

Station	Coordinates (m)		
	X	Y	Z
Brussels	4027833.0	307023.7	4919536.9
Trondheim	2814799.3	516821.8	5681073.4
Geldingaholt	2625128.9	−969339.4	5712459.0
Wettzell	4075543.2	931834.9	4801602.7
Toulouse	4627844.0	119867.1	4372987.6

The values of the range-rate biases recovered from the 6-arc solution are plotted for each pass in Fig. 22.13. The biases for Brussels, Trondheim, and Toulouse fall into the lower narrow band with a mean slope of about 2.55×10^{-7} m/s^2, which corresponds to a frequency drift of 0.024 Hz per day. This value agrees nicely with the actual satellite oscillator drift of about 0.02 Hz per day (Anderle 1980). The range-rate biases for passes over Geldingaholt constitute the upper band with an offset of about 5 m/s and a larger slope, indicating a considerable ground receiver oscillator drift. The four points outside these two bands refer to passes over Wettzell; their irregular distribution is probably due to instrumental problems or data processing errors.

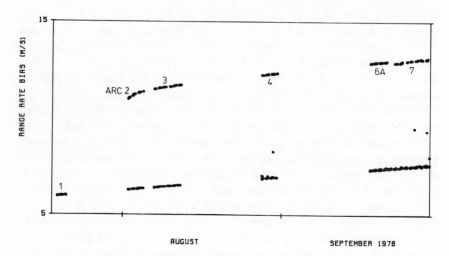

Fig. 22.13 – The recovered range-rate biases for all doppler passes over Brussels, Toulouse, Trondheim (lower band), Geldingaholt (upper band), and Wettzell (lower band and four points outside the two bands).

22.9 CONCLUSIONS

Laser range measurements acquired during 256 passes over 11 ground stations were processed. Seven data arcs were formed with lengths of 3 to 6 days. From these data arcs very accurate orbits were determined and values for the satellite's drag coefficient and solar reflectivity were recovered. The sensitivity of the orbit solution and parameter estimation to the applied gravity and atmospheric drag model was investigated. For the gravity field the NASA GEM-9, GEM-10B, PGS-S3, and PGS-S4 models were used. The atmospheric drag was modelled by constant and time-varying cross-sectional areas and adjustable drag coefficients. Two laser data arcs were used to tune some PGS-S4 gravity model coefficients such that an improved orbital fit was obtained. Finally, doppler data acquired during 540 passes over 5 European doppler stations were processed simultaneously with the laser data to position the doppler stations relative to the global PGS-S4 coordinate system.

The main conclusions from this study are:

- The newly-developed PGS-S3 and PGS-S4 tailored gravity models yield significantly more-accurate orbits than the general-purpose GEM-9 and GEM-10B models.

- Applying the PGS-S4 gravity model and an appropriate atmospheric drag model the orbit of SEASAT has been reconstituted from 3- to 6-day arcs of laser range tracking data with an rms error of about 0.8 m in radial direction and about 3 m along track.

- The dynamic model errors and not the tracking data precision limit the accuracy of a long-arc orbit solution. The orbit errors arise mainly from inaccuracies of the gravity models and, to a somewhat lesser extent, of the models for the surface forces, in particular atmospheric drag.
- To attain a radial orbital accuracy of 10 cm, as was a goal in the SEASAT mission, a considerable effort will be required in further improving the gravity model. To develop such an advanced model many more altimeter data have to be included, and an accurate modelling of tidal and surface forces, like atmospheric drag and lift, and radiation pressure, will be necessary.
- An accurate orbit determination of SEASAT requires a precise modelling of the satellite's time-varying cross-sectional area and the application of adjustable time-varying drag coefficients.
- In the current atmospheric models the density fluctuations due to short-period solar and geomagnetic disturbances are modellled relatively poorly. The orbital effects of these model errors can be removed effectively by applying a multiple drag coefficient model.
- Although PGS-S4 definitely is the best available SEASAT gravity model, orbits computed with this model show a distinct sinusoidal trend in the along-track position error component. Solving for a limited number of gravity field coefficients has resulted in a modified PGS-S4M model that yields a considerable reduction of the along-track orbit errors.
- The addition of the European doppler data to the global laser tracking data hardly affected the orbital accuracy.
- The coordinates of 5 European doppler stations were determined by simultaneously processing six laser and doppler data arcs. The accuracy of these coordinates is estimated to be about 1 m.
- For the determination of very accurate orbits from tracking data arcs of several days, a better global distribution of laser stations (or any other high-precision tracking system) and an extensive use of altimeter data are more important than the improvement of the current precision of the laser measurements.

REFERENCES

Amberg, L., Hauck, H., Herzberger, K. & Schluter, W., Final results of SEATOC and NORSDOC, to be published in: *Validation of SEASAT altimetry using ground truth in the North Sea region,* Deutsche Geod. Komm., Munich, in preparation.

Anderle, D., (1980) Naval Surface Weapons Center, Dahlgren, SEASAT frequencies, priv. comm., September.

Anon., (1981) *ESA remote sensing satellite no. 1,* Technical mission description, document APP(81) 4, ESA.

Cutting, E., Born, G. H., & Frautnick, J. C., (1978) Orbit analysis for SEASAT, *J. Astronaut. Sc.*, **26** 315–342, October.

Cutting, E., & Pounder, E., (1978) SEASAT opens new phase in earth observations, *Aeronautics and Astronautics*, 42–50, June.

Goad, C. C., Douglas, B. C., & Agreen, R. W., (1980) On the use of satellite altimeter data for radial ephemeris improvement, *J. Astronaut. Sc.* **28** 419–428, October.

Jacchia, L. G., (1971) Revised static models of the thermosphere and exosphere with empirical temperature profiles, *Smithson. Astroph. Observ. Spec. Report* 322.

Latimer, J. H., (1979) Smithsonian Astrophysical Observatory, Letter to all investigators using laser data from Orroral, September.

Lelgemann, D., (1980) Inst. fur Angew. Geodasie, Frankfurt, SEASAT doppler data from Brussels, Geldingaholt, Toulouse, Trondheim and Wettzell, priv. comm., May.

Lerch, F. J., Klosko, S. M., Laubscher, R. E. & Wagner, C. A. (1979) Gravity model improvement using GEOS-3 (GEM 9 and 10), *J. Geophys. Res.*, **84** 3897–3916, July.

Lerch, F. J., Marsh, J. G., Klosko, S. M. & Williamson, R. G., (1981), *Gravity model improvement for SEASAT,* paper submitted to J. Geophys. Res., February.

Lerch, F. J., Putney, B. H., Wagner, C. A., & Klosko, S. M., (1980) *Goddard earth models for oceanographic applications* (GEM 10B and 10C), paper presented at Marine geodesy symposium, Miami, September.

Lerch, F. J., Wagner, C. A., Klosko, S. M., Belott, R. P., Laubscher, R. E., & Taylor, W. A., (1978) *Gravity model improvement using GEOS-3 altimetry* (GEM 10A and 10B), paper presented at spring annual meeting of American Geophysical Union, Miami, April.

Marsh, J. G., (1980) NASA Goddard Space Flight Center, NASA modified New Orleans station coordinates, priv. comm., February.

Marsh, J. G. & Williamson, R. G., (1980) Precision orbit analyses in support of the SEASAT altimeter experiment, *J. Astronaut. Sc.,* **28** 345–369, October.

Martin, T. V., (1978) *Geodyn descriptive summary, report contract no. NAS 5-22849,* Washington Analytical Services Center, Washington.

Schutz, B. E. & Tapley, B. D., (1980) Orbit accuracy assessment for SEASAT, *J. Astronaut. Sc.,* **28** 371–390, October.

Tapley, B. D., & Born, G. H., (1980) The SEASAT precision orbit determination experiment, *J. Astronaut. Sc.,* **28** 315–326, October.

Wakker, K. F., & Ambrosius, B. A. C., (1981) Accurate orbit determinations from laser range observations of LAGEOS, STARLETTE and GEOS-3, *ESA SP-160,* 25–35, August.

Wakker, K. F., & Ambrosius, B. A. C., (1981) *Some results of numerical experiments on the computation of the Kootwijk and Wettzell satellite laser ranging stations coordinates,* paper presented at the 4th LAGEOS working group meeting, Goddard Space Flight Center, September; also *Memorandum M-409,* Delft Univ. Techn., Dept. Aerospace Eng.

Wakker, K. F., & Ambrosius, B. A. C., (1982) *Summary of orbit computations from laser range observations of LAGEOS, STARLETTE, GEOS-3 and SEASAT,* paper presented at ESOC workshop on orbit determination for ERS-1 altimeter mission, ESOC, January; also: *Report LR-342,* Delft Univ. Techn., Dept. Aerospace Eng.

Wilson, P., & Aardoom, L., (1982) *SEASAT tracking over Europe,* paper presented at SURGE meeting, London, April.

Chapter 23

Gravity field investigation in the North Sea

R. RUMMEL, G. L. STRANG VAN HEES & H. W. VERSLUIJS

Afdeling der Geodesie, Delft University of Technology, Thijsseweg 11,
2629 JA Delft, The Netherlands

23.1 INTRODUCTION

The GEOS-3 mission has already demonstrated the usefulness of satellite altimetry for solid earth research. In the sequel the current status of our investigation of the gravity field in the North Sea, or more precisely the Dutch part of the North Sea, will be described. The study was carried out within the frame of the SEASAT Users Research Group of Europe (SURGE). The data to be analysed are, first, SEASAT altimeter measurements and, second, shipborne gravimetry. Whereas the internal measurement precision for GEOS-3 was ± 50 cm, the SEA-SAT altimeter precision appears to have exceeded 10 cm (see e.g. Marsh *et al.* (1982)).

One may ask how far satellite altimetry can contribute to gravity field investigations since it provides a topographic as opposed to an equipotential reference surface. Oceanographers require an equipotential reference surface which is stationary and physically defined. But the accuracy they require for the detection of ocean currents is better than 10 cm (see, for instance, Nat. Acad. Sc. 1979). Put another way, if the oceanographer wishes to monitor the absolute value of that part of the altimeter's signal provided by ocean currents and tides he must know the time-invariant geoidal surface to a very high (< 10 cm) precision. One can invert this reasoning. If an accurate profile of the sea surface can be obtained from a precise altimeter, then the geodesists can derive the equipotential surface which is the geoid by subtracting out the ocean effects, and this may also be done to a 10 cm accuracy in certain areas of the global oceans. Such a geoidal surface, if known globally, would serve as an excellent reference for gravity field investigations.

A different reasoning applies when the altimeter-derived sea surface is considered a topographic surface. The determination of the boundary surface of the earth, which is the main aim of geodesy, is then achieved automatically over that part of the globe covered by ocean. The free, unknown boundary surface of physical geodesy is then fixed and the introduction of a normal gravity field together with the known gravity field allows direct computation of gravity disturbances in the place of anomalies (see, for example, Hotine 1969 p. 314). The process can be taken a step further by combining an adjustment of gravity ratios as obtained from sea gravimetry with geocentric distance ratios (Baarda 1979). By calculating these ratios at a number of datum points then, in principle, the height datum problem might be approached — that is, potential differences between coastal points on different continents could be computed. In practice, the unreliability of altimetry in coastal areas and the incompleteness of the data coverage preclude such an approach at present.

23.2 GEOPHYSICAL SITUATION

For a description of the geophysical situation in the North Sea we rely on Pegrum et al. (1975) and Ziegler (1978). Three main factors are expected to characterize the structure of the gravity field in this area:

— a heavy sediment layer accumulated in this depositional basin,
— fault-bounded troughs or grabens, and
— remnants of former orogenic activity.

The North Sea area represents a geosyncline (that is, a sediment-filled basin) where the sediment rock reaches a thickness of more than 6000 m. Areas like that are favourable sites of oil and gas accumulation. In addition, there is the response of the lithosphere under a growing load of sediments: the role of geosynclines in the development of plate boundaries and in orogenic processes are areas of active research in solid earth physics.

About 200 million years ago the Atlantic opened from south to north and the North-American lithospheric plate gradually receded from the Euro-Asian plate. The separation went hand in hand with crustal thinning and an extensive faulting activity accompanied by vulcanism. In the North Sea, a large system of failed spreading zones was produced. Under continuous tension, and partial collapse along the separation line, troughs or grabens developed and filled with sediments. According to Pegrum et al. (1975) the Southern North Sea Trough is somewhat less defined and more difficult to identify than more northerly systems.

Finally, the area was almost certainly influenced by mountain-building activity. Remnants of the Caledonian phase are to be found not only in the mountains of Scotland and Norway but in the roots between them buried in the North Sea. The Hercynian phase produced mountains mainly south of Britain from southwest Ireland to central Europe, but affected the North Sea in the form of faulting. A separate position is taken by the London-Brabant Massif which is of earlier (possibly Cambrian) origin, (Ziegler 1978). Summarizing, one can expect that owing to underlying graben structures, mountain remnants, and faulting, the gravity field should show inhomogeneities, probably damped through the thick sediment layer.

23.3 GRAVITY SURVEY

From August to October 1979 a sea gravimetry survey was carried out in the southern part of the North Sea by the Dutch navy vessel *Buyskes*. The survey network was strictly designed according to geodetic requirements. The measurements grid covers the southern part of the North Sea (see Fig. 23.1) with a profile spacing of about 20 km. Each of the 300 cross-over points, where theroretically identical gravity values should be obtained, serves as a control point for the

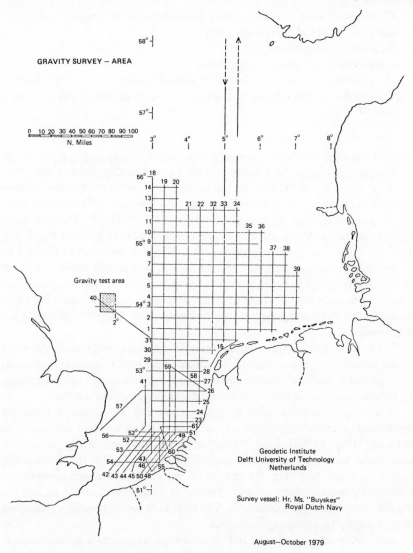

Fig. 23.1 − Profile grid of the sea gravimetry survey

adjustment process. The network is linked by land gravimetry to the International Gravity Standardization Network (IGSN-71) system at Den Helder (Netherlands), Ipswich (Great Britain), and Stavanger (Norway). The shipborne gravimeter measurements were performed with the Kss-5 system (Bodenseewerke), an improved Askania Gss-2 system, mounted on a stabilized platform (Anschütz). The sea gravimeter can be seen as an analytic, local level inertial system sensing only in the vertical direction. Consequently, the error sources are that of an inertial system.

Three critical error components may reach values up to 5 mgal:
— uncertainties in the *Eötvös-correction*,
— the cross-coupling effect, and
— non-gravitational vertical accelerations.

The *Eötvös-correction* models the vertical component of the Coriolis acceleration, due to the velocity of the ship relative to an earth-fixed coordinate system. It is

$$dg = 2 \omega \cos \phi \ v_{EW} ,$$

with ω the angular velocity of the earth, latitude ϕ, and east-west component of the velocity v_{EW}. Since the velocity is derived from the ship position, the accuracy of dg is strongly dependent on the navigational accuracy. For the standard deviation $\sigma(dg)$ to be better than 1 mgal the ship position difference has to be known to be better than 30 m inside a time interval of 10 min. This may cause problems in the world oceans, but should not be a difficulty in the North Sea, where the Hifix-6 navigation system, with a standard deviation of about 3 m, was available in the southern part and the Pulse-8 system with about ± 30 m for the rest. But lane slips in the hyperbolic pattern, a strictly systematic gross error, occur and are sometimes difficult to detect. Improved navigation with the forthcoming Global Positioning System, for example, could practically eliminate this error source.

The *cross-coupling effect* (see for example, Groten (1980)) is, in the Kss-5 system, separately computed from the measured horizontal accelerations acting on the platform, and can be applied afterwards as correction to the measurements. The accuracy level varies from 0 to 5 mgal depending upon the weather.

In principle, *non-gravitational accelerations* can be separated from the gravitational accelerations only via accurate gradiometer measurements, as shown by Moritz (1968). Thus, technological developments in the gradiometer sector could lead to a considerable improvement here, too. At the moment one can only use the fact that the spectra of gravitational and non-gravitational accelerations cover slightly different frequency ranges, the former mainly the low, the latter the high frequencies. The high ones are eliminated by damping and numerically by averaging techniques.

After preprocessing, the 32 000 observations underwent a combined least-squares adjustment and smoothing as described by Strang van Hess (1980). For the *a priori* variance-covariance model at the cross-over points, the measured cross-coupling effect could be used as an information source because it perfectly reflects the weather conditions. The adjustment yielded:

$$\sigma \text{ (cross-over point)} = \pm 1.7 \text{ mgal } (1 \text{ mgal} = 10^{-5} \text{ ms}^{-2})$$

and

$$\sigma \text{ (single observation)} = \pm 1.2 \text{ mgal}.$$

From the adjusted gravity observations free air anomalies were computed, ranging from −40 to +30 mgal with a maximum variation of 40 mgal over a distance of 20 km. The free air anomaly map is shown in Fig. 23.2.

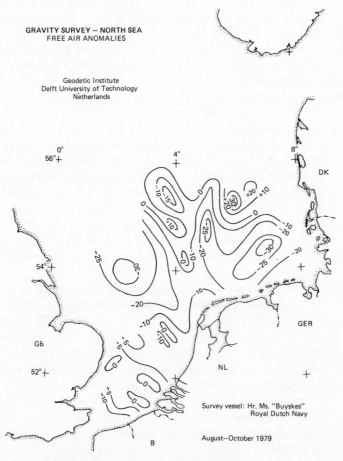

Fig. 23.2 − Free-air gravity anomaly map (contour interval : 5 mgal; gravity measurements linked to the IGSN-71; geodetic reference system 1967)

23.4 SATELLITE ALTIMETRY

As previously noted, the internal precision of the SEASAT altimeter, according to careful analyses by Marsh *et al.* (1982), is better than 10 cm. This does not imply that the sea surface height (that is, the separation of the instantaneous sea surface from a chosen reference ellipsoid) can be measured to the same precision. Even with optimal tracking and the best possible orbit determination, radial orbit uncertainties of the order of 70 cm remain. But in the same way that the accuracy of the gravity surveys may be improved by analysing and adjusting the cross-over points, so a similar technique applies to the satellite tracks. At a cross-over point the stationary part of the sea surface should be identical, so that deviations may be attributed to random altimeter errors, measurement bias, and timing errors as well as to unmodelled sea surface topography. As far as possible sea surface heights are corrected for topography using tidal models such as those of the UK Institute of Oceanographic Sciences used by Brennecke & Lelgemann (Chapter 24).

The processing and adjustment principles described by Rummel & Rapp (1977) have been applied to all North Sea altimeter data available to us. The main steps of the adjustment are

(i) elimination of blunders

(ii) identification of the cross-over points

(iii) least-squares adjustment.

By fitting to each altimeter track a low-order polynomial (degree zero to two) the adjustment procedure finds a synthesis between minimal cross-over discrepancies and optimal fit relative to geoid heights computed from potential coefficients. In this case the Goddard Earth Model 10 up to degree 30 was chosen (see Lerch *et al.* 1976). A consequence of any such adjustment procedure is that long-wavelength features in the data can be expected to be disturbed, but, considering the 70 cm r.m.s. radial orbit accuracy found by Marsh *et al.* (1982) this seems the price one has to pay. All details of the processing are given by Versluijs (1981).

The most objective internal quality measure for the adjusted altimeter data is probably the *a posteriori* standard deviation of the adjusted cross-over point discrepancies. From the global adjustment of almost 4 000 000 SEASAT sea-surface heights Rowlands (1981) obtained a value of ± 0.28 m, while for the eastern North Pacific Marsh *et al.* (1982) were able to reduce this value to ± 0.12 m. The cross-over discrepancy standard deviations of our computations with a polynomial fit, of degree zero (constant) and one, are:

number of arcs	:16
number of observations	:1417
number of cross-over points	:31
a priori	$\sigma = \pm 2.84$ m
a posteriori (zero degree)	$\sigma = \pm 0.23$ m(1)
a posteriori (1st degree)	$\sigma = \pm 0.08$ m(2)

Although both adjustments — constant bias (1) as well as 1st-order polynomial (2) — yielded satisfactory results, their differences show that the results
have to be treated with caution. The small number of cross-over points is responsible for a large part of this uncertainty; on the other hand the result may indicate
that in the commonly chosen adjustment procedure a factor of arbitrariness may
enter, especially when one wants to analyse sea surface phenonema with an
extension that is close to the shortest resolvable period of the polynomial
model used. The two pairs of frozen orbits in the data set (1158 & 1201 and
1163 & 1206) show excellent agreement after adjustment on the 10 cm level.

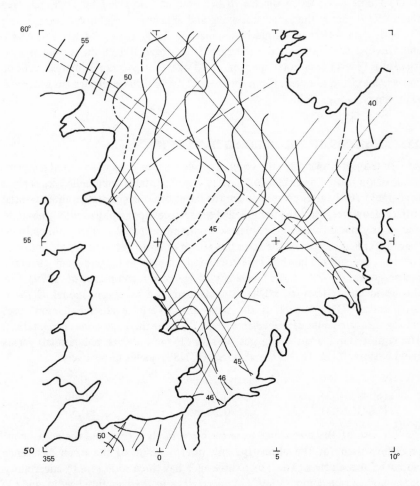

Fig. 23.3 — Estimated stationary sea surface heights (contour interval : 1 m)
————— ground tracks SEASAT-1
— — — ground tracks GEOS-3

The adjusted sea surface heights refer to a standard ellipsoid with

major axis $a = 6\ 378\ 140$ m, and
flattening $f = 1/293.257.$

The r.m.s. value of the adjusted sea surface heights minus the computed GEM 10 geoid heights up to degree 30, is ± 1.50 m for case (1) and ± 1.20 m for case (2).

Finally, the adjusted SEASAT data set has been merged with the adjusted GEOS-3 data available in the North Sea area and provided by Dr Rapp, (see Rapp 1979). Since the same processing and adjustment philosphy is employed for both data sets they should be compatible. Large offsets between SEASAT and GEOS-3 data sets, on the average of the order of 1.26 m, as observed by Rowlands (1981) from a comparison in 10 different geographical areas, do not appear to exist in our case. From the merged data set the contour map is derived as displayed in Fig. 23.3.

23.5 ANALYSIS OF THE DERIVED DATA SETS

As pointed out previously, the computed sea surface does not present a gravity information source, in principle, because it still contains unmodelled sea surface topography. Only after correction for the latter can one arrive at an equipotential surface. However, since the unmodelled sea surface topography is small, one could also simply not distinguish between sea and equipotential surface, and consider their difference as part of the error budget. In order to get some insight into this problem, sea surface heights were compared with geoid heights from the gravimetric geoid GGNS2 of the University of Hannover, (Monka *et al.* 1979). For this geoid an accuracy of ±0.40 m is claimed. A more recent model, GGNS3, computed by Dr. Wenzel of the same group (private communication) uses gravity measurements described in section 23.3 for the inner zone computation. The comparison sea surface height, version (1) (zero degree polynomial) versus geoid heights, Δ (SS–GGNS2) and Δ (SS–GGNS3), yields respectively:

	mean difference	r.m.s. difference
Δ (SS-GGNS2)	−0.18 m	±0.46
Δ (SS-GGNS3)	−0.20 m	±0.56

Because of the zero-order term problem in the geoid and the adjustment procedure used for the altimetry, any interpretation of the mean difference would be meaningless. The r.m.s. difference has three sources: (1) uncertainty of the sea surface computation, (2) uncertainty in the computed geoid, and (3) the r.m.s. signal of the unmodelled sea surface topography. Profile spacing of only 20 km for the gravity survey was chosen in order to achieve a gravity accuracy of about 1 mgal which, together with high-quality gravity material outside the North Sea, would produce a geoid height difference with a standard deviation

— of 10 cm for 20 km distance
— and of 60 cm for about 500 km, and
— a point geoid height with a standard deviation of 40–60 cm.

The error model has been described by Christodoulidis (1976). Considering the uncertainties in the sea surface and in the geoid computation, it is difficult to judge which one of the three contributions is dominant. This is a strong argument in favour of the approach followed by Wunsch (1981), who links directly the sea surface and equipotential surface to an ocean circulation model. For a comparison of the information content of the computed sea (or 'approximate equipotential') surface with the measured gravity one can choose between two alternatives. Either one compares the sea surface with an equipotential surface computed from the gravity material as described above, or a gravity field approximation is computed from altimetery and compared with the measured gravity. The reasons that the latter approach is preferred, despite its inherent instability, are twofold:

(1) Since the quality of the gravity data of the outer zone (that is, outside the gravity survey area) influences the accuracy of the computed geoid heights, the comparison of geoid heights derived from gravimetry with those derived from altimetry may not yield representative results.

(2) The gravity field analysis should form a base for geophysical interpretations. Gravity anomalies imply lateral mass inhomogeneities or, in other words, density anomalies. If one makes a comparison between the gravity anomaly field and a density layer in the spectral domain in terms of spherical harmonic coefficients it is

$$\mu_{nm} = \frac{n + \frac{1}{2}}{n - 1} g_{nm}$$

where μ_{nm} and g_{nm} represent the spherical harmonic coefficients of degree n and order m of the density layer and the gravity anomaly field respectively. For larger n the relation becomes one-to-one. The corresponding relations for the

the disturbing potential coefficient t_{nm} $\left(= \dfrac{R}{n - 1} g_{nm} \right)$ is

$$\mu_{nm} = \frac{n + \frac{1}{2}}{R} t_{nm}.$$

Dividing by the mean gravity G, this relation holds good for a geoid height coefficient. Since the same equations are also valid for the (white) noise contained in the estimated coefficients, one concludes that gravity anomaly and density layer are on the same information level. The computation of a density layer from geoid heights is, however, unstable, because of the amplification of the error components for increasing n. Thus, in terms, of the geophysical inference it is more meaningful to compare measured gravity anomalies with those derived from altimetry, in order to get an insight into the stability behaviour of the latter. For a comparison of this type a small test area with good altimeter coverage was selected. Gravity anomalies were then computed along a gravimetry line from the adjusted altimeter results and compared with the measured *in situ* gravity anomalies. The method employed was least-squares collocation (see Rummel & Rapp, (1977)). As expected, the amplitude of the altimeter gravity anomaly profile is drastically damped (see Fig. 23.4). On the other hand, the gravity profile derived from altimetry shows the same wavelengths as the measured gravity profile with a resolution of about 40 km. This result compares well with the experience of Eren (1980) who made an extensive

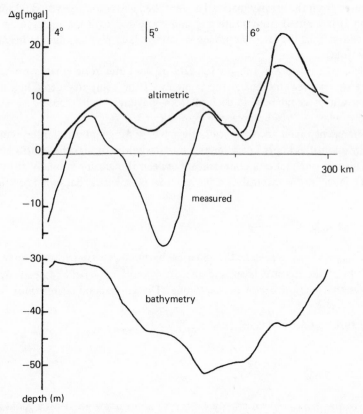

Fig. 23.4 – Comparison of a measured gravity profile (free air) at latitude 55°.5 with one derived from altimetry

analysis of this type with GEOS-3 data. Further, more detailed analysis of the sea surface heights as derived from SEASAT and GEOS-3 will be carried out using a new estimation concept, described in Rummel (1982) and tested with GEOS-3 data in the Atlantic ocean.

A purely qualitative comparison of the main geological features of the area with the computed sea surface (Fig. 23.6) and with the gravity anomaly map (Fig. 23.5) shows significant correlations. It remains to be seen how independent measurements of altimetry and sea gravimetry can best be combined, in relation to their overall accuracy and sampling rate, to provide constraints to geophysical

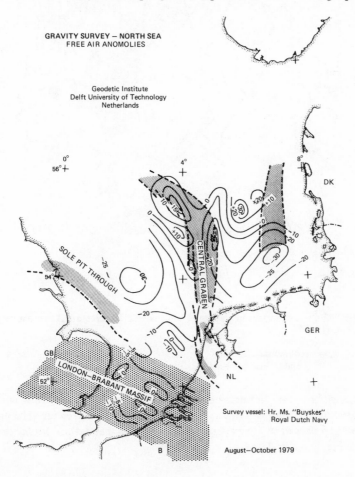

Fig. 23.5 — Free-air gravity anomalies and main geological features, as taken from (Pegrum *et al.* 1975)
- - - - - - faults
░░░░░░ sedimentary troughs
░░░░░░ London-Brabant Massif

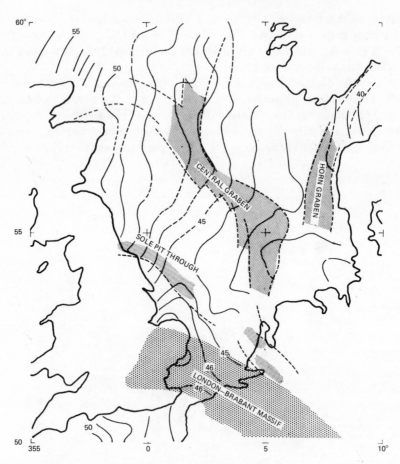

Fig. 23.6 – Sea surface heights and main gelogical features, as taken from Pegrum *et al.* (1975).

------ faults
▨▨▨▨ sedimentary troughs
▦▦▦▦ London-Brabant Massif

hypotheses, however, McKenzie *et al.* (1980) demonstrated the use of combined altimetry, gravity, and bathymetry in modelling mantle convection. The problem must be approached by investigating the solution spaces defined by the two data sets in a highly under-determined mass model.

ACKNOWLEDGEMENTS

This work was partly carried out in the framework of the SFB-78, Munich, who also provided the SEASAT data. Additional thanks go to Dr Rapp, Columbus, for making available the GEOS-3 data, and Dr Wenzel, Hanover for providing his gravimetric geoid.

REFERENCES

Baarda, W. (1980) *Mathematical models,* OEEPE publication office, 11, 73–101, Frankfurt.

Brennecke, J., Lelgemann, D., Torge, W., Wenzel, H. -G. (1982) *Validation of SEASAT-1 Altimetry using groundtruth in the North-Sea region,* in press.

Christodoulidis, D. C. (1976) *On the realization of a 10 cm relative oceanic geoid.* Dept. Geodetic Science, 247, The Ohio State University, Columbus.

Eren, K. (1980) *Spectral analysis of GEOS-3 altimeter data and frequency domain collocation,* Dept. Geodetic Science, 297, The Ohio State University, Columbus.

Groten, E. (1980) *Geodesy and the earth's gravity field,* Volume 2, Dümmler, Bonn.

Hotine, M. (1969) *Mathematical geodesy,* ESSA-Monograph, 2, U.S. Dept. Commerce, Washington D.C.

Lerch, F. J., Klosko, S. M., Laubscher, R. E., & Wagner, C. A. (1976) *Gravity model improvement using GEOS-3 (GEM9 & 10),* NASA-Goddard Space Flight Center, X-927-77-246, Greenbelt.

Marsh, J. G., Cheney, R. E., Martin, T. V., & McCarthy, J. J. (1982) Computation of a Precise Mean Sea Surface in the Eastern North Pacific Using SEASAT Altimetry, *EOS-Transactions* 63 9, 178–179.

McKenzie, D., Watts, A., Parsons, B., & Roufosse, M. (1980) Platform of mantle convection beneath the Pacific Ocean, *Nature* 288 4, 442–446.

Monka, F. M., Torge, W., Weber, G., & Wenzel, H. -G. (1979) *Improved vertical deflection and geoid determination in the North Sea region,* Wissensch. Arbeiten der Fachrichtung Vermessungswesen, 94, Universität Hannover.

Moritz, H. (1968) *Kinematical geodesy,* Deutsche Geodätische Kommission, A-59, Munich.

Panel of Gravity Field & Sea Level Requirements (1979) *Applications of a dedicated gravitational satellite mission,* National Academy of Science, Washington, D.C.

Pegrum, R. M., Rees, G. & Naylor, D. (1975) Geology of the North-West European Continental Shelf Vol 2, Graham Trotman Dudley, London.

Rapp, R. H. (1979) GEOS-3 Data processing for the recovery of geoid undulations and gravity anomalies, *Journal Geophysical Research* 84 B8, 3784–3792.

Rowlands, D. (1981) *The adjustment of SEASAT altimeter data on a global basis for geoid and sea surface height determinations,* Dept. Geodetic Science, 325, The Ohio State University, Columbus.

Rummel, R. (1982) Gravity parameter estimation from large and densely spaced homogeneous data sets, *Bolletino di Geodesia e Scienze Affini*, 41, 2, 149–160.

Rummel, R., Rapp, R. H. (1977) Undulation and anomaly estimation using GEOS-3 altimeter data without precise satellite orbits, *Bulletin Geodesique* 51, 1, 73–88.

Strang van Hees, G. L. (1983) Gravity survey in the North Sea, *Marine Geodesy* (in press).

Versluijs, H. W. (1981) The analysis of SEASAT-A data in the North Sea, thesis, afdeling der geodesie, Delft University of Technology, Delft.

Wunsch, C. (1981) An interim relative sea surface for the North Atlantic Ocean, *Marine Geodesy, 5*, 2, 103–119.

Ziegler, P. A. (1978) North West Europe: tectonics and basin development, *Geologie en Mijnbouw,* 57, 4, 589–626.

Chapter 24

The altimetric geoid in the North Sea

J. BRENNECKE and D. LELGEMANN
Institute für Angewandte Geodase, Frankfurt am Main, F. R. Germany

24.1 INTRODUCTION

If altimeter data are to be used to monitor oceanographic variability then variations in the level of the sea surface must be determined to about 10 cm. Spectral analysis of the output of SEASAT's altimeter indicates that random errors were of the order of ± 6 cm, so that future altimeter missions may be planned to provide useful information on ocean circulation.

The principal objectives of the SEASAT North Sea Altimeter Project were:
(i) to estimate the precision of mean sea surface heights by adjusting the altimeter profiles;
(ii) determination of the accuracy of these mean sea surface heights through comparison with a gravimetric geoid.

The North Sea appeared to be one of the best European sites for this work since a programme of extensive gravimetric surveys was considered to have provided a reliable model of the geoid over the area (Monka *et al.* 1979). However, to determine the gravimetric geoid and the mean sea surface to an accuracy better than 10 cm requires an extensive and coordinated programme of research. In this respect the North Sea experiment was conceived as a pilot study to be extended eventually to the eastern Atlantic seaboard.

24.2 DETERMINATION OF THE MEAN SEA SURFACE LEVEL FROM THE ALTIMETER DATA

For the computation of mean sea level, altimeter data from two periods of the SEASAT mission were provided by the Jet Propulsion Laboratory (one per second mean values corrected for instrumental and atmospheric effects). These data covered:

- 17 passes over different ground tracks when the satellite operated in the Cambridge orbit;
- 24 passes over 3 different ground tracks when the satellite followed the 3-day Bermuda frozen orbit pattern.

Data distribution is shown in Fig. 24.1 where the 3-day repeat pattern is marked by the letter B. Gross errors were filtered out of the data.

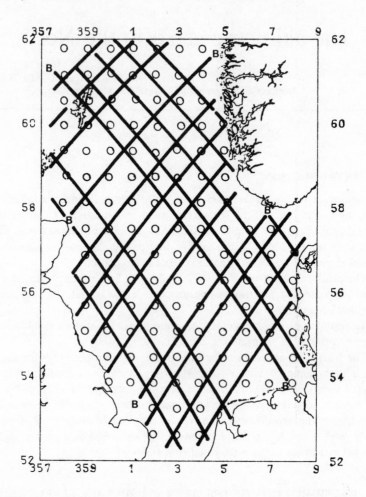

Fig. 24.1 – Distribution of SEASAT data and knot points (Bermuda Orbit passes are marked by the letter B)

Since only very sparse tracking data were gathered at the European tracking stations during July-August, the global SEASAT ephemerides data (GSFC Code 900 orbit as provided on the JPL geophysical records) were used for this period. But for the 24 Bermuda Orbit passes, short arc ephemerides were computed using Doppler tracking data from five European tracking stations.

An approximation to the time-dependent sea surface variations (tides, barotropic and wind effects etc.) in the North Sea, English Channel, and the

Eastern Atlantic was computed by the Institute of Oceanographic Sciences, Bidston, (Alcock & Cartwright 1982). A regional oceanographic model together with empirical corrections from ground truth observations provided a very high accuracy for the dynamic sea surface topography corrections. The high quality of these corrections was decisive in achieving the overall accuracy of the final result.

The fundamental relation in altimetry is

$$A = H - (\tau + \zeta),$$

where A = the processed and corrected altimeter measurement,

H = the ellipsoidal height of the spacecraft,
τ = time-dependent sea surface topography,
ζ = mean sea surface height,
which can be parameterised as follows.

Owing to the availability of the IOS data, τ could be considered known to sufficient accuracy. H was split into a pre-computed part H° (from satellite ephemerides) and a residual part dH (orbital error). ζ was separated into a quantity ζ° (computed from 15 × 15 spherical harmonics of the GEM 10B model) and an (unknown) residual part $\delta\zeta$. Thus, an amplified observation equation for an adjustment of reduced mean sea surface heights $\delta\zeta$ was derived as:

$$(A - H^{\circ}{}_k + \tau + \zeta^{\circ}) + \epsilon = A - A^{\circ}{}_k + \epsilon = dH(a_{jk}) - \delta\zeta(b_i) \qquad (24.1)$$

where b_i and a_{jk} are parameters of the approximation model
$i = 1, I$ is the number of knots of the harmonic splines
$j = 1,2$ are the bias and tilt parameters.
$k = 1, K$ is the number of satellite passes.

The errors dH in the satellite ephemerides which are predominantly due to errors in the gravity field used for the orbit computations (Tapley & Born 1980) have a long wavelength character. Since we restrict the computations to regional geoid determinations, then the error in the SEASAT ephemerides can be modelled as a tilt and a bias. This was the model used for the repeated Bermuda Orbit passes, whereas for the Cambridge Orbit passes we found modelling of only the bias to be sufficient. In order to relate the height datum to the system of European satellite tracking stations a constraint on the orbit bias parameter $(\Sigma_k a_k = 0)$ had to be added.

By making certain assumptions an effective procedure to determine the unknown orbit error parameters is to adjust crossing arcs so that the squares of differences at subsatellite intersections are minimised (Rummel & Rapp 1977,

Mather *et al.* 1977). This method uses the fact that by taking the difference of values at intersection points P (derived from equation (24.1)) the unknown time-independent sea surface height ζ drops out. Pseudo observations $A(P)$ must be generated at the crossovers P by using a suitable interpolation procedure.

For the final determination of the mean sea surface in the validation area we did not use crossover adjustment but the more rigorous direct solution of the system (24.1). However, to compare the results of both methods we also computed a solution using the crossover adjustment technique. The pseudo-observations at the crossovers have been generated using different interpolation procedures; among them the so-called Gauβ-Filter was considered as a very effective simple method yielding the best results.

Before the final adjustment the *a priori* r.m.s. value of the 167 crossover differences was ± 1.92 m, which reduced after crossover adjustment to ± 0.103 m. The rigorous adjustment method gave an r.m.s. value of ± 0.105 m for the crossover differences.

The r.m.s. value of the precision estimates of the orbit error parameters of the adjustment is nearly the same for both methods, namely ± 38 mm for the bias coefficients and ± 0.13 mm/km for the tilt coefficients. The r.m.s. value of the differences of the computed orbit error parameters obtained with the two different methods is ± 30 mm for bias and ± 0.16 mm/km for tilt. The present results confirm the use of the stepwise adjustment down to subdecimetre accuracy.

The mean sea surface (altimetric geoid) was approximated by an interpolation series with harmonic base functions $K(P,Q_i)$,

$$\zeta(P) = \sum_{i=1}^{I} b_i K(P,Q_i). \tag{24.2}$$

Several aspects may affect the choice of the parameters of the base functions $K(P,Q_i)$. The degree of smoothness, the rate of convergence and the degree of stability can be measured by suitably defined norms. From extensive numerical investigations we have found the following criteria to be important:
(a) the position of the poles should be chosen to provide a smooth function generating smooth derivatives;
(b) the rate of convergence may be estimated by the size of the error norm of the prediction functional, one factor in the expression for the upper bound of the absolute prediction error;
(c) the position of the knots Q should be chosen to provide a stable interpolation function without artificial oscillations between the knots by a reasonable restriction on the vector norm of the interpolation coefficients;
(d) the function values b_i in the knots may be determined to minimise the Euclidean norm of the residuals of the measurements, namely by conventional adjustment. Considering these points the following kernel function was used (Lelgemann 1981):

$$K(P,Q) = 2/ \sqrt{[R_Q/r_P)^2 + (R_Q/r_P) \cos \psi + 1]} \qquad (24.3)$$

where Q denotes the knots of the interpolation series, r_P the geocentric radius of a point P outside or on the geoid, R_Q the geocentric radius of the pole associated with the knot Q, ψ the spherical distance between Q and P. If the set of knots forms a (quasi-) equidistant grid on a sphere with radius $r_Q = r_P$ then one common radius R_Q may be chosen as a function of the knot distance $\delta\psi$. Numerical investigations have shown that this radius R_Q has to be bounded to be $\delta\psi \leqslant r_Q - R_Q \leqslant 3 \, \delta\psi$.

As a result of test computations to obtain a stable interpolation procedure, $R_Q = r_Q - 2\delta\psi$ was chosen for a smooth and oscillation-free approximation of the mean sea surface in the North Sea region.

Regarding the choice of $\delta\psi$, the geoid resolution is limited mainly by the density of the distribution of altimeter observations and the ratio of signal (geoid) to noise power. The maximum degree of resolution which is allowed by the approximation model is a function of the knot point interval; the corresponding maximum frequency n_{max} of a spectral series is

$$n_{max} = \frac{180°}{\delta\psi°} = \frac{20\,000 \text{ km}}{\delta\psi \text{ km}}$$

Shorter wavelengths cannot be discriminated, owing to aliasing effects.

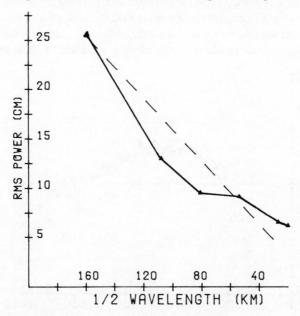

Fig. 24.2 – R.M.S. spectral power of residuals beyond max. frequency of resolution (Dashed line corresponds to ½ of Kaula's rule)

The total power of measurement noise, model erros, and geoidal variations beyond the maximum frequency is given approximately by the r.m.s. value of residuals σ_0 resulting from the adjustment. Fig. 24.2 displays the value of σ_0 as a function of the resolution of the approximation function as found from the adjustment of sea surface height profiles. Since the altimeter noise is about 5–8 cm (Born *et al.* 1979), it is concluded that half wavelengths as short as 30 km could easily be resolved from SEASAT data as far as profiles are concerned.

24.3 COMPUTATION RESULTS

If time-dependent effects such as tides are not taken into account the resulting geoid function will be contaminated by systematic errors. Such errors have a similar spectral distribution as the signal to be resolved with amplitudes clearly exceeding the noise level of the SEASAT altimeter data. However, if a reasonable number of collinear passes are included in the adjustment then the time-dependent effects will be at least partly filtered out and shown up in residuals of the adjustment.

Residuals with amplitudes up to 2 metres were obtained when using the data of 24 Bermuda Orbit passes without tidal reduction, reflecting the unmodelled tide effects. In Fig. 24.3a the residuals of a typical pass are shown. The amplitudes could not be reduced by using the globally modelled tides from the JPL–geophysical records. A comparison of those large residuals with the IOS sea surface topography data (see Fig. 24.3b) has shown a very strong correlation. When the IOS data were used to correct the altimeter data for tidal effects, the r.m.s. value of the residuals reduced to ±8 cm (see Fig. 24.3c, for example); only the noise and small time-dependent errors remained. A comparison of these residuals and

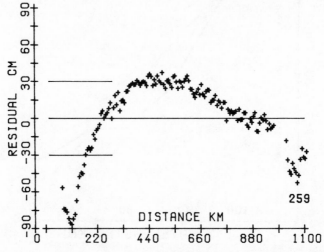

Fig. 24.3a – Residuals of a typical pass from an adjustment of repeated Bermuda Orbit passes (without considering the time-dependent sea surface topography).

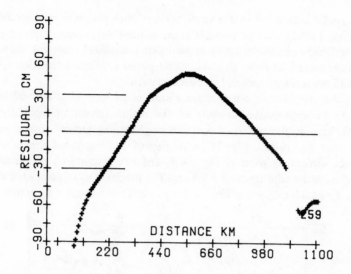

Fig. 24.3b – IOS time-dependent sea surface topography (for same pass as in Fig. 24.3a).

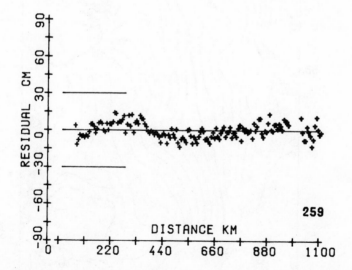

Fig. 24.3c – Residuals as in Fig. 24.3a (IOS data are taken to correct for time-dependent ocean effects).

the significant wave heights has shown that they are partly due to the so-called wave heights effect (Born *et al.* 1980); therefore consideration of this effect slightly reduces the residuals further.

It can be concluded that a verification of high-precision tidal models, e.g. Schwiderski (1981), may be possible at the sub-decimetre level, if altimeter data from a sufficient number of collinear passes are available. In any case the analysis of the residuals of repeated Bermuda Orbit passes confirmed the high accuracy of the IOS sea surface topography approximation.

For the two-dimensional solution a knot point interval of $\delta\psi\sim65$ km was chosen to be appropriate (in spite of the higher resolution capability along profiles), because the distances between neighbouring ground tracks were not shorter than 80 km (see Fig. 24.1). A map of the computed altimetric geoid (mean sea surface) is given in Fig. 24.4. The r.m.s. values σ_0 of the residuals resulting from the adjustment is ± 8.8 cm. The precision σ_ζ of computed altimetric geoid heights is mapped in Fig. 24.5; σ_ζ ranges from about ± 4 cm to ± 11 cm.

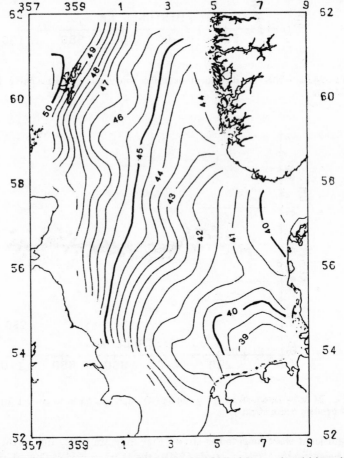

Fig. 24.4 – Mean sea surface computed from SEASAT altimeter data (altimetric geoid in m).

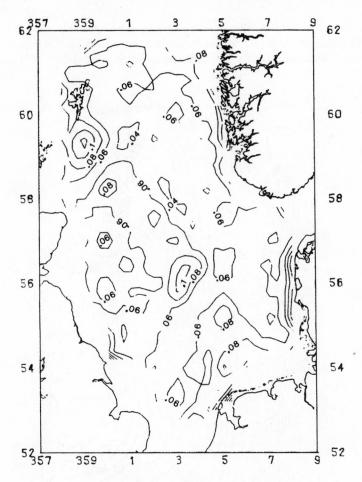

Fig. 24.5 – Standard error of the altimetric geoidal undulations (in m).

The value of σ_ζ is clearly a function of the distribution of available altimeter data. At two larger gaps (around $\Phi = 59°$, $\lambda = 359°$ and $\Phi = 56.3°$, $\lambda = 3.5°$) σ_ζ exceeds ± 10 cm; beyond the area where data were available the error estimate σ_ζ increases strongly up to several decimetres. The precision estimate σ_ζ is confirmed by the r.m.s. value of crossover differences of ± 10.5 cm.

24.4 COMPARISON AND DISCUSSION OF THE RESULTS

Because the sea surface topography is estimated to be less than 10 cm in the North Sea, the mean sea surface computed from SEASAT altimetry can be compared directly with the gravimetric geoid GGNS3 (Fig. 24.6), (Torge *et al.*

1982). The comparison has been carried out using 983 points located in a 12′ X 20′ grid (the gravimetric geoid has been directly evaluated in these points; the SEASAT-1 mean sea surface has been interpolated from a 0.6° X 1.0° knot grid).

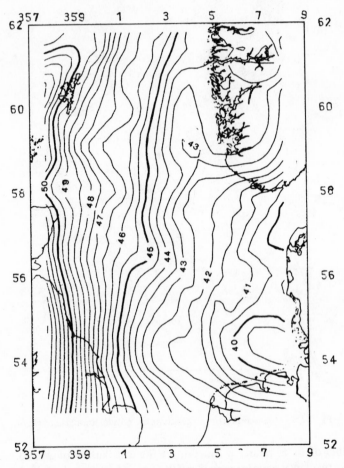

Fig. 24.6 – Gravimetric geoid GGNS 3 in the North Sea region.

The SEASAT mean sea surface heights may have an unknown constant bias due to the constraints used in the adjustment of orbital bias parameters. There may also exist residual tilts in the mean sea surface heights due to remaining systematic tilts of the GSFC 900 orbital data. The gravimetric geoid clearly has a constant bias depending on the choice of the major semi-axis of the reference ellipsoid; tilt errors in the gravimetric geoid can be produced by any remaining low degree errors of the gravity anomalies and the GEM9 spherical harmonic model.

We have therefore applied a three-parameter bias fit to the differences between mean sea surface heights and the gravimetric geoid, resulting in a constant bias of −46 cm, a NS tilt of 0.28″ and a EW tilt of 0.38″ (SEASAT minus GGNS 3); the maximum tilt of 0.47″ occurs in the azimuth 54°. The computed bias parameters are rather small and within the assumed error level of the altimetric and the gravimetric geoid.

The bias and tilt reduced differences are shown in Fig. 24.7; the r.m.s. value is ± 24 cm. With the exception of three small areas marked A, B, C in Fig. 24.7, the bias reduced differences are small and compatible with the precision estimate of the SEASAT-1 altimetry adjustment (see Fig. 24.5). They are also well below the accuracy estimates of the gravimetric geoid.

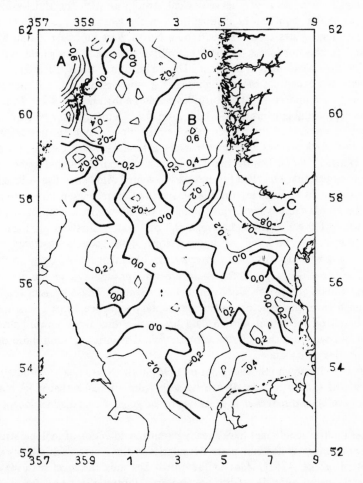

Fig. 24.7 – Differences between SEASAT altimetric geoid and gravimetric geoid GGNS 3, bias and tilt reduced.

The differences in the three marked areas — A, B, C — exceed the normal range of the differences; but in these areas neither larger residuals nor larger crossover discrepencies occur in the SEASAT-1 altimetry. Because of the good agrement of the SEASAT altimeter geoid with one computed from GEOS-3 data it is supposed that these differences are due to shortcomings in the gravity data used for the geoid computation.

24.5 CORRELATION WITH GEOLOGICAL STRUCTURES

A distinct correlation of the geological structures with the short-wavelength features of the geoid can be distinguished in the North Sea area. The North Sea occupies large parts of the intracratonic North West European Basin, which is flanked to the north-east by the Precambrian Fennoscandian shield and to the west by the Caledonides of Scotland. The sea bottom of the North Sea is relatively flat with the exception of a trough near the Norwegian coast, but the sedimentary sequence underlying the North Sea reaches a thickness of 9 km. Indeed, the metamorphic and igneous rocks which form the basement of the North Sea basin cannot be mapped in the deeper parts by reflection seismic profiles; insufficient refraction data are available to construct a basinwide top basement structure map (Ziegler 1977). During the Triassic a north-south oriented graben system was developed that breached the east-west oriented Mid North Sea — Ringkobing Fyn high; main features are the Viking and the Central graben, the Horn graben, and the Horda fault system which parallels the Norwegian coast. The crystalline basement of the Fennoscandian high and of the east Shetland platform results in an uprising of the geoid. An upward slope of the geoid is associated with the Ringkobing Fyn high formed by Precambrian basement rocks.

The local uprisings of the geoid are very pronounced in the area of the central graben and in the Moray Firth Fault System. Those uprisings cannot be explained from the structure of the sedimentary sequence. Indeed, first information based on refraction seismic data has shown that those upward slopes are correlated with uprisings of the Mohorovicic discontinuity, and more detailed investigations are planned.

It is well known that the major oil fields in the North Sea are situated in the area of the graben system. It can be seen from Fig. 24.8 that most major oil fields have been discovered in areas where the geoid shows distinct local upward slopes.

Altimetric geoid maps have already permitted detection of geologic structures such as fracture zones and plate boundaries in the North Pacific (for example by Marsh et al. 1982). Also in the North Sea area the local features of the altimetric geoid provide additional geologic information, but for a deeper interpretation a higher data resolution is required.

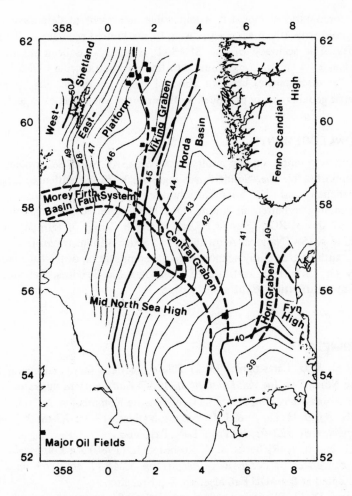

Fig. 24.8 – Correlation with geological structures.

24.6 CONCLUSIONS

The evaluation of SEASAT altimetry in the North Sea project has shown that the precision of mean sea surface determination for regional areas using SEASAT altimetry is better than ± 10 cm, which agrees with the noise level of the instrument. The condition for obtaining such a high precision is an excellent tidal model and a high quality altimetry adjustment model. A verification of high-precision tidal models is possible at the sub-decimetre level, if SEASAT altimeter data from a sufficient number of collinear passes are available.

The validation of the accuracy of SEASAT altimetry was considerably hampered by the limited accuracy of the gravimetric ground truth; even with a refined computational model it is impossible at present to determine a gravimetric geoid with 10-cm accuracy, owing to the limited accuracy of the gravity anomalies. The accuracy of the SEASAT altimetry has been shown to be better than ± 24 cm by comparison with the gravimetric geoid GGNS 3; a large part of the discrepancies between SEASAT mean sea surface heights and the gravimetric geoid is clearly produced by the errors of the gravimetric geoid.

ACKNOWLEDGEMENTS

This work is a report on the results of the SEASAT North Sea project. It has been supported by the 'Sonderforschungsbereich 78 (Satellitengeodaesie) der TU Muenchen' and the 'Sonderforschungsbereich 149 (Vermessungs − und Fernerkundungsverfahren an Kuesten und Meeren) der Universitaet Hannover' as part of the SURGE (SEASAT User Research Group of Europe) SEASAT proposal to NASA (National Aeronautics and Space Administration).

The authors greatefully acknowledge the computation of the tidal reduction data by Dr Cartwright, Institute of Oceanographic Sciences, Bidston, as an important contribution to this investigation.

REFERENCES

Alcock, G. & D. Cartwright, (1982) Tide corrections for the altimeter data in the North Sea. In: *Validation of SEASAT-1 altimetry using ground truth in the North Sea region,* Deutsche Geodaetische Kommission, B, 263.

Borne, G. H., D. H. Lame, & J. C. Wilkinson (Eds), (1979) *GOASEX Workshop Report. Rep. 622-01,* Jet Prop. Lab., Pasadena.

Born, G. H., M. A. Richards, & G. Rosburough, (1980) *An Empirical estimate of the effect of ocean wave heights on SEASAT altimeter data.* Paper presented at the AGU Fall Meeting, San Francisco.

Lelgemann, D., (1981) On numerical properties of interpolation with harmonic kernel functions. *Manuscripta Geodaetica* 6 157-191.

Marsh, J. G., Cheyney, R. E., Martin, T. V. & McCarthy, J. J., (1982) Computation of a precise linear sea surface in the Eastern North Pacific using SEASAT altimetry. *EOS* 63 no.9.

Mather, R. S., R. Coleman, C. Rizos, & B. Hirsch, (1977) A preliminary analysis of GEOS-3 altimeter data in the Tasman and Coral Sea. *Austr. Journal of Geod. Photogr., Surveying,* 26 27-46.

Monka, F. M., W. Torge, G. Weber, & H. G. Wenzel, (1979) Improved vertical deflection and geoid determination in the North Sea Region. *Wiss. Arb. der Fachr. Vermessungswesen der Univ. Hannover* No. 94.

Ronai, P., (1979) SEASAT: Interim Geophysical Data Record (IGDR), *Users' Handbook, Initial Version Altimeter, JPL-Document 622-97* Pasadena, Cal.

Rummel, R. & R. Rapp, (1977) Undulation and Anomaly Estimation using GEOS-3 Altimeter data without precise satellite Orbits. *Bull Geodesique,* 51.

Schwiderski, E. W., (1979) *Global ocean tides.* Naval Surface Weapons Center, NSWC TR 79-414.

Tapley, B. D. & G. H. Born, (1980) The SEASAT precision orbit determination experiment. *Journal of Astronautical Sciences* **XXVIII** No. 4, 315-326.

Torge, W., G. Weber, & H. G. Wenzel, (1982) Gravimetric geoid determination for the SEASAT-1 North Sea Experiment, DGK, B, 263.

Ziegler, P. A., (1977) Geology of hydrocarbon provinces of the North Sea. *Geojournal, Akademische, Verlagsgesellschaft, Wiesbaden.*

Chapter 25

Wave measurements with the SEASAT radar altimeter – A Review

D. J. WEBB
Institute of Oceanographic Sciences, Wormley, Godalming, Surrey, UK

25.1 INTRODUCTION

The design and operation of the SEASAT radar altimeter has been described by MacArthur (1976, 1978). The algorithms used in processing the data are given by Hancock, Forsythe, & Lorell (1980). Aspects of the mechanical and electrical behaviour of the instrument are discussed by Townsend (1980), Lipa & Barrick (1981) and also in SEASAT (1980).

The accurate measurement of distance and wave-height using a radar altimeter, depends on the use of a very short radar pulse. It is also necessary to get sufficient power into the pulse for it to be detected. For SEASAT this combination of power and resolution was obtained by using a long (3.2 μs) swept-frequency pulse. The returned signal was correlated with a replica of the transmitted signal, and Fourier transformed to give the effective returned pulse from a transmitted pulse of only 3.2 ns duration.

Because of random interference between the returns from different specular reflectors on the sea surface, each returned pulse is very noisy. However, when the signal is averaged over a large number of pulses a smooth curve may be obtained. This shows initially a steep linear increase in the power returned, followed by a short plateau region and then a gradual decay in the returned signal. Wave-height and distance measurements are made using the linear rise section. If the sea is calm, the rise-time is roughly equal to the width of the transmitted pulse. If waves are present, then the rise-time is increased by the time it takes the pulse to travel from the crests of the waves down to the troughs and back again.

For SEASAT the slope of the linear rise section was found by using a number of sample gates centred on the point where the returned pulse reached its half-power level. When large ocean waves were being measured, the length of the linear rise section encompassed more gates, and this should have given a more accurate estimate of the slope and wave-height. The effect of the specular noise was reduced by making averages over many radar pulses. The satellite transmitted 1000 pulses per second, and in Webb (1981) it was found that by averaging over 20 s of data it was possible to reduce the statistical fluctuations to very low levels. The noise is considered further in Challenor's (1983) paper (this volume).

SEASAT's on-board processor converted the slope measurement into an estimate of wave-height. This was transmitted to the ground, where two corrections were made, one to allow for the tilt of the satellite and a second to correct for an error in the Fourier transform algorithm. The wave-height conversion and correction tables were prepared before the satellite was launched, and so far as is known in the work reported here no additional empirical corrections have been made to the satellite measurements.

25.2 CALIBRATION TESTS

Unfortunately, SEASAT operated for only about 100 days, during which time the ground experiments were concerned primarily with the calibration of the instruments. In the US such studies have been reported by Tapley *et al.* (1979) and by Fedor & Brown (1982). In Europe work has been reported by Queffeulou, Braun, & Brossier (1981) and by Webb (1981). Tapley *et al.* compared wave-heights calculated by the on-board algorithm and an algorithm developed by Fedor, with wave-heights measured by an aircraft altimeter, a laser profilometer, the GEOS–3 altimeter, and computer hindcasts based on surface winds and spot measurements of wave-height. Their work showed that the standard deviation between the SEASAT altimeter and the other measurements was 50 cm.

Fedor & Brown (1982) used 51 buoy measurements to test the altimeter performance. Their work revealed a possible small bias with the satellite overestimating the wave-height by 28 cm. Above 2 m this bias may increase to 50 cm. The r.m.s. difference between the altimeter and the buoy measurements was 38 cm.

Queffeulou, Braun, & Brossier (1981) compared 163 ships' observations of waves with the satellite measurements. The standard deviation between the two sets of measurements is 87 cm, a figure which reflects the larger errors involved in ship observations. For waveheights of between 2 and 4 m the satellite appears to overestimate the wave-height by 19 cm, the probability of this or a larger figure occurring by chance being only 14%. In the rest of the data, wave-heights of up to 8 m were observed, but no statistically significant bias was found.

Webb (1981) reported on eight comparisons between the satellite and a pitch-roll buoy carried out during the JASIN experiment in the north Atlantic near Rockall. The wave-heights measured were in the range of 1–2 m, and the ratio of the satellite measurement to the buoy measurement was 0.96 ± 0.04. There was some evidence that for wave-heights near 1 m, the Fourier transform correction to the wave-height was too large and resulted in an underestimate of the wave-height measured by the altimeter.

If SEASAT had continued working, further tests would have undoubtedly been carried out to investigate these small discrepancies in more detail. However, the errors are small enough so that, as far as making use of the present SEASAT data is concerned, it seems reasonable to assume that the data are as good as those obtained from ocean surface measurements.

25.3 RESEARCH STUDIES

Research on ocean waves is conveniently split into two main categories. The first category contains studies concerned with the wave climate of the oceans and how this varies during the year. The second category is concerned with the physics of the wave-field, and how the waves grow, propagate, and eventually decay. At the moment, studies of the wave climate are economically the most important, because they are required in the design of ocean structures such as oil-drilling platforms.

An important feature of the SEASAT radar altimeter was that it gave regular worldwide measuremets of wave-height and so could be used to build up statistics on wave climate covering the whole world. Unfortunately its operational lifetime was short, but Chelton, Hussey, & Parke (1981) have combined all the data received to produce an average wave-height map for the world during the period of operation. They also used other data from the altimeter to produce maps of wind speed.

Their wave-height map is probably the first reliable one showing winter storm conditions in the southern hemisphere. Waves with an average wave-height of over 5.5 m were found in a large region south of the Indian Ocean. Elsewhere, regions of interest include an elongated area of high waves in the Arabian Sea. This is produced by the East African Jet, a wind which is part of the monsoon system and which turns out to sea along the north coast of Somalia.

North of New Zealand a region of low wave-height was observed, apparently a result of New Zealand blocking swell waves propagating north from the Antarctic storms belt. In the northern hemisphere low wave-heights were also seen on the western side of each ocean. This behaviour might be expected in the latitudes of the westerlies, but it also appears to occur in the north-easterly trade wind regions.

Another, more detailed, study of the wave climate of the southern hemisphere has been carried out by N. Mognard (1983) and her co-workers (chapter 26, this volume). Their work shows that the region with maximum winds and waves is found in the south Atlantic in June, but then moves eastwards and reaches an area south of New Zealand in September and October. Their work also used measurements of windspeed to estimate how the wave-field was partitioned into the local wind, sea, and swell.

Queffeulou, Braun, & Brossier (1981) compared SEASAT measured waves with the output from wave forecast models. The work is significant because it indicates one way in which altimeter measurements of wave-height can be used to study the physics of ocean wave generation, propagation, and decay. If the physics used in the wave forecast model is wrong this will show up as a discrepancy between the model and observations. In the present case the work confirmed two suspected errors in the meteorological side of the forecast model. These were, a tendency to underestimate wind speed in the model and a tendency to underestimate the speed at which atmospheric low-pressure systems moved. The latter error meant that the predicted position of the highest waves was wrong.

It is to be hoped that further work will be done with the SEASAT data to learn more about the wave climate and the physics of the ocean wave-field. It would be interesting to see how the waves of the Arabian Sea vary with the wind field and whether they are affected by any of the strong currents which are found in the area. Another interesting study would be to look at the trade wind regions and see how much of the average wave-field is due to distant swell. Possibly a model could be produced for the decay of this swell.

25.4 FUTURE IDEAS

The SEASAT radar altimeter has proved to be a successful instrument, and as a result a number of similar instruments are likely to be launched in the next ten years. As an example, one will be included in the European Space Agency satellite ERS-1, planned to be launched in 1986.

25.4.1 Directional spectra

One improvement that could be usefully incorporated into the altimeter is to make use of the phase of the returned signal as well as its amplitude. Clifford & Barrick (1978) showed that if the phase is correlated it gives a structure function from which the one-dimensional wave-height spectrum in the direction of satellite motion can be calculated. To obtain a two-dimensional spectrum, correlations at right angles to the satellite track are required. These could be obtained with a number of receivers at suitable spacing (25–200 m), or with a single receiver rotating about the main satellite at a similar distance.

25.4.2 Wave climate

As has been mentioned above, the main economic use of the radar altimeter measurements will be in studies of wave climate. In this work possibly the most important quantity is the estimate of the extreme 50-year return wave-height, as this is the figure used in engineering design studies.

At present the data required to obtain this estimate come from wave-measuring buoys. Storms last typically from six to twelve hours, and so in order that the largest waves in each storm be measured, the buoys usually record for 15 minutes every three hours. During the 15 minutes, the buoy encounters about 200 waves, and this gives a measurement accuracy of about 7 per cent. The length of time that records have to be taken to build up good long-term statistics depends on the year-to-year variation in the wave climate. Unfortunately this can be large, and it is usually recommended that at least seven years' data be obtained. As an example of the year-to-year variations, P. Challenor (private communication) has shown that the 50-year return wave for Seven Stones (Land's End, UK) is 15 m when based on seven years data taken between 1968 and 1977, but if' based on the data of 1968 alone it is only 11 m.

In choosing the orbit of a particular satellite, one has to compromise between the horizontal spacing of sub-satellite tracks on the Earth and the time between observations at a particular point. The compromise used in SEASAT, a 3-day repeat orbit with a 1000 km operation between tracks at the equator, seems to be a reasonable one. Using the 1968 Seven Stones data, P. Challenor (private communication) found that if these data were sampled, as if from a satellite with one up-pass and one down-pass in the vicinity every three days, then although many of the storm events were missed, the predicted 50-year return was often as good as that obtained with 3-hourly data. In most subsamples the 50-year wave was predicted to be near 11 m and only one sub-sample was seriously in error, giving 8 m. If sampling were continued for a number of years, the chance of such a low value being predicted in error would be much reduced.

Two measurements every three days corresponds to a point on the orbit where the northward-going and southward-going satellite tracks intersect. In the deep ocean where Challenor (1983) shows that wave climate is not expected to vary much over 1000 km, interpolation can be used to give the wave-field at intermediate points. Unfortunately, the economically most important regions are near coasts, and here the wave climate is likely to change appreciably over distances much smaller thatn the distances between satellite tracks. Thus in the North Sea there is only one satellite path going diagonally northwest through the region, and two shorter ones going southwest. The data from such tracks would be insufficient to produce a usable wave climate map of the region. In the English Channel there is even less data available.

One possible solution to this spatial problem may be to use a precessing orbit, like the 17-day orbit used by SEASAT during part of its mission. This, however, will require the development of new techniques to predict the long-term wave climate using data from spatialy distinct points.

25.5 CONCLUSIONS

The radar altimeter carried on SEASAT has proved to be a useful instrument for measuring waves. It may suffer from some small systematic errors, but for most purposes the systematic errors are small enought that they can be neglected.

The instrument has produced some interesting data on the wave climate of the ocean, and has shown how the climate changes from month to month. The data have also been used to check the performance of wave forecast models.

It is to be hoped that similar instruments will be launched in the future. In addition to its uses as an altimeter, the main justification for such an instrument would be the collection of good wave climate statistics for the whole world. In addition it may be possible to obtain wave-directional spectra with such an instrument. It should also be possible to learn more about the physics of the ocean wave-field.

REFERENCES

Challenor, P. (1983) (in this volume).

Chelton, D. B., Hussey, K. J. & Parke, M. E. (1981) Global satellite measurements of water vapour, wind speed and wave-height. *Nature* **294** (5841) 529-532.

Clifford, S. F. & Barrick, D. E. (1978) Remote sensing of sea state by analysis of backscattered microwave phase fluctuations. *IEEE Transactions on Antennas and Propagation,* **AP-26** (5) 699-705.

Fedor, L. S. & Brown, G. S. (1982) *Journal of Geophysical Research* (in the press).

Hancock, D. W., Forsythe, F. G. & Lorell, J. (1980) SEASAT altimeter sensor file algorithms. *IEEE Journal of Ocean Engineering* **OE-5** 93-99.

Lipa, B. J. & Barrick, D. E. (1981) Ocean surface height-slope probability density function from SEASAT altimeter echo. *Journal of Geophysical Research* **86** (C11) 10921-10930.

MacArthur, J. L. (1976) Design of the SEASAT-A radar altimeter. *Oceans 76 conference record,* 10B-1 – 10B-8.

MacArthur, J. L. (1978) SEASAT, *a radar altimeter design description.* Applied Physics Laboratory Report SDO-5232, John Hopkins University, Baltimore.

Mognard, N. M. (in this volume)

Queffeulou, P., Braun, A. & Brossier, C. (1981) A comparison of SEASAT-derived wave-height with surface data. In: *Oceanography from space,* ed J. F. R. Gower. Plenum Press, New York.

SEASAT (1980) *SEASAT Gulf of Alaska workshop II reported.* NASA report no 622-107. Jet Propulsion Laboratory, Pasadena.

Tapeley, B. D. & 17 others (1979) SEASAT altimeter calibration: Initial results. *Science* **204** 1410-1412.

Townsend, W. F. (1980) An initial assessment of the performance achieved by the SEASAT-1 radar altimeter. *IEEE Journal of Ocean Engineering* **OE-5** 80-92.

Webb, D. J. (1981) A comparison of SEASAT-1 altimeter measurements of wave height with measurements made by a pitch-roll buoy. *Journal of Geophysical Research* **86** (C7) 6394-6398.

Chapter 26

Swell propagation in the North Atlantic ocean using SEASAT altimeter

NELLY M. MOGNARD

Groupe de Recherches de Geodesie Spatiak, CNES, 18 Ave. E. Belin, Toulose, France

26.1 INTRODUCTION

During the lifetime of the SEASAT satellite, from 26 June to 10 October, 1978, extensive along-track measurements of surface wind speed and significant wave height were obtained by the radar altimeter over the world oceans. It is with the data acquired by the GEOS–3 radar altimeter, from April 1975 to early 1979, that several algorithms to compute significant wave heights ($H_{1/3}$) were developed and that comparisons with sea truth data were made (Mognard 1977, Fedor & Barrick, 1978, Rufenach & Alpers 1978, Gower 1979). It is also with the GEOS–3 altimeter that the possibility to deduce wind velocity from the variations of the reflected power along the satellite track was empirically and simultaneously discovered by Brown (1979) and Mognard & Lago (1979). The SEASAT altimeter measurements of $H_{1/3}$ and wind speed have been compared to sea truth data during the two experiments that occurred during SEASAT's lifetime: GOASEX (Tapley *et al.* 1979) and JASIN (Webb 1981). Comparisons of SEASAT altimeter inferred estimates of significant wave height and surface wind speed with buoy data yield a mean difference of 0.07 m and a standard deviation of 0.29 m over a range of 0.5 to 5 m significant wave height, and a mean difference of −0.25 m/s and a standard deviation of 1.6 m/s for wind speed ranging from 1 to 10 m/s (Fedor & Brown 1982).

Significant wave heights ($H_{1/3}$) and wind speed measurements along the altimeter tracks can be used to determine sea-state parameters such as: minimum swell height, maximum wind waves, and wave spectra (Mognard *et al.* 1982). In this paper, a method to compute minimum swell heights from the altimeter measurements is presented. This method is used to compute a global mean swell map for the three moths of SEASAT lifetime (Fig. 26.1). This map is the first picture of a mean swell field over the world ocean ever obtained. Swell propagation patterns are analysed in the North Atlantic Ocean during the last period of SEASAT's lifetime in September-October 1978 when the North Atlantic Ocean was beginning to be influenced by deep atmospheric depressions originating in the Arctic regions.

26.2 COMPUTATION OF MINIMUM SWELL HEIGHT

An estimate of the minimum swell height along the altimeter track can be obtained using the altimeter $H_{1/3}$ and wind speed measurements and a sea state model for a fully developed sea. The Pierson & Moskowitz (1964) general expression for a fully developed sea is:

$$S(\omega) = \frac{\alpha g^2}{\omega^5} \, \exp\left[-\beta \left(\frac{\omega_0}{\omega} \right)^4 \right] \tag{26.1}$$

where ω is the pulsation;

$$\omega_0 = \frac{g}{U_{19.5}}$$

with g the gravity acceleration and $U_{19.5}$ the wind speed in (m/s) measured on board ships at 19.5 m above mean sea level.

α and β are two constants empirically adjusted to:

$$\alpha = 8.1 \times 10^{-3}$$
$$\beta = 0.74.$$

The zero order estimator of the spectral energy density is expressed by the relation:

$$m_0 = \int_0^\infty S(\omega) \, d(\omega) \tag{26.2}$$

From the relation (26.2), an expression relating the wind speed as measured by ships U_{195} and $H_{1/3}$ for a fully developed sea $(H_{1/3})_{FD.}$ can be obtained from equation (26.1) using the relation:

$$\omega_0 = \frac{g}{U_{19.5}}$$

For a fully developed sea the following expression is written:

$$(H_{1/3})_{F.D.} = 0.022 \, U^2_{19.5} \tag{26.3}$$

Using the altimeter wind speed measurement this equation can be used to estimate the significant wave height of the fully developed sea. However, the algorithm used to compute the wind speed from the altimeter estimate is at 10 m above mean sea level (Brown, 1979). In order to adjust the two wind speeds to the same reference level, a logarithmic model can be used to describe the boundary layer. The following relation is thus obtained:

$$U_{19.5} = 1.08 \, U_{10} \tag{26.4}$$

$H_{1/3}$ as deduced from equations (26.3) and (26.4) is representative of fully developed waves, which are the maximum limit for wind waves. The relation between the zero order estimator of the spectral energy density m_0, and $H_{1/3}$ is expressed by Longuet-Higgins (1952):

$$H_{1/3} = 4\sqrt{m_0} \tag{26.5}$$

where m_0 is representative of the wave energy (it is the wave energy when multiplied by ρg, where ρ is the water density). From $H_{1/3}$ we can thus determine the wave energy. Two wave energies can be computed: $E_{F.D.}$ corresponding to fully developed waves and E_{ALT} deduced from the altimeter $H_{1/3}$ measurement. The energy $E_{F.D.}$ corresponding to the fully developed sea is obtained using the altimeter wind speed measurement U_{10} after computing the equivalent ship measurement $U_{19.5}$ equation (26.4), and using the relationship given by equation (26.3). We thus obtain:

$$(H_{1/3})_{F.D.} = 0.025\, U_{10}^{2}. \tag{26.6}$$

The following relationships can then be written:

$$E_{F.D.} = k\, U_{10}^{4} \tag{26.7}$$

$$E_{ALT} = l\, (H_{1/3})^{2} \tag{26.8}$$

where $k \cong 4.10^{-5}$ and $l = 0.062$.

When the energy E_{ALT} is higher than $E_{F.D.}$, the residual energy E_s comes from the presence of swell. When the frequency domain of wind wave and swell are separated, which means that the nonlinear interactions can be neglected, the following relation can be written:

$$E_{ALT} = E_{F.D.} + E_S \tag{26.9}$$

where E_S is the swell energy.

In the limit case of a fully developed sea, the residual energy E_S is an estimate of the swell energy.

For wind waves being generated, E_s is a minimum estimate of the swell energy. The significant swell height $(S_{1/3})$ is deduced from the residual energy E_S using equation (26.5):

$$S_{1/3} = 4\,(E_S)^{\frac{1}{2}} \tag{26.10}$$

$S_{1/3}$ is an estimate of the significant swell height in the limit case of a fully developed sea or a minimum limit of the swell height in a more usual case.

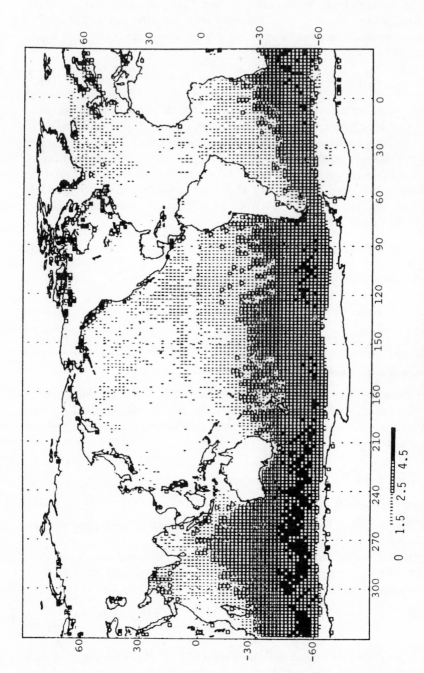

Fig. 26.1 — Field of mean significant swell heights in metres deduced from SEASAT altimeter measurements for the period of 7 July to 10 October 1978.

There are cases when the actual swell height is higher than the computed height $S_{1/3}$. This is often the case when the satellite covers a region where a storm is forming.

This method can be used to obtain the first global mean swell field for the three months of SEASAT lifetime.

26.3 SWELL CLIMATOLOGY

Global maps of mean wind speed and $H_{1/3}$ have been obtained from the processing of the ensemble of the three months of SEASAT altimeter data (Chelton *et al.* 1981). These are the first global pictures from space of surface wind speed and $H_{1/3}$ and present the first accurate picture of global wave conditions ever obtained. They exhibit a rather striking zonal banding of both fields, with the highest wind speed and $H_{1/3}$ located in the southern high-latitude ocean where it was winter. Comparable features are found of the global map of mean swell height where the highest swells are located in the Southern Ocean (Fig. 26.1). A feature that appears in the global mean wave field, and that is even more striking in the mean swell map, it that the smallest waves are found on the western borders of the oceanic basins. An explanation could be found if it is assumed that the wave feature is primarily produced by the swell field (Mognard, 1982).

26.4 SWELL PROPAGATION IN THE NORTH ATLANTIC OCEAN

Swell events in the North Atlantic Ocean are mostly found at the end of the SEASAT satellite lifetime in September–October 1978. In order to use the maximum satellite coverage to analyse swell propagation, the ensemble of the data acquired over the North Atlantic Ocean by the SEASAT altimeter for periods of 3 consecutive days has been processed. Successive 3-day swell maps have been obtained and are shown in Fig. 26.2 for the period, 19 September–9 October 1978. The analysis of the successive maps shows swell families propagation and attenuation. The storms that have generated the observed swells can be located using the $H_{1/3}$ and wind speed altimeter measurements, whose maxima and locations are reported in Table 26.1. The atmospheric low pressure corresponding to the centre of the cyclones and their locations are deduced from the weather maps over the North Atlantic Ocean and reported in Table 26.2 along with the maximum wave heights estimation and location given by the prediction model DSA5 (Gelci & Chavy, 1978). During the period analysed, the maximum swell height measured by the altimeter is 7 m. The two swell patterns observed on the first swell map, September 19 to 20 (Fig. 26.2a) have receded in the next 3-day period where another swell pattern appears south of Greenland and Iceland (Fig. 26.2b). The swell pattern moved and was reinforced on the next 3-day period (Fig. 26.2c) and then reached the British Isles where the energy was

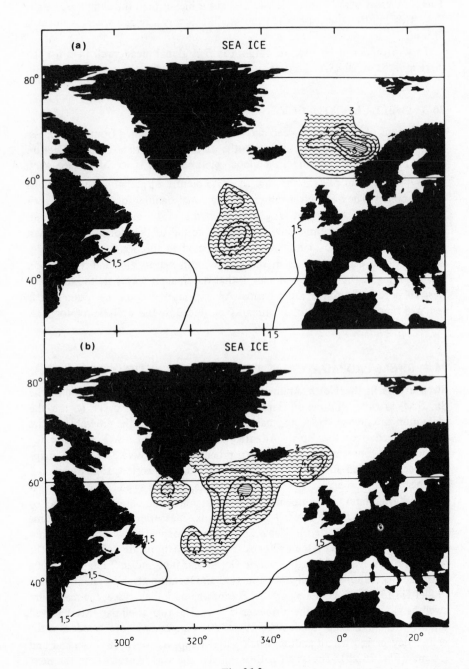

Fig. 26.2

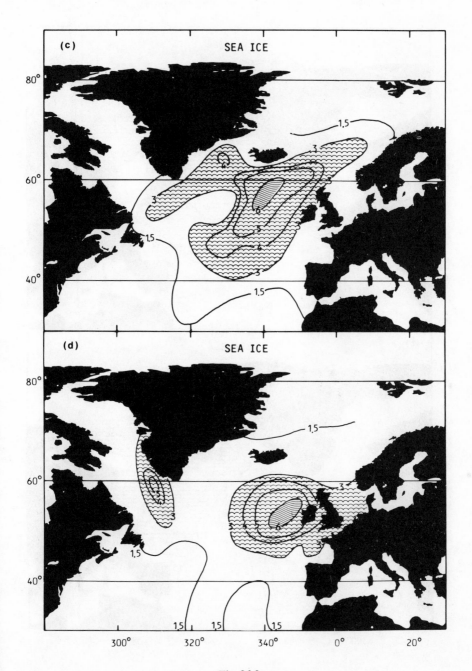

Fig. 26.2

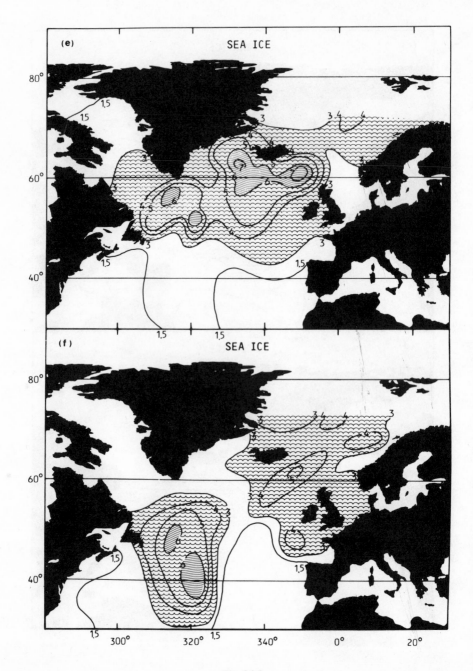

Fig. 26.2

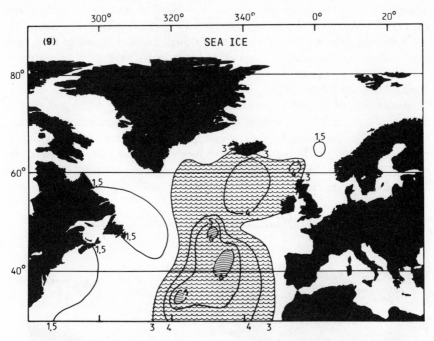

Fig. 26.2 – North Atlantic Ocean swell maps deduced from SEASAT altimeter measurements for the periods: (a) September 19 to 21, (b) September 22 to 24, (c) September 25 to 27, (d) September 28 to 30, (e) October 1 to 3, (f) October 4 to 6, (g) October 7 to 9, 1978.

Table 26.1

Storm location in the North Atlantic Ocean as deduced from maximum $H_{1/3}$ and U_{10} altimeter measurements for the period 19 September to 9 October 1978

| | LOCATION | | | |
Dates	Latitude (°N)	Longitude (°E)	U_{10} (m/s)	$H_{1/3}$ (m)
24 September	57/58	329/331	17	8/9
29 September	56/54	348/351	17	8/9
30 September	54/60	305/311	20/25	8/10
01 October	59/61	319/321	17	7/8
02 October	57/60	320/330	15/17	7/8
05 October	50/53	317/320	16/18	9

attenuated along the coasts (Fig. 26.2d). New swell families were generated on the next 3-day period (Fig. 26.2e) covering most of the North Atlantic Ocean. In the following 3-day period (Fig. 26.2f) the swell families moved to form two swell patterns which were the result of respectively eastward and southeastward motions. During the last 3-day period before SEASAT stopped working, the swell pattern was located in the middle of the North Atlantic Ocean (Fig. 26.2g).

The motion of the swell families and their generation are influenced by the motion of atmospheric depressions whose locations and magnitude are reported in Table 26.2. The location of the maximum swell amplitude is usually found near the location of the maximum wave height obtained with the prediction model DSA5 (Table 26.2). Fig. 26.3 shows the spatial and temporal evolution of the swell families as defined by the 3 m contour lines between September 19 and 28.

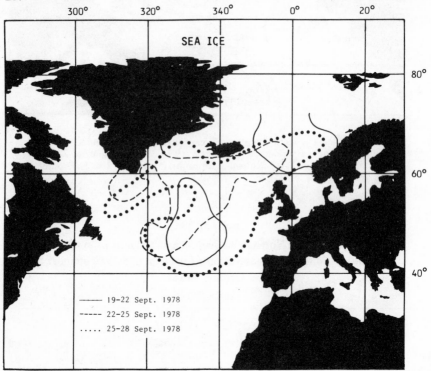

Fig. 26.3 – Spatial evolution of 3 m swell isopleths during the period 19 to 28 September 1978 in the North Atlantic Ocean as deduced from SEASAT altimeter measurements.

It would of course be most interesting to study swell patterns and swell propagation during the winter season. Since the SEASAT lifetime was during the austral winter, it was in the southern hemisphere oceans that the most striking and dynamic swell propagation was observed (Mognard 1982).

Table 26.2

Location of depression centres and wave height maxima as deduced respectively
from weather maps and from the DSA5 maps for the period 19 September to
9 October 1978

Dates	Atmospheric pressure (m bar)	Latitude (°N)	Longitude (°E)	$H_{1/10}$† (m) DSA 5	Latitude (°N)	Longitude (°E)
19 September	1 000 "Hope"	40	318	5	35	323
	995	57	303			
20 September	1 000 "Hope"	47	329	7	52	308
September	1 000	70	350	6	60	360
21 September	990	70	360	6	55	340
	1 000	58	330	5	42	330
22 September	985	58	303	8	50	325
	995	55	325			
23 September	970	61	332	10	54	332
24 September	980	62	328	11	54	332
25 September	985	63	332	11	57	346
				6	34	335
26 September	995	64	347	8	56	348
				5	40	340
27 September	995	63	358	6	55	350
				6	50	328
28 September	995	57	335	8	54	342
				6	57	320
29 September	990	60	355	8	53	352
				6	62	355
				5	40	308
30 September	980	55	298	10	53	318
				6	49	355
01 October	975	63	317	11	57	332
02 October	985	68	337	9	56	332
03 October	980	64	342	8	70	350
04 October	985	67	355	8	60	360
				8	48	310
	990	50	315	6	45	320
05 October	985	50	320	7	55	347
				5	63	328
06 October	990	60	337	6	50	315
	990	47	330	7	62	335
	995	58	337	6	44	338
07 October	985	52	335	5	45	305
				7	43	344
				6	60	340
08 October	995	55	342	7	45	345
	995	48	341	6	57	340
				6	45	315
				4	27	325
09 October	1 000	55	348	6	55	317
				5	47	340
				4	30	325

†$H_{1/3} = 0.787\, H_{1/10}$

26.5 CONCLUSION

In the North Atlantic Ocean the data acquired by the SEASAT satellite were during the summer, when the air-sea interactions are not very strong. However, the three-month SEASAT lifetime was sufficient to demonstrate the ability of microwave sensors to monitor efficiently and accurately sea-state phenomena from space. The radar altimeter is a very promising sensor which can estimate several important sea-state parameters such as significant wave heights, minimum swell height, maximum wind waves, and, of course, surface wind speed. Its ability to acquire data in the midst of a storm is needed to study air-sea interactions. An extensive mapping of swell routes and a monthly or seasonal climatology of wind speed, $H_{1/3}$, and swell can only be achieved with an altimeter in space working continuously for several years.

REFERENCES

Brown, G. S. (1979) Estimation of surface wind speed using satellite-borne radar measurements at normal incidence, *J. Geophys. Res.* **84** NB8, 3974–3978.

Chelton, D. B., Hussey, K. J. & Parke, M. E. (1981) Global satellite measurements of water vapour, wind speed, and wave height, *Nature* **294** 529–532.

Fedor, L. S. & Barrick, D. E. (1978) Measurements of ocean wave heights with satellite radar altimeter, *EOS Trans. AGU,* **59** (9) 843–847.

Fedor, L. S. & Brown, G. S. (1982) Waveheight and wind speed measurements from the SEASAT radar altimeter, *J. Geophys. Res.* **87** NC5 3254–3260.

Gelci, R. & Chavy, P. (1978) Sept ans de prevision numerique de l'etat de la mer par le modele DSA5, *Note de l'EERM,* No. 406.

Gower, J. F. R., Measurements of ocean surface wave height using GEOS-3 satellite radar altimeter data, (1979) *Remote Sensing Environ.* **8** 79.

Longuet-Higgins, M. S. (1952) On the statistical distribution of the heights of sea waves, *J. Mar. Res.,* **11** (3) 245–266.

Mognard, N. M. (1977) Utilisation des donnees de GEOS-3 pour la mise en evidence des variation de l'etat de la mer en fonction des conditions meteorologiues, *Journees de Teledection* 77 GDTA 111–118.

Mognard, N. M. and Lago, B. (1979) The computation of wind speed and wave heights from GEOS-3 data, *J. Geophys. Res.* **84** B8, 3979–3986.

Mognard, N. M. (1982) Apport de l'altimetrie-radar sur satellite a la determination de l'etat de la mar, *These de Doctorat d'Etat,* No. 1043, Universite Paul Sabatier de Toulouse (Sciences).

Mognard, N. M., Campbell, W. J., Cheney, R. E., Marsh, J. G. & Ross, D. B. (1982) Southern ocean waves and winds derived from SEASAT altimeter measurements, to be published in *Wave Dynamics and Radio Probing of the Ocean Surface.*

Pierson, W. J. & Moskowitz, L. (1964) A proposed spectral form for fully developed wind seas based on the similarity theory of S.A. Kitaigorodoskii, *J. Geophys. Res.*, **69** (24) 5181-5190.

Rufenach, C. L. & Alpers, W. R. (1978) Measurement of ocean wave heights using the Geos-3 altimeter, *J. Geophys. Res.* **83** (C10) 5011-5018.

Tapley, B. D., Born, G. H., Hagar, H. H., Lorell, J., Parke, M. E., Diamante, J. M., Douglas, B. C., Goad, C. C., Kolenkiewicz, R., Marsh, J. G., Martin, C. F., Smith, S. L.. III, Townsend, W. F., Whitehead, J. A., Byrne, H. M., Fedor, L. S., Hammond, D. C. & Mognard, N. M. (1979) SEASAT altimeter calibration: initial results, *Science,* **204** (4400), 1410-1412.

Webb, D. J. (1981) A comparison of SEASAT-1 altimeter measurement made by a pitch-roll buoy, *J. Geophys. Res.* **86** C7 6394-6398.

Chapter 27

Use of ocean skewness measurements in calculating the accuracy of altimeter height measurements

G. WHITE

Space Department, Royal Aircraft Establishment, Farnborough, Hants

27.1 INTRODUCTION

Experience with SEASAT's radar altimeter has shown that information about current variations, mesoscale eddies and tides can be determined from a suitable analysis of the height measurements. Further, very accurate significant wave-height values can be obtained from examination of the leading edge of an altimeter backscattered pulse. Of great interest, primarily because of their effect on the Earth's climate, are the large-scale ocean current patterns. In theory, it is possible to measure such currents by the slope they introduce into the sea surface as a result of the Coriolis force. In practice it is difficult to determine this slope for two reasons. First, its gradient is very small and, secondly, the small gradient is often completely swamped by the much larger slope of the geoid.

To calculate current patterns from altimeter height measurements three quantities must be known as accurately as possible: the geoid height, the altimeter geocentric distance, and the distance between the altimeter and mean sea level at the point in question. The work presented here is concerned with determining how accurately the distance between the SEASAT altimeter and mean sea level can be measured by utilising SEASAT altimeter height measurements. To answer this question an understanding of how the altimeter performed a height measurement must be gained, and the sources of error appreciated. Also, a realistic quantitative description of a mean backscattered pulse power-envelope, hereafter called a wave form, must be available so that the point on the leading edge of this envelope, to which the transit time is measured, can be identified.

27.2 DESIGN AND OPERATION OF THE SEASAT ALTIMETER

The design of the SEASAT altimeter is shown in Fig. 27.1. In practice this instrument measured the transit time of a short-duration pulse by simply timing the interval between the transmission of a pulse's leading edge and the reception of a backscattered pulse's leading edge. The measurement must then be corrected for various instrumental effects and propagation delays introduced by the ionosphere and troposphere (Ronal 1979). It is assumed here that such corrections have been applied very precisely (resulting r.m.s. noise contribution around a few centimetres).

The system (MacArthur 1976) illustrated in Fig. 27.1 employs a pulse compression technique in order to synthesise the very short-duration pulse necessary for the attainment of a height precision around the 10 cm level. In the design of the altimeter system the transit time measurement was set to correspond to the time interval between transmission of the leading edge of a pulse and reception of the half power point on the leading edge of the back-scattered pulse. This point was selected because it produced a transit time value independent of the $H_{1/3}$ (significant wave-height) value. However, as will be shown later, this is only true if the quantitative description for the backscattered pulse is that described by Miller & Brown (1974). In this model the half power point represents the point at which a signal first arrives from mean sea level. Consequently the transit time measurement is that obtained by ranging to mean sea level.

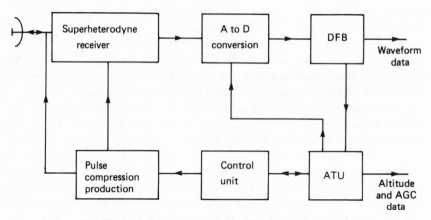

Fig. 27.1 – Block schematic of the SEASAT altimeter system.

To track to this point an ATU (Adaptive Tracking Unit) was employed which consisted of a microcomputer for summing the outputs of the DFB (Digital Filter Bank) and a feedback loop. In practice the DFB gate outputs were combined to form three outputs as shown in Fig. 27.2. Summing the DFB gates in this manner is equivalent to using three contiguous gates of greater width.

For the case of the SEASAT altimeter the gate widths used could have one of five values. The actual value used depended on the magnitude of $H_{1/3}$. After synthesising one of the five gates the quantity I where $I(T) = M(T) - \alpha L(T+W)$ was calculated. Here M is the middle gate output, L the late gate output, while the value of α determines which point on the pulse leading edge will be tracked for a given L and M. The variable T is the displacement of the centre of the middle gate from the origin of the time axis measured in seconds, and W is the gate width. When the altimeter was tracking correctly $I = 0$.

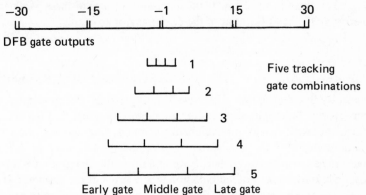

Fig. 27.2 – Formation of the five tracking gates, and their relationship to the DFB.

27.3 HEIGHT ERRORS

With this altimeter design there are two major error sources referred to as the tracking and discriminator error. The first is a result of the variation in the number of specular reflectors contained in successive footprints. An expression for evaluating the value of this error was derived by Dorley *et al.* (1974).

A plot of $I(T)$ versus T is shown in Fig. 27.3 and is known as the discriminator curve. As $T \neq 0$ when $I = 0$ then T must be determined so that the transit time measurement can be made to represent the round trip time

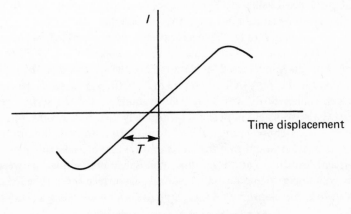

Fig. 27.3 – Typical discriminator curve.

to mean sea level. This value of T can be determined only if the discriminator curve is known quantitatively, which in turn depends on having a quantitative description of a waveform. As will be shown, to determine T two parameters must be calculated, and errors in the determined values of these parameters cause an error in T. This is the basis of the discriminator error.

27.4 REALISTIC QUANTITATIVE DESCRIPTION OF A WAVEFORM

A quantitative description of a waveform is given by the consecutive convolutions of four functions, these being the impulse response of a flat sea, the pulse shape, the wave-height probability distribution, and the instrument transfer function. Owing to the small footprint size and the fact that the attitude error angle (angle between the boresight axis and the nadir direction) was less than ½°, essentially only backscattered radiation was received at the altimeter. It has been shown rigorously by Kodis (1966) that those surface elements on the ocean surface which are normal to the impinging EM (electromagnetic) wave will produce the dominant contribution to the backscattered component of reflected radiation. These surface elements or facets reflect specularly. An aroused sea surface will have water elevations both above and below mean sea level. Consequently, the leading edge of a backscattered pulse will be smeared out in time compared to the flat sea case as signals arrive both before and after the time of reception of a signal from mean sea level. Each level within a sea surface will contribute power to the backscattered pulse, the actual magnitude depending on the number of specular reflectors at a particular level that are illuminated by the altimeter footprint. Consequently the wave-height-slope pdf (probability density function) is required to determine the relevant numbers. This function written as $P(Z; \mathrm{D}x, \mathrm{D}y)$ depends on Z – the surface elevation – and $\mathrm{D}x$ and $\mathrm{D}y$, these being orthogonal measurements of surface slope. The function $P(Z; \mathrm{D}x, \mathrm{D}y)$ gives the normalised value of the numbers of surface elements likely to occur at a particular surface elevation for a specified slope value.

For the altimeter only the backscattered component of radiation is of interest, so $\mathrm{D}x = \mathrm{D}y = 0$. Consequently the functional form of $P(Z; 0, 0)$ is required. For the model used in determining the quantitative description of a waveform for the SEASAT altimeter, $P(Z; 0, 0)$ was taken to be a normal distribution (Miller & Brown 1974). Observations of the sea surface generally show that the wave-height pdf cannot be normally distributed; otherwise the sea surface elevations above mean sea level would mirror those below. In practice, however, it is often found that the crests of waves are considerably peakier than the adjacent troughs. To explain this, non-linear interactions between wave motions combining to produce an observed sea surface are invoked. Longuet-Higgins (1963) has derived $P(Z; \mathrm{D}x, \mathrm{D}y)$ for a long-crested sea ($\mathrm{D}y = 0$) and finds that:

$$P(Z;Dx) = (2\pi \sigma_z^2 \sigma_{\Delta x}^2)^{-\frac{1}{2}} \{1 + (\lambda/6)((Z/\sigma_z)^3 - 3Z/\sigma_z)$$
$$+ (\lambda_1 Z/Z\sigma_z)((\Delta x/\sigma_{\Delta x})^2 - 1)\} \exp\{-\frac{1}{2}((Z/\sigma_z)^2 + (\Delta x/\Delta^2 \sigma_{\Delta x}^2))\},$$

where σ_z and $\sigma_{\Delta x}$ are the standard deviations of the surface elevations and slopes, and λ and λ_1 are the skewness coefficients defined by $\lambda = \mu_{30}/\sigma_z^2$ and $\lambda_1 = \mu_{12}/\sigma_z \mu_{02}$, where μ_{mn} is a moment of the distribution $P(Z;Dx)$ given by $\langle Z^m D x^n \rangle$.

Using this description for backscatter radiation, then:

$$P(Z;0,0) =$$
$$= (2\pi \sigma_z^2)^{-\frac{1}{2}} \{1 + (\lambda/6)(((Z/\sigma_z)^3 - 3Z/\sigma_z) - (\lambda_1 Z/2\sigma_z))\} \exp(-\frac{1}{2}(Z/\sigma_z^2)).$$

Work by Jackson (1979) has shown that to a reasonable approximation $\lambda_1 = 2\lambda$, hence:

$$P(Z;0,0) =$$
$$= P(Z) = (2\pi \sigma_z^2)^{-\frac{1}{2}} \{1 + (\lambda/6)((Z/\sigma_z)^3 - 9Z/\sigma_z)\} \exp\{-Z^2/2\sigma_z^2\}.$$

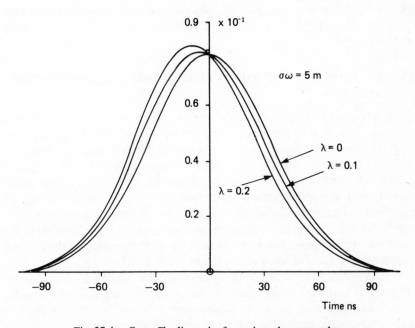

Fig. 27.4 – Gram-Charlier series for various skewness values.

This expression for the wave-height pdf is used in the convolution process mentioned above. The form of this curve is plotted in Fig. 27.4 for $\lambda = 0, 0.1$, and 0.2. It is seen that when $\lambda \neq 0$ the resulting distribution is skewed and consequently the sea surface elevation will be a symmetric about the mean. Furthermore, the mean of the distribution is located at $-\lambda \sigma_\omega$ metres. This offset from zero, zero being the value of the mean for a normal ($\lambda = 0$) wave-height pdf, is known as the wave bias. Consequently in refining the altimeter height measurement, λ must be determined so that the magnitude of this bias can be calculated.

To obtain a quantitative description of a mean waveform P, the following expression must be evaluated:

$$P = I_R * P_S * W_n * I_T.$$

Here I_R is the flat sea impulse response, P_S the pulse slope distribution, W_n the wave-height pdf, and I_T the instrument transfer function. $*$ denotes a convolution operation. The convolution operations can be carried out in any order. In evaluating this expression, $P_S * W_n$ was evaluated first, forming T, then $T * I_T$ giving R and finally $R * I_R$. The final convolution integral $R * I_R$ reduced to six integrals, five of which were found to be related to the remaining one by a suitable reduction formula. For further details see White. The resulting expression for P shows that $P = f(\sigma_\omega, \lambda, E_R, \alpha_i)$ where σ_ω is the r.m.s. wave-height, λ the skewness of the wave-height pdf, E_R, the attitude error angle, and α_i various instrument variables. All the α_i are known to high accuracy; only λ, σ_ω, and E_R have to be found.

27.5 DETERMINATION OF THE SKEWNESS AND R.M.S. WAVEHEIGHT VARIABLES

As a consequence of employing a more realistic description of the wave-height pdf in the derivation of the quantitative description of a waveform it is now possible to answer the question about the altimeter height accuracy. To refine the SEASAT altimeter height values for each waveform the value of λ, E_R, and σ_ω must be found. E_R can be calculated from the pitch, roll, and yaw angles obtained from the spacecraft's attitude control system, while λ and σ_ω are determined by fitting the theoretical description of a waveform to actual received waveforms. This fitting takes place in the least squares sense employing the variables λ, σ_ω and time. The latter enters because the position of the received waveform with respect to the origin of the time axis is not known.

Having found λ, σ_ω and E_R, the value of T corresponding to the condition $I(T) = 0$ can be found by using the discriminator curve and finding where the central linear portion of this curve cuts the time axis (see Fig. 27.3). In developing the quantitative description of a waveform the origin of the time variable has been translated to the time at which backscattered radiation from specular points lying at mean sea level is first received where $\lambda = 0$, that is, for a wave-height pdf that is normally distributed. If $\lambda \neq 0$ then this origin will not correspond to the time at which a mean sea level return first arrives. Instead the point preceding this by $2\lambda\sigma_\omega/C$ seconds will be the desired one. The factor $2\lambda\sigma_\omega/C$ arises owing to the wave bias effect. Consequently the transit time must be decreased by $2\lambda\sigma_\omega/C$ seconds. Fig. 27.5 illustrates the leading edge of a waveform for $\sigma_\omega = 5$, $\lambda = 0$ and Fig. 27.6 the wave bias corrections.

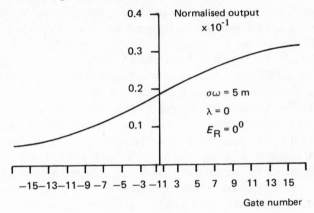

Fig. 27.5 – Waveform leading edge.

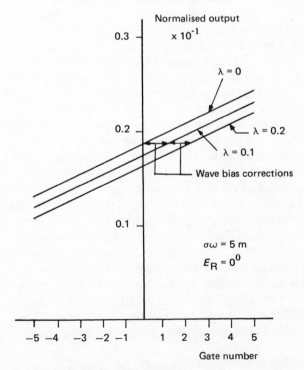

Fig. 27.6 – Waveform leading edge position as a function of skewness.

To answer the question how accurately can the satellite measure the distance to mean sea level, the error in the magnitude of T and the wave bias must first be found. This uncertainty arises owing to the method of calculating σ_ω, λ, and T as discussed below. Finally, the effect of tracker noise must be added in.

This question was answered by adopting a modelling process. Initially a modelled waveform was produced using specified values of σ_ω and λ, E_R being set to 0. Random noise was added to this waveform such that a SNR $= 10$ resulted (this was the typical SNR for SEASAT waveforms). σ_ω, λ, and T were calculated by a differential least squares technique using these modelled waveforms. This technique was utilised because the functional form relating the waveform to the variables σ_ω, λ, and T is highly non-linear. The process is iterative in that the solutions obtained in one iteration are used as the starting values in the next. At first the procedure diverged, and to overcome this difficulty a diagonal weighing matrix of dimension 3×3 was introduced. In using this matrix the variables σ_ω and λ were weighted so as to prevent large changes in their magnitude between successive iterations. After some trials a set of weighting values was found which, if decreased, caused the procedure to diverge, and if increased, to converge.

Unfortunately, introducing this weighting matrix also caused a high correlation between the fitted variables, so that the standard deviation obtained from the covariance matrix was meaningless. To derive realistic standard deviations the following procedure was adopted. Initially a value of λ in the range $0 < \lambda < 0.5$ and σ_ω in the range $0.25 < \sigma_\omega < 5$ were selected, a modelled waveform produced, and random noise added to attain a SNR $= 10$. For this waveform a fit for σ_ω and λ was obtained 100 times by taking a set of 100 σ_ω and λ values as the initial guess values for the differential least squares procedure. Each of these guess values was derived by adding a noise component to the true σ_ω and λ value. For σ_ω the magnitude of this component was such that the guess value of σ_ω lay within 10% of the true value. For λ the guess value was taken to lie within the range of 0 to 0.5. Starting with each of these guess values a prediction of the true value of σ_ω and λ was obtained for each of the 100 sets of guess σ_ω and λ values. Having obtained these estimates of the true values the standard deviation of the predicted σ_ω and λ could be calculated. This procedure was repeated for further modelled noisy waveforms.

Finally, to arrive at a figure for the accuracy of a height measurement, since the variables were highly correlated, the corrections mentioned above were carried out for each set of predicted σ_ω and λ values. Consequently for each true σ_ω and λ value 100 corrections to the height value were calculated. Then the standard deviation of the height correction was obtained. Fig. 27.7 illustrates the resulting height accuracy obtained by this simulation procedure as a function of σ_ω.

One further point should be noted — the curve illustrated in Fig. 27.7 was found to be virtually independent of λ and E_R.

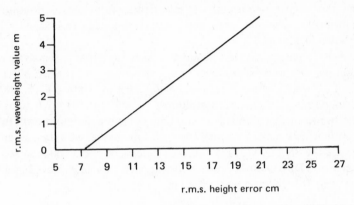

Fig. 27.7 — Accuracy of height measurements derived from SEASAT altimeter
data.

27.6 CONCLUSIONS AND RECOMMENDATIONS FOR FURTHER WORK

The modelling discussed above has shown that the height accuracy — that is, the
accuracy in determining the satellite-to-mean sea level distance — is dependent
on the value of σ_ω. For low sea state values ($\sigma_\omega < 2\pi$) the height error is
< 12 cm.

The predictions shown in Fig. 27.7 are more suspect for the more aroused
sea states, $\sigma_\omega > 3$ m. As the sea surface becomes more perturbed owing to
an increased wind speed, the non-linear interaction between wave components
increases, and consequently some effect on the wave-height pdf is to be
expected. Regarding future work, two areas of endeavour are: (a) verification
of the quantitative description of the waveheight pdf given above, and (b) a
comparison of the wavelength pdf from altimeter and wave buoy measurements.
This comparison will verify whether the specular point height pdf is the same as
the wave-height pdf.

27.7 IMPROVEMENT OF AN ALTIMETER MEASUREMENT OF OCEAN
SKEWNESS

One difficulty in using the skewness values (derived from SEASAT's waveform
for wave bias and oceanographic parameter estimation) is the error bar on
the skewness value determined by the differential least squares procedure. This
error is often large, percentage-wise very large when σ_ω and λ are both small.
Consequently consideration must be given as to how this error on both variables
can be reduced. A reduction may be attained by the following means:
(a) Increase the width of the DFB so that more of a mean waveform's trailing
    edge is observed.

(b) Scan the altimeter footprint – obtain returns from a nadir footprint and footprints lying within a ring concentric with a nadir (sub-satellite point), as shown in Fig. 27.8.

Further, to ensure that the sampled area is such that λ remains constant over it, an increase in the pulse repetition frequency (over the SEASAT case) by a factor of 10 would be needed. Employing a DFB of width 500 ns, 400 ns of which is used in sampling a mean waveform's trailing edge, then the radius of the footprint shown in Fig. 27.6 is ~ 11 km. Employing the pattern shown, then the area over n which remain constant is ~60 km square. Sampling seven areas will increase the SNR by $\sqrt{7}$, and a further increase of ~$\sqrt{3}$ will arise according to the increased DFB width. Consequently the SNR may be improved by a factor of $\sqrt{21}$ ~ 4.5. This will be a most useful increase.

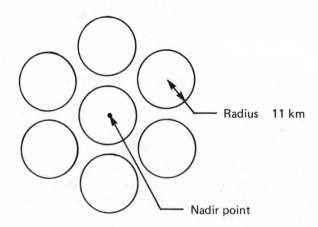

Fig. 27.8 – Footprint illumination pattern.

REFERENCES

Dorley, *et al.* (1974) *Study of radar pulse compression for high resolution satellite alitmetry.* NASA Report CR-137474.

Jackson, F. C. (1979) The reflection of impulses from a non-linear random sea. *J. Geophys. Res.* **84**, 4939–4943.

Kodis, R. D. (1966) A note on the theory of scattering from an irregular surface. *IEEE trans. Antennas Prop.* **14**, 77–82.

Longuet-Higgins, M. S. (1963) The effect of non-linearities on statistical distributions in the theory of sea waves. *J. Fluid Mech.* **17**, 459–480.

MacArthur, J. L. (1976) Design of the SEASAT-A radar altimeter. MTS–IEEE, *Oceans '76*, 10B1–10B8.

Miller, L. S. & Brown, G. S. (1974) *Engineering studies related to the GEOS–C radar altimeter.* NASA Report CR-137462.

Ronal, P. (1979) *SEASAT — Interim geophysical data record (IGDR) users'
Handbook — altimeter.* Jet Propoulsion Laboratory, 15 April.

White, G. C. The accuracy of SEASAT-1 altimeter attitude measurements under
various sea state conditions. RAE Technical Report. In press.

Chapter 28

Spatial variation of significant wave-height

P. G. CHALLENOR
Institute of Oceanographic Sciences, Wormley, Godalming, Surrey, UK

28.1 INTRODUCTION

Although work has been done on the shorter scales, both temporal and spatial, associated with surface waves, especially in situations of wave growth, e.g. Stewart & Teague (1980), little effort has been put into investigating the larger scale variations of significant wave height (H_s) in either time or space. What work has been done has concentrated on the persistence of sea states over time, a subject of obvious commercial application. Variation of H_s in space has not so far received any attention; indeed without remote sensing it would be difficult to conceive how such a study could be achieved. The problem of spatial variation of a sea state is of considerable practical interest. While it may be reasonable to assume H_s is stationary in time, allowing for a seasonal cycle, it would be obviously incorrect to assume stationarity in space over the entire globe. How far then is it reasonable to assume spatial stationarity? Can measurements made at a point be used to infer return values etc. at another location 100 km away? Or 500 or 1000 km away? What is the optimum spacing for the grid of a wave-hindcasting model? The altimeter data from SEASAT give a suitable database to enable a start to be made on answering questions such as these.

28.2 THE DATA USED

If we are to attempt to answer questions such as those posed above, a series of measurements is needed that are made at the same location at different times, so as to be able to remove temporal variation; therefore a satellite repeat orbit has to be used. This cuts down the amount of available data from the altimeter since SEASAT only started its three-day repeat on the 16 September and ceased operation on the 10 October, giving a usable series of approximately 24 days or 8 passes. In order to gain some insight into spatial variability *per se* it is best to remove all complications of variable fetch and consider a track in the open ocean. The track used here is shown in Fig. 28.1; the satellite is travelling south. This track has the added advantage of passing over OWS *Lima*, which is fitted with a Shipborne Wave Recorder enabling us to estimate the temporal variability in this region. Owing to the speed of the satellite it is assumed that the data are contemporaneous.

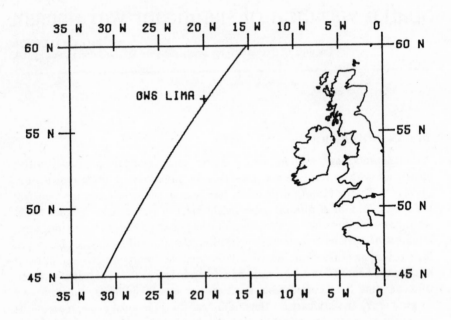

Fig. 28.1 – The satellite track. SEASAT is travelling south along this track.

Significant wave-height along the eight passes is shown in Fig. 28.2. The left-hand end is the most northerly point on the track, and each graph is labelled with the day number of the pass. There is considerable variation in the values of H_s from pass to pass with what appear to be two storms passing through the south of the track on days 259 and 280 (16 September and 7 October). Fig. 28.3 shows the mean and variance. The mean is reasonably flat, implying that average wave conditions are probably constant over the region. This is supported by the contour plot of the mean of all the altimeter data given by Chelton, Hassey, & Parke (1981). The variance, however, has a large peak in the south which is probably associated with two storms mentioned above. Unfortunately, eight passes are not nearly enough to give us any definite information on the variance, or for that matter on the mean, and although it is possible that wave conditions are in fact more variable along the southern half of the track, there is really insufficient evidence to come to any conclusion.

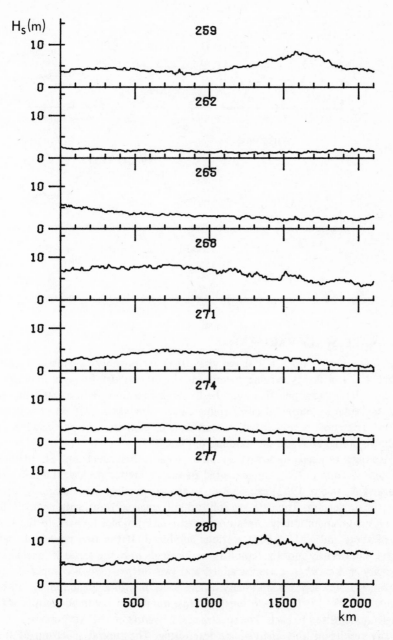

Fig. 28.2 – Along-track significant wave-height (passes are identified by Julian day)

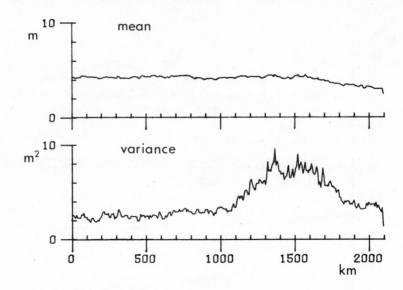

Fig. 28.3 – Mean and variance of significant wave-height. along the track

28.3 SMALL-SCALE VARIATION

An examination of Fig. 28.2 shows that there are possibly two distinct scales present; one is a slowly varying 'mean', and the other a high-frequency component. In order to examine this hypothesis the large-scale variation was removed from the signal by means of cubic splines. Each pass was divided into five equal sections, and cubic splines fitted by least squares to these. The cubic spline fits are shown in Fig. 28.4 and the residuals in Fig. 28. 5.

Are these residuals stationary and, if so, what is their spectrum? The method of Ozaki & Tong (1975), implemented in the computer package TIMSAC-78 (Kitagawa & Akaike 1982) can be used to analyse possibly non-stationary series and estimate spectra. The data are split up into small sections, within which the series is assumed stationary. An autoregressive (AR) model is fitted to the first two of these and also to both sections combined. If the two models do not explain significantly more variation than the single one, the series is considered stationary and another section is added and two further models computed and compared. If the series is stationary a large section with a single model will be produced, while if it is non-stationary a large number of the small sections with a single model fitted to each. The spectra are derived from the AR models.

The results of this analysis are interesting. The typical spectrum of this short-term variation is either flat or U-shaped, showing that the cubic splines are not a perfect low-pass filter and that there is considerable variation caused by instrument noise and sampling variation, from data point to data point.

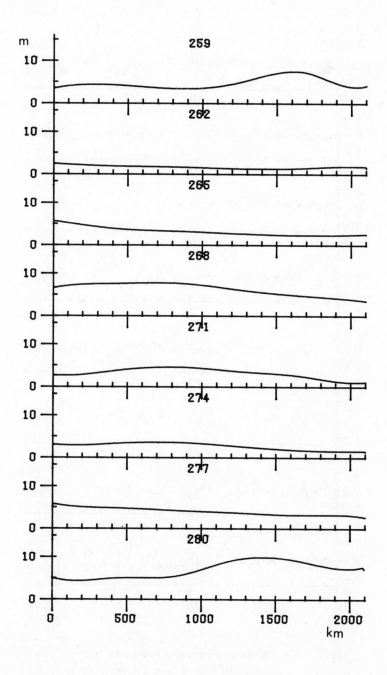

Fig. 28.4 – Least squares cubic spline fits to the data shown in Fig. 28.2

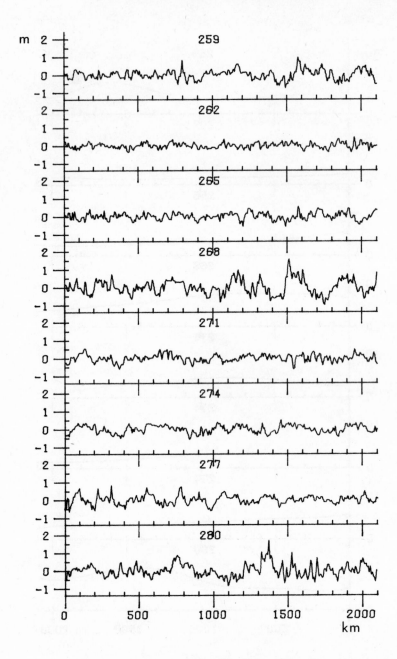

Fig. 28.5 – Residual variation after fitting cubic splines

The series are, in general, stationary, however they do have non-stationary portions. These may well be joining different stationary sections and be areas where the properties of the H_s series are in the process of changing. The portions of the pass that are stationary change from pass to pass, and are not related to the knots in the cubic splines so do not appear to be an artefact of the analysis. Although in certain of the spectra there are peaks indicating some structure, these are transitory phenomena and, with the proviso that the residuals may be organised over short distances, it seems reasonable to regard the high wavenumber variation as noise, which is stationary for long distances but not over the whole series.

28.4 LARGE-SCALE VARIATION

Eight passes are not enough to make definite statements about the large-scale variations, apart from what was said above about the mean and variance. If there were more data it would be possible to do a principal component analysis (sometimes called empirical orthogonal function analysis) and see what, if any, underlying patterns are present in the data. However, with the small amount of data this is simply not feasible.

28.5 COMPARISON WITH *LIMA* DATA

One problem with satellite data is the relatively infrequent sampling which is a grave disadvantage if we want to estimate return values or some similar parameter; however, if we could somehow link the spatial and temporal scales we could use the along-track information in the same way as we now make use of data taken in time. The *Lima* data gives an opportunity to investigate this possibility. OWS *Lima* is situated at 57°N 20°W, directly under the track considered here, and is equipped with a Shipborne Wave Recorder (Tucker 1958). Significant wave height is estimated from a 15-minute record taken every 3 hours. Fig. 28.6 shows the data for period day 259 to 280 (16 September to 7 October) for 1975 to 1979, and the mean. All the inferences given here are of course only relevant to this period, as there is considerable seasonal variation in the distribution of H_s. Unfortunately a large proportion of the data is missing.

The temporal and spatial scales can be linked by considering the group velocity, v_g. If k is the wavenumber this is given by

$$v_g = \tfrac{1}{2}\left(\frac{g}{k}\right)^{\frac{1}{2}}$$

but, by the dispersion relationship (for deep water),

$$\left(\frac{2\pi}{t}\right)^2 = gk$$

where t is the wave period.

Then $v_g \cong 0.78t$ ms^{-1}
If we suppose $t \cong 9$ s
$v_g \cong 7$ ms^{-1}
      $\cong 7$ km per 15 min.,

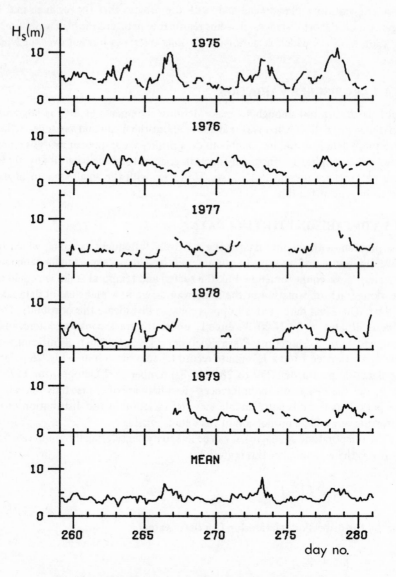

Fig. 28.6 — Significant wave-height measured at OWS *Lima*

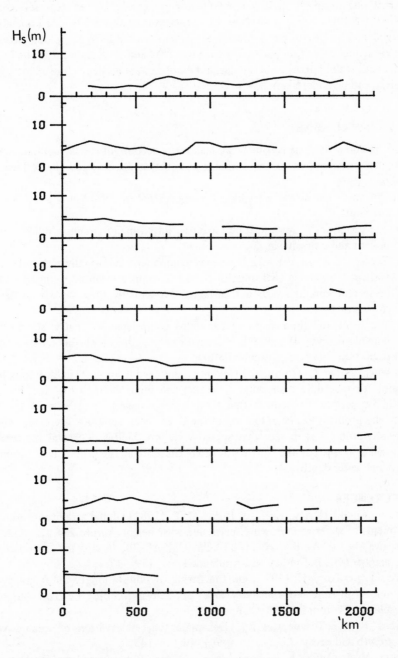

Fig. 28.7 − *Lima* 1976 data plotted on a scale of 1 km = 2 minutes as described in section 5.

so that one footprint for the altimeter is approximately equal to one record from the SBWR, and if we assume two minutes is equivalent to one kilometre we will not be far wrong. A plot of the *Lima* data for 1976 using this scale is shown in Fig. 28.7. Comparing it with Figs. 28.2 and 28.4 shows that the time and length scales appear to be similar, but more data are needed before a thorough analysis can be undertaken.

28.6 CONCLUSIONS

The above analysis is somewhat shallow on account of the lack of data, and is of necessity limited to the autumn; however, it has revealed the following points:

(1) The mean significant wave-height along a track of 2100 km appears to be constant.
(2) The variations in space can be separated into a high-frequency 'noise' component and a slowly varying component.
(3) The spectrum of the high-frequency component is normally either flat or U-shaped, implying that most of these variations are caused by instrument noise and sampling variation, although occasionally these signals do have some structure.
(4) The space and time scales appear to be compatible, so that it should be possible to use along-track information to supplement the rather poor resolution in time of altimeter data.

Where do we go from here? Obviously the need is for more data. It may be possible to take data from adjacent tracks and lump them all together, but it would be necessary to ensure that there was no spatial trend between them. Once the problem has been tackled in the open ocean the more interesting, and difficult, problem is to see what spatial variation there is in coastal and semi-enclosed seas, such as the North Sea, where there are all the problems of varying fetch and water depth.

REFERENCES

Chelton, D. B., Hassey, K. J., & Parke, M. E. (1981) Global satellite measurements of water vapour, wind speed, and wave height. *Nature* **294** 529–532.
Kitagawa, G., & Akaike, H. (1982) On TIMSAC-78. In *Applied time series analysis II* D. F. Findley ed., Academic Press, 499–547.
Ozaki, T., & Tong, H. (1975) On the fitting of non-stationary antoregressive models in time series analysis, *Proc. 8th Hawaii Internat. Con. on System science,* Western Periodic Company.
Stewart, R. H. & Teague, C. (1980) Dekameter radar observations of ocean wave growth and decay. *J. Phys. Oceanogr.* **10** 128–143.
Tucker, M. J. (1958) A shipborne wave recorder. *Trans. Inst. Naval Architects* **98** 236–50.

PART 5

Scanning Multichannel Microwave Radiometer

Chapter 29

The scanning multichannel microwave radiometer – an assessment

P. K. TAYLOR
Institute of Oceanographic Sciences, Wormley, Godalming, Surrey, UK

29.1 INTRODUCTION

The primary purpose of the Scanning Multichannel Microwave Radiometer (SMMR) carried by SEASAT was to measure sea surface temperature, and secondly surface wind (Born, Lame, & Rygh 1981). However, the SMMR also provided estimates of atmospheric water vapour and liquid water, and rain rate. A similar instrument, at present operating on the NIMBUS 7 satellite, has additionally been used in sea ice studies (e.g. Gloersen *et al.* 1981). Francis, Thomas & Windsor (1983) describe algorithm design and testing for the NIMBUS 7 SMMR; this chapter will assess the performance of the SEASAT instrument. The Joint Air Sea Interaction experiment, JASIN (Pollard, Guymer & Taylor 1983), held in the eastern North Atlantic during a period, July to September 1978, when SEASAT was operating, represented a major European contribution to the SMMR evaluation. Examples will be shown of the combined use of SMMR and JASIN data.

29.2 METHOD OF OPERATION

The SMMR (Gloersen & Barath 1977, Njoku, Stacey, & Barath 1980b) measured microwave radiation from the Earth at frequencies of 6.63, 10.69, 18.0, 21.0, and 37.0 GHz for both horizontal and vertical polarisations.

The SEASAT SMMR scanned an approximately 600 km wide swath to the right of the satellite, this being the major difference from that on NIMBUS 7 which scanned to both sides. The physical principles which allow the microwave measurements to be converted into estimates of geophysical quantities have been described by Wilheit (1978), Swift (1980), Wilheit, Chang & Milman (1980), and others. Essentially the emission of microwave radiation by the ocean surface depends on such characteristics as surface roughness, foam, and sea surface temperature. The radiation reaching the satellite represents that emitted by the sea modified by emission, absorption, and scattering by atmospheric constituents such as water vapour, cloud liquid water, and rain. Radiation at different microwave frequencies and polarisations is differently affected by the various oceanic and atmospheric characteristics. Thus, measurement at several frequencies should, theoretically, allow evaluation of these geophysical properties. In practice incomplete knowledge of the physical equations, the large number of unknowns, and the low signal-to-noise ratios make evaluation difficult. A number of geophysical retrieval algorithms have been proposed (section 29.3); Table 29.1 shows the channels used by the algorithm chosen for SMMR geophysical data production (JPL 1981b).

Table 29.1

Geophysical quantities determined by SEASAT, channels used (Chester algorith, section 29.3.1) and retrieval grid (3 dB limit, km)

Parameter	Channels used GHz	Retrieval grid
(1) SST	6.6 V & H + vapour & cloud corrections (see 3 & 4)	150 × 150
(2) Surface wind	10.7H + SST, vapour & cloud corrections. . (see 1, 3, 4)	85 × 85
(3) Water vapour	18V & H, 21V & H	54 × 54
(4) Cloud liquid water	18V & H, 21V & H, 37V & H	54 × 54
(5) Rain rate	18H, 37H	54 × 54

All channels, representing wavelengths between 8 and 46 mm, used the same 790 mm antenna. The maximum spatial resolution therefore varied with frequency and the results were determined on a number of nested grids (Wilheit 1978). Land has a high emissivity, approximately twice that of a water surface, and the atmospheric water amounts are not retrieved over land. Further, the presence of land within an antenna pattern sidelobe can lead to serious errors if not properly compensated (section 29.4.2).

The conversion of the satellite data to geophysical parameters was performed in three major stages (Njoku *et al.* 1980b). The first stage (Swanson & Riley 1980) was to determine the microwave radiance (expressed in terms of black body temperature) at the antenna. Secondly, Antenna Pattern Corrections (APCs) for the antenna sidelobe pattern, polarisation coupling, Faraday rotation, and other effects (Njoku 1980, Njoku, Christensen, & Cofield 1980a) were applied to give microwave brightness temperatures, T_B's. The T_B's represent the input to the final stage, the geophysical retrieval algorithms. Unfortunately, early evaluation attempts showed that biases existed in the T_B's possibly caused by calibration errors (Gloersen 1981, JPL 1981a). For the SEASAT SMMR these biases have been empirically evaluated by applying a chosen geophysical retrieval algorithm to a subset of the data. The biases required to minimise the difference between those results and *in situ* measurements have been calculated and incorporated in the algorithm. The modified algorithm has then been tested against the 'withheld' portion of the dataset and the results evaluated at a number of workshops (JPL 1979a-c, 1980a-b, 1981a-b). The results have been summarised by Lipes *et al.* (1979) and Lipes (1982) and will be discussed below for each of the retrieved geophysical parameters.

29.3 GEOPHYSICAL RETRIEVAL ALGORITHMS

29.3.1 'Linear regression algorithms'

In the Wilheit algorithm (Wilheit & Chang, 1980, Wilheit, Chang, & Milman 1980) regression coefficients, previously calculated from an ensemble of possible geophysical models, are applied to functions of the T_B's.

The Chester algorithm (Chester 1980b) was developed from that of Wilheit. Chester attempted to minimise spurious correlations between the derived values by heavily weighting that channel which was most sensitive to each required parameter. This algorithm has been criticised for containing a large number of empirical corrections (Wentz 1981). However, with certain modifications (Chester 1981c) it has been chosen for the SMMR geophysical data production since, at the time of that decision, it was considered to give the best results (JPL 1981b).

29.3.2 'Least squares algorithms'

The Wentz algorithm (JPL 1979a) uses a least squares fitting technique to derive the geophysical parameters from a theoretically derived function for the T_B's observed by the SMMR (Wentz 1982). The algorithm had been developed (JPL 1980a, Wentz 1980) by, for example, incorporation of an improved sea surface emissivity versus wind relationship (Wentz, Christensen, & Richardson 1982). The Wentz algorithm generally gave better results than the Chester algorithm and possessed a sounder physical basis (JPL 1981b). However, these advantages were not considered sufficient to warrant changing from the Chester algorithm for geophysical data production.

A modification of the Wentz algorithm was produced by Bierman (JPL 1980a) who used the same geophysical model but a different solution technique, designed to take advantage of the greater number of independent measurements available in the higher frequency channels. However, this was found to degrade the SST retrievals (JPL 1980a p. 5.54).

29.3.3 Other algorithms

Early sea surface temperature estimates reported by Hofer, Njoku & Waters (1981) and Njoku & Hofer (1981) used a linear regression algorithm (Hofer &

Njoku 1981). Rozenkrantz (1981a,b) has developed a Fourier analysis algorithm which has the advantage of a more rigorous treatment of the variations in spatial resolution between the different channels.

In the SMMR-SASS algorithm (JPL 1980a) the SMMR atmospheric water channels were used to correct SASS backscatter coefficients (Moore *et al.* 1982) and the SASS winds then used to aid separation of the SST and wind effects in the SMMR. Initial analyses emphasised the importance of good wind estimates in determining SST from the SMMR (JPL 1980a).

29.4 STATUS OF GEOPHYSICAL RETRIEVALS

29.4.1 Introduction
The status of the geophysical algorithms for many puposes became fixed in April 1981 when the Chester algorithm was confirmed for geophysical data record (GDR) production. This section will review the status of the parameter retrievals at that time.

29.4.2 Sea surface temperature
Early estimates of SST for the Pacific (Hofer *et al.* 1981, JPL 1979c) showed standard deviations of 1.5°C or better when compared with 'surface truth' However, SMMR SST determinations for the JASIN area showed biases and scatter both amounting to a few degrees, with significant differences between ascending and descending spacecraft passes (JPL 1980a). Possible causes were radio frequency interference (RFI) and land contamination in the antenna side-lobes. Subsequently Chester & Wind (1980) found, for the Pacific, that retrievals became poor for distances less than 600 km from land. Schacher, Jarrell & Lucas (1981) confirmed that the JASIN estimates degraded with increasing proximity to Europe.

A summary of comparisons between current SMMR algorithms is given in Table 29.2. Excluding categories (h), for which the algorithm is not stated, and (j), which represents a single ship track in a region of high oceanic variability, the mean biases and mean standard deviations about the bias are:

$$-0.14 \pm 1.1°C \text{ (Chester algorithm)}$$
$$-0.07 \pm 1.0°C \text{ (Wentz algorithm)}$$

Since these differences include the errors inherent in using spot *in situ* measure-ments to represent sea surface temperature over a region of ocean, SMMR accuracies of better than 1°C are suggested (JPL 1981a). However, to achieve these accuracies a number of important conditions must be imposed on the SMMR data. The presence of sunglint, RFI, or significant rain degrades the retrieval. During the daytime 1° C errors may occur at the swath edges because of residual Faraday effects. Most serious is the requirement that SMMR estimates must be at least 600 km away from large land masses, thus preventing measurements in such areas as the Norwegian Sea (e.g. Gloersen *et al.* 1981) or the JASIN area.

Table 29.2

Summary of SMMR and 'surface truth' (S/T) comparisons for SST, °C. The mean bias and standard deviation, σ, about that bias are shown together with the number, n, of comparisons. The results are taken from JPL (1980b, 1981b).

Difference SMMR-S/T	σ °C	n	SMMR algorithm	Surface truth Source	References
(a) −0.2	0.9	49	C	OWS Papa, buoy & XBTs	Chester (1980b)
(b) 0.03	0.8	85	W	XBTs	Wentz (1980)
(c) −0.03	0.9	168	W	National Marine Fisheries Services map	
(d) −0.5	1.5	203	C	Unscreened XBTs	Chester (1981a)
(e) −0.2	1.4	203	W		
(f) −0.08	1.0	142	C	XBTs	Wentz (1981)
(g) −0.08	0.9	142	W		
(h) −1.4	1.3	∿40	?	Ships	Liu (1981)
(i) 0.22	0.8	−	C	Time-averaged gridded ship & SMMR data	Bernstein (1981)
(j) 1.4	1.3	43	C	Ship: S. A. Agulhas	Gerber, Chester & Wind (1981)

The GDR production for the entire SMMR data set will allow SST evaluations over a far greater range of geophysical conditions. There is the possibility that the algorithm will perform less well for SST below 7°C (Gerber, Chester, & Wind 1981) or for rough sea conditions (Chester 1981a); the accuracy might have changed with time (Chester & Wind 1980). However, when assessing the results the projected use of SMMR data should be considered. Because of the sparsity of ship observations and cloud contamination of satellite IR measurements, SST distributions produced by meteorological forecasting offices are usually composites or averages over several days. For such purposes, the large number of SST estimates produced by SMMR means that the rejection of some points, because of RFI or rain, is, provided those points can be identified, of little significance. Bernstein (1981) has shown that SMMR can produce large-scale maps of both SST and SST anomaly from climatological mean, suitable for climate research. Further, the combined use of conventional and SMMR estimates should enable any long-term SMMR calibration changes to be monitored, and Gautier (1981) has shown that SMMR-derived SST gradients are potentially more accurate than the absolute values. Thus, although further work is needed, experience with the SEASAT SMMR suggests that such an instrument is capable of providing SST information over much of the world's oceans to useful accuracy for weather forecasting and climate research purposes.

29.4.3 Sea surface wind speed
Evaluations using the latest algorithm versions are shown in Table 29.3 which includes some comparisons with wind speeds from the SEASAT Scatterometer, SASS (see Pierson 1983, Guymer 1983 for SASS description and assessment). Mean differences from 'surface truth' (excluding SASS) are

$$1.4 \pm 2.1 \text{ m/s Chester algorithm}$$
$$0.8 \pm 1.9 \text{ m/s Wentz algorithm}$$

where some of the scatter is attributable to surface truth 'errors'. These retrievals degrade within 500 km of land (Lipes 1981), owing to sunglint (Cardone 1980, Chester 1980) or rain (Cardone & Chester 1981). The Chester algorithm gives poor results for large liquid water amounts and a variable bias (Wentz, 1981) which may be due to an *ad hoc* SST correction (JPL 1981b p. A5).

Table 29.3
Summary of SMMR and 'surface truth' comparisons for wind speed, m/s, including SMMR/SASS comparisons. The mean bias and standard deviation, σ, about that bias are shown together with the number, n, of comparisons. The algorithms are Chester (C) or Wentz (W). Results from JPL (1980b, 1981b).

Difference SMMR-S/T	σ m/s	n	SMMR algorithm	Surface truth source	Reference
(a) 0.2	1.2	69	W	SASS winds	Wentz (1980)
(b) 0.7	2.3	–	C	S/T fields	Chester (1980c)
(c) 1.3	2.3	–	W		
(d) 0.7	1.1	–	?	A/c	Ross & Lawson (1980)
(e) 0.3	1.4	17	W	OWS Papa & buoy measurements	Wentz (1981)
(f) 2.0	1.9	17	C		
(g) 1-2	<2	–	W & C	SASS	Cardone & Chester (1981)

Early evaluations using the Wentz algorithm had shown a difference, SMMR-S/T, of 0.6 ± 2.6 m/s for Gulf of Alaska Experiment (GOASEX) passes. However, equivalent JASIN values were 2.6 ± 1.9 m/s (JPL 1980a). Chester (1980) argues that this offset is not due to RFI as suggested by Born *et al.* (1981). Allan & Guymer (1982) suggest that the calibration of the SMMR winds changed on or near rev 900 (29 August), becoming smaller for the period thereafter in which GOASEX occurred. Chester & Wind (1980) present SMMR/buoy comparisons which may also suggest a calibration change sometime between revs 820 and 1200.

Thus, while the SMMR wind speeds meet the design objectives (JPL 1981a) care will be needed in using GDR values, and there is scope for futher investigation. One difficulty is the sparsity of good 'surface truth', and this has led to investigations in which different SEASAT wind sensors are compared (e.g. Wentz, Cardone & Fedor 1982).

29.4.4 Atmospheric water

Katsaros, Taylor, Alishouse & Lipes (1981) review the evaluation of the atmospheric water estimates up to and including the SEASAT-JASIN Workshop (JPL 1980a). At that time the Wentz algorithm was, for the JASIN area, giving integrated atmospheric water vapour, q_v, estimates (Table 29.4) at least as accurately as the high quality JASIN radiosonde ascents (estimated accuracy ± 1.7 kg/m^2, Taylor, Katsaros, & Lipes 1981). A bias in SMMR estimates in the tropics was later removed by the use of more realistic air temperatures within the algorithm (Wentz 1980). In contrast, the Chester algorithm was found to underestimate midlatitude water vapour amounts (Alishouse 1981), and modification was required before GDR production (Chester 1981c). This modification may, however, degrade retrievals in the JASIN area (Wentz 1981).

Few evaluations of SMMR liquid water and rain rate estimates have been made, because of the lack of *in situ* measurements. Taylor, Guymer, Katsaros, & Lipes (1983) compared total atmospheric liquid water content q_1 from the Wentz (JASIN) algorithm to values derived from radiosonde profiles. For strato-cumulus clouds with $q_1 < 0.2$ kg/m^2 they found agreement to better than ± 0.05 kg/m^2. At that time Wentz and Wilheit q_1 estimates were in reasonable agreement up to about 1.5 kg/m^2 (Katsaros *et al.* 1981). Chester (1981c) noted that for increasing true liquid water amounts, corresponding to increasing rain rate, the Wilheit algorithm actually predicted decreasing liquid water values. For such cases Chester redefined q_1 to be a multiple of the rain rate. The q_1 estimates may require correction by a scale factor (Chester 1981a) or bias removal, possibly by comparison with cloud imagery such as that from GOES (Gautier 1980).

Most rain rate comparisons have been qualitative in nature. For JASIN the Wentz algorithm indicated rain for cases in which rain was widespread, that is, reported by several of the ships. Taylor *et al.* (1983) found agreement between the area of rain indicated by the SMMR and the 'present weather' reports from the ships. Where two quantitative *in situ* rainfall estimates were available the SMMR values agreed within a factor of two; however, in both cases the rain was light, less than 1 mm/hour (Katsaros *et al.* 1981, Taylor *et al.* 1983). The Wilheit and Chester algorithms have been found to predict rain more frequently than that of Wentz (Katsaros *et al.* 1981, Ross & Lawson 1980, Alishouse 1980, 1981); however, these algorithms still miss light or scattered showers. Heavier rain, particularly if convective, may be even more difficult to estimate accurately. Chester (1981b) has suggested that the SMMR rain rates might possibly be reasonable up to 10 mm/hour. However, because of the poor spatial resolution, liquid water and rain rate determinations in such conditions should be viewed with caution (e.g. Lovejoy & Austin 1980). Independent freezing level estimates could improve SMMR rain rate algorithms. (e.g. Wilheit 1978).

Table 29.4
Comparison of SMMR atmospheric water vapour and radiosonde values, kg/m^2. The algorithms are W* = Wentz (JASIN Workshop version, March 1980), W = Wentz, C = Chester, C* = ammended Chester. Results from JPL (1980b, 1981b).

Difference SMMR-S/T	σ kg/m^2	n	SMMR algorithm	Comparison Regions	Reference
1.2	1.6	19	W*	Midlatitudes (JASIN)	Katsaros *et al.* (1981)
−4.1	2.0	26	C	Midlatitudes (GOASEX)	Alishouse (1981)
0.0	2.0	26	C*	"	(See text)
4.9	3.7	6	W*	Tropics (early algorithm)	Katsaros *et al.* (1981)
0.0	3.4	35	W	Tropics (no rain)	Alishouse (1981)
−0.4	3.7	30	C	Tropics (no rain)	Alishouse (1981)
0.5	5.3	15	W	Tropics (incl. rain)	Alishouse (1981)
2.0	5.0	15	C	Tropics (incl. rain)	Alishouse (1981)

In assessing the quality of the SMMR atmospheric water estimates the potential applications must be considered. The water vapour estimates are excellent compared to *in situ* measurements. In addition to estimating SASS attenuation (Moore *et al.* 1981) q_v measurments are required for altimeter range correction. The JASIN radiosonde observations show that, for a given surface specific humidity, q_v varied by up to 15 kg/m^2. If a standard model atmosphere were used to determine the moist tropospheric range correction (e.g. Saastamoinen 1971) this variation would represent an error of up to 11 cm, which is large compared to the accuracy required for ocean circulation sensing (e.g. Mather, Rozos, &

Coleman 1979). In contrast the SMMR provided accurate measurements of the range correction (Tapley, Lundberg, & Born 1981).

In JASIN the q_v distributions from SMMR have also been used to position synoptic scale fronts even for a case where lack of frontal cloud prevented detection in visible and IR images (Taylor *et al.* 1981). The potential of SMMR q_v estimates in climate research has been demonstrated by Chelton, Hussey, & Parke (1981) who produced global q_v distributions.

The lack of *in situ* validation data for liquid water and rain rate estimates is in itself a measure of their uniqueness. However difficult precise calibration may be, the SMMR values still represent a considerable aid for meteorological analysis. This will be illustrated in the next section by an example from the JASIN experiment.

29.5 A JASIN CASE STUDY

On 5 September 1978 an occlusion approached the JASIN area from the southwest (Fig. 29.1). East-southeasterly winds of over 15 m/s were reported by ships in advance of the surface front. NOAA 5 infra-red images (Fig. 29.2) showed a

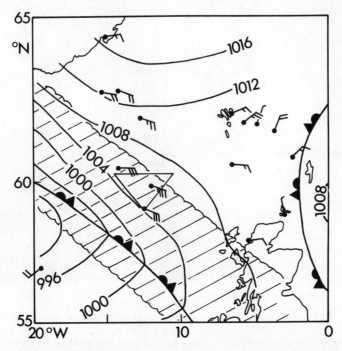

Fig. 29.1 – Surface analysis for 06 GMT 5 September 1978. Frontal positions derived by the UK. Meteorological Office are shown together with surface wind reports from ships. The high cloud band observed by NOAA 5 satellite is also indicated (see Fig. 29.2).

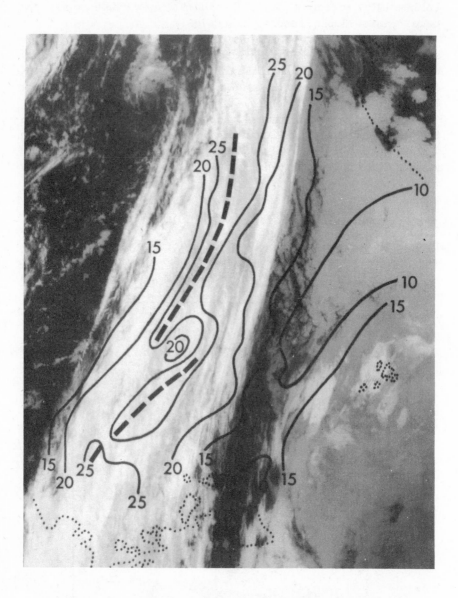

Fig. 29.2 – NOAA 5 infra-red satellite image at 1037 GMT 1978 (courtesy Univ. of Dundee) and SMMR integrated water vapour q_v distribution at 0813 GMT (kg/m^2). The frontal cloud stretched from Ireland and Scotland (lower left) to Iceland (upper right). The JASIN area was situated just above the break in the dashed line which marks the q_v maximum (Reprinted by permission from *Nature* **294** (5843) 737–739, Copyright © 1981 Macmillan Journals Limited; rotate the image 90° anticlockwise for comparison with other figures).

400 km band of high cloud associated with the front. SMMR integrated water vapour values showed a maximum (broken line in Fig. 29.2) along the cloud band of about 27 kg/m^2. The surface frontal position would be expected to lie along the rear edge of this maximum (Taylor *et al.* 1981). Allowing for frontal movement this suggests that the Meteorological Office analysis frontal position was some 50 to 100 km to the southwest of the true position, a discrepancy which is not surprising considering the lack of conventional reports in that region (Fig. 29.1).

The region of q_v maximum was also a region of high liquid water, q_l (Fig. 29.3). This showed an along-front variation on a scale of perhaps 300 km. Maxima in q_l were associated with SMMR rain rates of up to 0.8 mm/hour (Fig. 29.4). Similar q_l/rain rate maxima were observed on the SMMR pass 8 hours previously. Assuming that the same region of enhanced rain rate was observed on both passes, the displacement shown in Fig. 29.4 represents an along-front propagation of about 5 m/s. The banded nature of frontal precipitation has been the subject of much recent research (see reviews by Browning & Mason 1980, Hobbs 1981). Precipitation areas on scales above 100 km tend to propagate with a system velocity rather than the wind speed at any level (e.g. Harrold 1973) and have durations of order several hours (Austin & Houze 1972). Thus it is not inconceivable that the same precipitation region was observed on both passes.

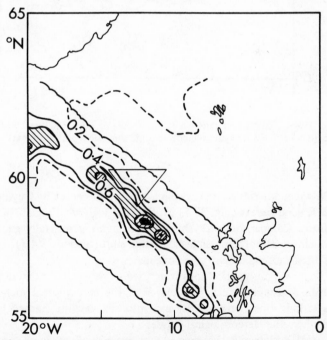

Fig. 29.3 – Atmospheric liquid water distribution (kg/m^2) at 0813 GMT, 5 September 1978. The limits of the high cloud band are also shown.

Quantitative rainfall observations were not available from the JASIN ships at this time; however, one ship reported drizzle at 0800 GMT, followed by intermittent rain until 1700 GMT, at the position shown in Fig. 29.4. This could correspond to the northward propagation of the main rainfall maximum.

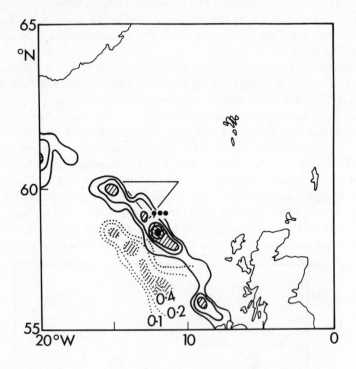

Fig. 29.4 – SMMR rain rate values (mm/hour) at 0004 GMT and 0813 GMT, 5 September 1978. The rain symbols indicate ship rain reports (see text).

SASS winds for this frontal system have been presented by Guymer (1982). The SMMR wind speeds (Fig. 29.5) confirm the high wind region values ahead of the front. Comparison with ship measurements is generally good for this period which was after the possible calibration shift (section 29.4.3).

This JASIN example illustrates three major points:

(1) The water vapour distributions were used to determine the frontal position to higher accuracy than possible from the conventional meteorological observations (including visible and infra-red satellite imagery) normally available to the Meteorological Office.

(2) The SMMR detected the presence of regions of enhanced precipitation probability within the frontal zone.

(3) The region of higher wind speeds ahead of the front was mapped by the
SMMR, and the wind speed values in this region determined.

If routinely available at, or soon after, the time of observation, such information could improve forecasts of frontal timing, precipitation, and winds for
systems approaching, for example, the British Isles from the Atlantic Ocean.
That at present such forecasts can be hampered by lack of reliable surface
observations has been stressed by Woodroffe (1981) in discussing the 1979
'Fastnet storm'. The strong winds associated with that storm occurred in a
belt probably of order 200 km wide and were therefore difficult to forecast
but detectable by SMMR.

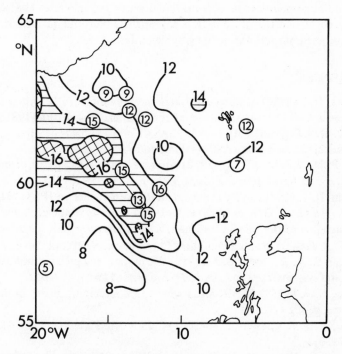

Fig. 29.5 – SMMR wind speed values (m/s) for 0813 GMT, 5 September 1978.
Ship observations are shown in circles.

29.6 CONCLUSIONS

The SMMR is capable of quantitative estimates of sea surface temperature, sea
surface wind speed, and total atmospheric water vapour. Under good conditions
accuracies of ± 1°C (SST), ± 2 m/s (winds), and ± 2 kg/m² (vapour) are attained;
however, SST and wind measurements need careful screening, and SST measurements in particular must be more than 600 km from land. The SMMR can also

produce liquid water and rain rate distributions which are in qualitative agreement with surface truth, but quantitative comparisons are lacking.

That the water vapour estimates should be most accurate is not surprising; not only is estimation relatively simple (Table 29.1), but the measurement techniques have been developed on the NIMBUS series of satellites. In comparison, SST and surface winds are new measurements, and future microwave radiometers can be designed to avoid some of the present restrictions. A larger antenna, for example, would allow measurement closer to land. Nevertheless, the SEASAT SMMR has provided data suitable both for climate research and for synoptic and subsynoptic meteorological analysis. The European based JASIN experiment contributed significantly to the calibration of the SEASAT SMMR, and has benefitted substantially from the SMMR derived distributions of surface wind speed and atmospheric water vapour, liquid water and rain rate. JASIN experience has confirmed that the SMMR is potentially a powerful quantitative instrument for meteorological and oceanographic analysis.

REFERENCES

Alishouse, J. C. (1980) Atmospheric water determinations, *in* JPL (1980b).
Alishouse, J. C. (1981) Atmospheric water determinations, *in* JPL (1981b).
Allan, T. D. & Guymer, T. H. (1982) SEASAT measurements of wind and waves on selected passes over JASIN (submitted to *Int. J. Rem. Sen.*).
Austin, P. M. & Houze, R. A. (1972) Analysis of the structure of precipitation patterns in New England, *J. Appl. Meteorol.* **11** (6) 926–935.
Bernstein, R. L. (1981) SST mapping in North-West Pacific, *in* JPL (1981b).
Born, G. H., Lame, D. B. & Rygh, P. J. (1981) A survey of the goals and accomplishments of the SEASAT mission, *in Oceanography from space* (edited by J. F. R. Gower), Plenum Publishing Corporation, New York, 3–14.
Browning, K. A. & Mason, B. J. (1980) Air motion and precipitation growth in frontal systems, *Pure and Applied Geophysics* **119** (3) 407–691.
Cardone, V. J. (1980) SMMR-SASS wind speed intercomparisons in the QE II storm, *in* JPL (1980b).
Cardone, V. J. & Chester, T. J. (1981) SMMR-SASS wind comparison, *in* JPL (1981b).
Chelton, D. B., Hussey, K. J. & Parke M. E. (1981) Global satellite measurements of water vapour, wind speed and wave height, *Nature* **294** (5841) 529–532.
Chester, T. J. (1980a) JASIN wind bias examination, *in* JPL (1980b).
Chester, T. J. (1980b) The Chester geophysical algorithm, *in* JPL (1980b).
Chester, T. J. (1980c) Wind field comparisons, *in* JPL (1980b).
Chester, T. J. (1981a) Geophysical parameter contour maps, *in* JPL (1981b).
Chester, T. J. (1981b) SMMR rain rate algorithm modifications, *in* JPL (1981b).
Chester, T. J. (1981c) Final changes in SMMR geophysical data production algorithm, *in* JPL (1981b).

Chester, T. J. & Wind, B. B. (1980) Spot comparisons from mini-workshop II data set, *in* JPL (1980b).

Francis, C. R., Thomas, D. P., & Windsor, E. P. L. (1983) The evaluation of SMMR retrieval algorithms (this volume).

Gautier, C. (1980) Comparison of SMMR liquid water retrievals with GEOS data, *in* JPL (1980b).

Gautier, C. (1981) Information content of the SEASAT SMMR brightness temperatures for sea surface temperature retrieval, *in Oceanography from space* (edited by J. F. R. Gower), Plenum Publishing Corporation, New York, 727-734.

Gerber, M., Chester, T. J. & Wind, B. B. (1981) SST off the South African coast *in* JPL (1981b).

Gloersen, P. & Barath, F. T. (1977) A scanning multichannel microwave radiometer for NIMBUS G and SEASAT A, *IEEE J. Oceanic Eng.* **OE-2** 172-178.

Gloersen, P. (1981) Passive microwave observations — an introduction, *in Oceanography from space* (edited by J. F. R. Gower), Plenum Publishing Corporation, New York, 663-664.

Gloersen, P., Campbell, W. J. & Cavalieri, D. (1981) Global maps of sea ice concentration, age and surface temperature derived from NIMBUS 7 scanning multichannel microwave radiometer data: A case study, *in Oceanography from space* (edited by J. F. R. Gower), Plenum Publishing Corporation, New York, 777-784.

Gloersen, P., Cavalieri, D., Crawford, J., Campbell, W. J., Farrelly, B., Johannessen, J., Johannessen, O. M., Svendsen, E. & Kloster, K. (1981) Evaluation of NIMBUS 7 SMMR sensor with airborne radiometers and surface observations in the Norwegian Sea. *in Oceanography from space* (edited by J. F. R. Gower), Plenum Publishing Corporation, New York, 717-726.

Guymer, T. H. (1983) Validation and applications of SASS over JASIN (this volume).

Harrold, T. W. (1973) Mechanisms influencing the distribution of precipitation within baroclinic disturbances, *Quart. J. Roy. Meteorol. Soc.* **99** (420) 232-251.

Hobbs, P. V. (1981) Mesoscale structures in midlatitude frontal systems, *in* Nowcasting: Mesoscale observations and short-range prediction, (*Proc. IAMAP symp.* 25-28 Aug. 1981, Hamburg), 29-36, European Space Agency, SP-165.

Hofer, R. & Njoku, E. G. (1981) Regression techniques for oceanographic parameter retrieval using space borne microwave radiometry, *IEEE trans. Geoscience and Remote Sensing* **GE-19** (4) 178-189.

Hofer, R., Njoku, E. G. & Waters, J. W. (1981) Microwave radiometric measurements of sea surface temperature from satellite: first results. *Science* **212** 1385.

JPL (1979a) SEASAT Gulf of Alaska Workshop Report, Vol. I, Panel Reports (Born, G. H., Wilkerson, J. C. & Lame, D. B., eds.). JPL Internal Document 622-101, April 1979.

JPL (1979b) SEASAT Scanning Multichannel Microwave Radiometer Mini-Workshop Report, (Lipes, R. G. & Born, G. H. eds.), JPL Internal Document 622-208, June 1979.

JPL (1979c) SEASAT Scanning Multichannel Microwave Radiometer Mini-Workshop II Report, (Lipes, R. G. & Born, G. H. eds.), JPL Internal Document 622-212, October 1979.

JPL (1980a) *SEASAT-JASIN Workshop Report,* Vol. I: Findings and Conclusions (Businger, J., Stewart, R. H., Guymer, T. H., Lame, D. B. & Born, G. H., eds.), JPL Publication 80-62, December 1980.

JPL (1980b) SMMR Mini-Workshop III, 26-27 August 1980, (Lipes, R. G. & Born, G. H. eds.), JPL Internal Document 622-224, December 1980.

JPL (1981a) SEASAT Data Utilisation Project, Final Report (Born, G. H., Held, D. N., Lame, D. B., Lipes, R. G., Montgomery, D. R., Rygh, P. J. & Scott, J. F.). JPL Internal Document 622-233, September 1981.

JPL (1981b) SMMR Mini-Workshop IV (Lipes, R. G., ed.), JPL Internal Document 622-234, December 1981.

Katsaros, K. B., Taylor, P. K., Alishouse, J. C. & Lipes, R. G. (1981) Quality of SEASAT SMMR (Scanning Multichannel Microwave Radiometer) atmospheric water determinations, *in Oceanography from space* (edited by J. F. R. Gower), Plenum Publishing Corporation, New York, 691-706.

Lipes, R. G., Bernstein, R. L., Cardone, V. J., Katsaros, K. B., Njoku, E. G., Riley, A. L., Ross, D. B., Swift, C. T. & Wentz, F. J. (1979) SEASAT Scanning Multichannel Microwave Radiometer: Results of the Gulf of Alaska Workshop, *Science* **204** 1415-1417.

Lipes, R. G. (1981) Conclusions and summary, *in* JPL (1981b).

Lipes, R. G. (1982) Description of SEASAT radiometer status and results, *J. Geophys. Res.* **87** (C5) 3385-3398.

Liu, C-T (1981) STT in the Eastern tropical Pacific, *in* JPL (1981b).

Lovejoy, S. & Austin, G. L., (1980) The estimation of rain from satellite borne radiometers, *Quart. J. Roy. Meteorol. Soc.* **106** 255-276.

Mather, R. S., Rozos, C. & Coleman, R., (1979) Remote sensing of surface ocean circulation with satellite altimetry, *Science* **205** 11-17.

Moore, R. K., Birrer, I. J., Bracalente, E. M., Dome, G. J. & Wentz, F. J. (1982) Evaluation of atmospheric attenuation from SMMR brightness temperature for the SEASAT Satellite Scatterometer, *J. Geophys. Res.* **87** (C5) 3337-3354.

Njoku, E. G. (1980) Antenna pattern correction procedures for the scanning multichannel microwave radiometer (SMMR), *Boundary Layer Meteorol.* **18** 79-98.

Njoku, E. G., Christensen, E. J. & Cofield, R. E. (1980a) The SEASAT Scanning Multichannel Radiometer (SMMR): antenna pattern corrections — development and implementation, *IEEE J. Oceanic Eng.* **OE-5** (2) 125-137.

Njoku, E. G., Stacey J. M. & Barath, F. T. (1980b) The SEASAT Scanning

Multichannel Microwave Radiometer (SMMR): instrument description and performance, IEEE *J. Oceanic Eng.* **OE-5**, (2), 100-115.

Njoku, E. G. & Hofer, R. (1981) SEASAT SMMR observations of ocean surface temperature and wind speed in the North Pacific, *in Oceanography from space* (edited by J. F. R. Gower) Plenum Publishing Corporation, New York, 673-681.

Pierson, W. (1983) Highlights of the SEASAT SASS program (this volume).

Pollard, R. T., Guymer, T. H., & Taylor, P. K. (1983) Summary of the JASIN 1978 field experiment *Phil. Trans. Roy. Soc.,* London **A308**, 221-230.

Rosenkrantz, P. W. (1981a) Inference of sea surface temperature, near surface wind, and atmospheric water by Fourier analysis of Scanning Multichannel Microwave Radiometer Data, *in Oceanography from space* (edited by J. F. R. Gower), Plenum Publishing Corporation, New York, 707-716.

Rosenkrantz, P. W. (1981b) Inversion of data from diffraction limited multi-wavelength remote sensors, 3, Scanning Multichannel Microwave Radiometer data, *Radio Science* (in press).

Ross, D. B. & Lawson, L. (1980) An evaluation of SMMR observations of surface windspeed during rev 1339, *in* JPL (1980b).

Saastamoinen, J. (1971) Use of artifical satellites for geodesy; atmospheric correction for the troposphere and stratosphere in radio ranging for satellites, *in Proc 3rd Intern. Symp., Am. Geophys. Union.*

Schacher, G. E., Jarrell, J. & Lucas, G. (1981) SMMR JASIN sea-surface temperature comparisons, *in* JPL (1981b).

Swanson, P. N. & Riley, A. L. (1980) The SEASAT Scanning Multichannel Microwave Radiometer (SMMR): radiometric calibration algorithm development and performance, IEEE, *J. Oceanic Eng.* **OE-5** 116-124.

Swift, C. T. (1980) Passive microwave remote sensing of the ocean — a review, *Boundary Layer Meteorol.* **18** 25-54.

Tapley, B. D., Lundberg, J. B. & Born, G. H. (1982) The SEASAT altimeter wet tropospheric range correction, *J. Geophys. Res.* **87** (C5) 3213-3220.

Taylor, P. K., Katsaros, K. B. & Lipes, R. G. (1981) Determinations by SEASAT of atmospheric water and synoptic fronts, *Nature* **294** 737-739.

Taylor, P. K., Guymer, T. H., Katsaros, K. B. & Lipes, R. G. (1983) Atmospheric water distributions determined by the SEASAT Multichannel Microwave Radiometer, in *Proc. Symp. Variations in the Global water budget,* 10-15 August 1981, Oxford, 93-106, D. Reidel, Dordrecht.

Wentz, F. J. (1980) Developments in the Wentz geophysical algorithm, *in* JPL (1980b).

Wentz, F. J. (1981) Comparison of SEASAT SMMR geophysical algorithms, *in* JPL (1981b).

Wentz, F. J. (1982) A model function for ocean microwave brightness temperatures, *J. Geophys., Res.* (submitted).

Wentz, F. J., Cardone, V. J. & Fedor, L. S. (1981) Intercomparison of wind speeds inferred by the SASS, Altimeter and SMMR, *J. Geophys. Res.* **87** (C5) 3378-3384.

Wentz, F. J., Christensen, E. J. & Richardson, K. A. (1982) SEASAT SMMR least squares geophysical algorithm (in preparation).

Wilheit, T. T. (1978) A review of applications of microwave radiometry to oceanography, *Boundary Layer Meteorol.* **13** 277-293.

Wilheit, T. T. & Chang, A. T. C. (1980) An algorithm for retrieval of ocean surface and atmospheric parameters from the observations of the Scanning Multichannel Microwave Radiometer (SMMR), *Radio Science* **15** 525-544.

Wilheit, T. T., Chang, A. T. C. & Milman, A. S. (1980) Atmospheric corrections to passive microwave observations of the ocean, *Boundary Layer Meteorol.* **18** (1) 65-78.

Woodroffe, A. (1981) The Fastnet storm — a forecasters viewpoint, *Met. Mag., Lond.* **110** (1311) 271-287.

Chapter 30

The evaluation of SMMR
retrieval algorithms

C. R. FRANCIS, D. P. THOMAS and E. P. L. WINDSOR
British Aerospace Dynamics Group, Bristol, UK

30.1 INTRODUCTION

In 1976 BAe (Dynamics Group) was accepted as a member of the Experiment Team for the SMMR on Nimbus-7, with the aim of developing a retrieval algorithm for sea surface parameters. This general aim was later concentrated on to the development of an iterative algorithm, as an alternative to the linear approach adopted for the routine production of geophysical data. The intention was to investigate the feasibility of such an approach and to evaluate its performance. This chapter describes the development of this algorithm.

30.2 OUTLINE

Microwave radiometers measure the natural microwave emissions from the ocean and atmosphere. In order to use these measurements we must know which geophysical parameters influence the emissions, and how, in a quantitative way. A model of the radiative transfer processes has been generated from information available in the literature, which estimates the observed microwave spectrum resulting from a given set of geophysical parameters. This model is used in an iterative algorithm, to estimate geophysical parameters, given the microwave measurements.

30.3 ENVIRONMENTAL MODEL

30.3.1 General description

The observed spectrum is formed from the four components shown in Fig. 30.1; i.e. radiation emitted from the sea and the atmosphere and received from the space background and which is reflected from the sea and absorbed in the atmosphere. The received power is expressed in the model as a brightness temperature, which is the product of physical temperature and emissivity. This proportionality between power and physical temperature is described by the Rayleigh-Jeans approximation, and allows the brightness temperature of each of the four components of Fig. 30.1 to be simply summed to find the overall brightness temperature.

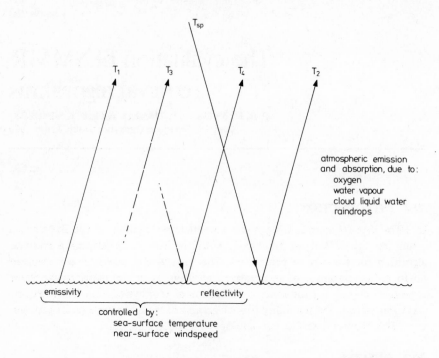

Fig. 30.1 – The four components of upwelling microwave radiation which are observed by a passive microwave radiometer above the atmosphere.

Emission and reflection from the sea are influenced by the sea-surface temperature and the surface windspeed (which controls surface roughness), while atmospheric absorption and emission is actually the combined effect of the relevant constituents:

— water vapour
— oxygen
— liquid water in clouds
— raindrops

In principle, then, all of these parameters are measurable by microwave radiometry.

The emissivity and reflectivity of the sea are found from a set of equations which combine analytical, numerical, and empirical methods, while the atmosphere is modelled as a series of slabs to allow numerical integration. The number of atompospheric calculations is minimised by careful accumulation of brightness temperatures and absorption coefficients, using the equality between absorptivity and emissivity to allow a single calculation for each constituent per layer.

30.3.2 Ocean model

The operations required to find the radiation from the sea are shown in Fig. 30.2. This radiation is given by the product of sea-surface temperature, T_S, and the emissivity ϵ.

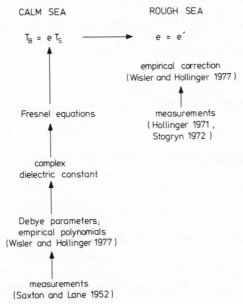

CALM SEA   ROUGH SEA

$T_B = e\, T_S$ ⟶ $e = e'$

empirical correction
(Wisler and Hollinger 1977)

Fresnel equations   measurements
(Hollinger 1971,
Stogryn 1972)

complex
dielectric constant

Debye parameters;
empirical polynomials
(Wisler and Hollinger 1977)

measurements
(Saxton and Lane 1952)

Fig. 30.2 — The series of operations required to find the brightness temperature of the sea.

Emissivity is a function of the complex dielectric constant of sea water, whilst the dielectric constant is, in turn, a function of the Debye parameters — the static dielectric constant, the relaxation time, and the ionic conductivity — of sea water. There are therefore essentially three stages to the process of calculating the sea surface emissivity: finding values for the Debye parameters, calculating the dielectric constant, and thus finding a value for the emissivity.

30.3.2.1 *Debye parameters*

Measurements of E_s, the static dielectric constant, the dielectric constant for very high frequencies E_∞, the relaxation time t, and the ionic conductivity σ have not been published for sea water, but Saxton & Lane (1952) have tabulated these quantities as functions of temperature and salinity for aqueous sodium chloride solutions. Although ocean water has many types of salts in solution, 85.6% of the total dissolved salts are sodium and chloride ions. Since NaCl is the principal salt in sea water, one may expect that a reasonable approximation for sea water may be obtained by assuming that the results for NaCl solutions can be used without modification.

Wisler & Hollinger (1977) have fitted a second-order polynomial in salinity and temperature by the method of least squares to the experimentally determined values of E_s, t and σ given by Saxton & Lane (1952).

$$E_s = 88.00 - 4.399 \times 10^{-1}\, S + 1.710 \times 10^{-3}\, S^2 - 4.035 \times 10^{-1}\, T$$
$$+\, 8.065 \times 10^{-4}\, T^2 + 6.170 \times 10^{-3}\, ST - 8.910 \times 10^{-5}\, S^2 T$$
$$-\, 6.934 \times 10^{-5}\, ST^2 + 1.439 \times 10^{-6}\, S^2 T^2$$

$$t = (18.70 - 7.924 \times 10^{-2}\, S + 6.350 \times 10^{-4}\, S^2 - 5.489 \times 10^{-1}\, T$$
$$+\, 5.758 \times 10^{-3}\, T^2 + 1.889 \times 10^{-3}\, ST - 7.209 \times 10^{-6}\, S^2 T$$
$$-\, 5.299 \times 10^{-7}\, ST^2 - 2.101 \times 10^{-7}\, S^2 T^2) \times 10^{-12}$$

$$\sigma = (7.788 \times 10^{-3}\, S - 1.672 \times 10^{-6}\, S^2 - 8.570 \times 10^{-15}\, T$$
$$+\, 2.996 \times 10^{-16}\, T^2 + 4.059 \times 10^{-4}\, ST - 3.215 \times 10^{-6}\, S^2 T$$
$$-\, 1.423 \times 10^{-6}\, ST^2 + 3.299 \times 10^{-8}\, S^2 T^2) \times 10^{11}$$

where S is salinity (°/oo) and T is the sea surface temperature (°C). In the present work a value for salinity of 34°/oo is assumed. This is a valid assumption, particularly since the SMMR measurement frequencies are insensitive to changes in salinity value. Saxton & Lane suggest that a constant value of 4.9 be taken for E_∞.

30.3.2.2 *Complex dielectric constant*

A polar liquid such as water has a complex dielectric constant E in which the imaginary part represents losses due to ionic conductivity and other effects. The Debye parameters evaluated in section 30.3.2.1 may be used to find the real and imaginary parts of the dielectric constant through the Debye theory of polar liquids. If the complex dielectric constant E is expressed:

$$E = E' - jE''$$

then the real and imaginary parts, E' and E'' may be approximated by

$$E' = \frac{E_s - E_\infty}{1 + (2\pi v t)^2} + E_\infty$$

$$E'' = \frac{(E_s - E_\infty)\, 2\pi v t}{1 + (2\pi v t)^2} + \frac{2\sigma}{v}$$

where v is the frequency.

30.3.2.3 *Emissivity*

The emissivity of the smooth surface of the undisturbed sea is calculated using the Fresnel formulae:

$$\epsilon_v = 1 - \left| \frac{E \cos \theta - \sqrt{(E - \sin^2 \theta)}}{E \cos \theta + \sqrt{(E - \sin^2 \theta)}} \right|^2$$

$$\epsilon_h = 1 - \left| \frac{\cos \theta - \sqrt{(E - \sin^2 \theta)}}{\cos \theta + \sqrt{(E - \sin^2 \theta)}} \right|^2$$

where ϵ_v and ϵ_h are the vertically and horizontally polarised emissivities and θ is the angle of incidence.

It is therefore possible to obtain values for the emissivity of a calm sea, given the sea surface temperature, by carrying out the series of steps described in sections 30.3.2.1, 30.3.2.2, and 30.3.2.3.

30.3.2.4 *The wind-driven sea surface*

The real sea surface is not smooth however, and it is therefore necessary to express the effect of wind-induced roughness on the emissivity. The development of this expression follows closely the treatment by Wisler & Hollinger (1977).

Wind speed acts on the sea surface in two ways. The first is due to the roughness of the water surface due to waves and the second to the increased coverage of foam on the surface which also distorts the fine structure of the surface. Since roughness which can influence the emissivity has dimensions of the same order as (or smaller than) the microwave wavelengths, then the wind-driven waves which cause this roughness are the capillary waves, which respond rapidly to changes in the local wind fields.

If roughness effects are not too severe, Wisler & Hollinger suggest that the specular solution derived above should be combined with an empirical expression for the increase in emission due to surface roughness. They propose an empirical model, based on measurement of the ocean roughness effect from an ocean tower by Hollinger (1971) and a survey of published measurements of the sea foam effect by Stogryn (1972).

The effective reflection coefficient for polarisation mode p, r_p, is given as:

$$r_p = r_{rp} (1 - f) + r_{fp} f$$

There are two components; for the foam-covered fraction, where the fractional foam coverage is f, the reflection coefficient is r_{fp} whilst for the non-foam-covered fraction $(1 - f)$, the reflection coefficient is r_{rp}.

If the reflection coefficient of a specular calm sea is r_{sp} the reflection coefficient r_{rp} of sea without foam is given by:

$$r_{\text{rp}} = r_{\text{sp}} - \frac{\Delta T_{\text{rp}}}{T_{\text{s}}}$$

where T_{s} is the sea surface temperature and ΔT_{rp} is given by curves fitted to the measurements of Hollinger (1971) for both vertical and horizontal polarisation modes:

$$\Delta T_{\text{rv}} = W\,(a + b\,\exp{(c\theta)})\,\sqrt{\nu}$$

$$\Delta T_{\text{rh}} = W\,(d + e\theta^2)\,\sqrt{\nu}$$

in which θ is the incidence angle, and the frequency, ν, is measured in GHz.

In these expressions the constants have the following values:

$a = 1.71 \times 10^{-1}$
$b = -2.09 \times 10^{-3}$
$c = 7.32 \times 10^{-2}$
$d = 1.15 \times 10^{-1}$
$e = 3.8 \times 10^{-5}$

The reflection coefficient of foam-covered sea, r_{fp}, is given by Stogryn (1972):

$$r_{\text{fp}} = 1 - \frac{(208 + 1.29\nu)}{288}\,G_{\text{p}}$$

where $G_{\text{V}} = 1 - 9.946 \times 10^{-4}\,\theta + 3.218 \times 10^{-5}\,\theta^2 - 1.18\ 7 \times 10^{-6}\,\theta^3$
$$+ 7 \times 10^{-20}\,\theta^{10}$$

and $G_{\text{H}} = 1 - 1.748 \times 10^{-3}\,\theta - 7.336 \times 10^{-5}\,\theta^2 + 1.044 \times 10^{-7}\,\theta^3.$

In the model, the fractional foam coverage, f, is a function of wind speed and is given by an expression due to Stogryn (1972):

$$f = 7.751 \times 10^{-6}\,W^{3.231}$$

where W is the wind speed in m/s

30.3.3 Atmospheric model

The principal components in the atmosphere contributing to both attenuation and emission are molecular oxygen, water vapour, and liquid water droplets in clouds and rain. Ice, as in cirrus clouds, is not a significant factor in the microwave frequency range.

Before considering the attenuation coefficients of each of these atmospheric constituents, the general case of absorption and emission in the atmosphere will be addressed.

For the case of a sea surface with no intervening atmosphere and with no background space temperature, the black-body temperature which would be measured by a satellite-borne passive microwave radiometer is simply:

$$T_B = \epsilon \, T_s$$

where ϵ is the sea surface emissivity,
and T_s is the sea surface thermodynamic temperature.

If an atmosphere with opacity τ is introduced, then we must consider three components of brightness temperature. These are the sea surface component and the upwelling and downwelling atmospheric emission components. To complete the physical picture, the space background temperature reflected by the sea must also be included.

The equation of radiative transfer can thus be written as:

$$T_B = T_1 + T_2 + T_3 + T_4 \text{ (see fig. 30.1)}$$

In this equation, T_1, the sea surface component, is

$$T_1 = \epsilon T_s \exp(-\tau)$$

where τ is the one-way atmospheric attenuation.

T_2, the space component, is

$$T_2 = (1 - \epsilon) \, T_{sp} \exp(-2\tau)$$

where T_{sp} is the space temperature

T_3, the upwelling atmospheric component, is

$$T_3 = \int_0^\tau T(z) \exp(-\tau + \tau'(z)) d\tau'(z)$$

where $T(z)$ is the atmospheric thermodynamic temperature at altitude z

T_4, the downwelling atmospheric component, is

$$T_4 = (1 - \epsilon) \exp(-\tau) \int_0^\tau T(z) \exp(-\tau'(z) d\tau'(z)$$

The atmospheric attenuation referred to here is the slant-range opacity τ, and for a uniform atmosphere would be given by

$$\tau = \frac{\alpha d}{\cos \theta}$$

where α is the absorption coefficient
        d is the atmospheric depth
        θ is the incidence angle.

In the computer model numerical integration must replace the analytical expressions, and so the atmosphere is represented by a series of 100 uniform layers of equal depth. The variation of atmospheric pressure as a function of height is represented by:

$$P(z) = 1000 \times 10^{-0.0638z}$$

where P = pressure (mb)
           z = height (km).

Similarly the temperature is assumed to vary linearly with altitude, with a 6°/km lapse rate:

$$T(z) = T_0 - 6z$$

where T = temperature (°K)
           T_0 = temperature at sea level.

Although sea level atmospheric temperature can vary by several degrees above or below the sea surface temperature, the approximation that these temperatures are equal is made; it should be remembered that the sea surface temperature considered here is the temperature of a very thin (<1 mm) surface layer.

The density of water vapour, ρ, is assumed to decrease exponentially with a scale height, H, of 2.5 km, so that

$$\rho(z) = \rho_0 e^{-z/H}$$

where ρ_0 is the sea level density.

Clouds are assumed to have uniform density between two fixed levels; in the general model these levels are variable. Rain is assumed to be of constant density between sea level and any specified level.

In the version of the model used in the retrieval algorithm these limiting levels for cloud and rain have to be specified in advance to limit the number of variables, though they can be changed between retrievals.

The absorption coefficients of the relevant constituents will now be discussed.

30.3.3.1 *Water vapour*
Water vapour has a resonance feature at 22.235 GHz.
The expression given by Staelin (1966) and due to Barnett & Chung (1962) is used to find the absorption coefficient of water vapour. This expression is

$$\alpha_{H_2O} = 3.24 \times 10^{-4} \exp\left(-\frac{644}{T}\right) \frac{\nu^2 P_0}{T^{3.125}} \left(1 + 0.0147 \frac{\rho T}{\rho}\right)$$

$$\times \left\{ \frac{1}{(\nu - \nu_0)^2 + \Delta\nu^2} + \frac{1}{(\nu + \nu_0)^2 + \Delta\nu^2} \right\} + 2.55 \times 10^{-8} \frac{\rho\nu^2 \, \Delta\nu}{T^{1.5}} \times 10^5$$

where α_{H_2O} = absorption coefficient (nepers/km)

           ν = frequency (GHz)

           ν_0 = resonance frequency (22.235 GHz)

           T = temperature (°K)

           P = total pressure (mb)

           ρ = water vapour density (g m^{-3})

           $\Delta\nu$ = line width, given by

$$\Delta\nu = 2.58 \times 10^{-3} (P + 0.0147\,\rho T)\left(\frac{T}{318}\right)^{-0.625}$$

30.3.3.2 Oxygen

Oxygen has a series of resonances between 50 and 70 GHz which, at sea level pressure, merge to form a resonance band. The expression used to find the absorption coefficient is that used by Wisler & Hollinger (1977). In this expression resonances due to total angular momentum quantum number transitions are summed, using a given set of 40 molecular magnetic dipole transitions. Pressure line broadening and temperature dependence are also accounted for. The absorption coefficient is given by

$$\alpha_{O_2} = \frac{kP^2\nu^2}{T^2} \sum_N \Phi N \left(f_1 + f_2 + f_3 + f_4 + \frac{0.7W_b}{\nu^2 + (PW_b)^2}\right)$$

where α_{O_2} = absorption coefficient (nepers/km)

           k = constant

           ΦN = fractional population of state N

           $f_{1..4}$ = shape factors for the transition lines

           W_b = non-resonant line width

(Details of these paramters are given by Wisler & Hollinger.)

Summation is carried out over the odd rotational states from 1 to 39. High altitude effects (Zeeman splitting and Doppler broadening) are ignored.

30.3.3.3 Clouds

The expression given by Wisler & Hollinger (1977) is used:

$$\alpha_L = 0.0629 \frac{3E''}{(2 + E')^2 + E''^2} \rho_L \nu$$

where α_L = absorption coefficient of cloud liquid water

           E' and E'' are the real and imaginary parts of the relative

                dielectric constant of water.

           ρ_L = density of cloud liquid water (g m^{-3})

E' and E'' are given by the expressions in secion 30.3.2.2, though in deriving these quantities the salinity is assumed to be zero.

30.3.3.4 *Rain*

The opacity due to rainfall in the SMMR frequency range is found from an approximate expression developed by Gloersen & Barath (1977). This expression is:

$$\alpha_R = -a + (a^{1.2} + b^{1.2} R^{1.2})^{0.833}$$

where

α_R = absorption coefficient due to rain
a = $- 3.41 \times 10^{-2} + 5.55 \times 10^{-2}\lambda - 6.42 \times 10^{-3}\lambda^2$
b = $5.14 \times 10^{-2}\lambda^{-1.85}$
R = rain rate (mm/hr)

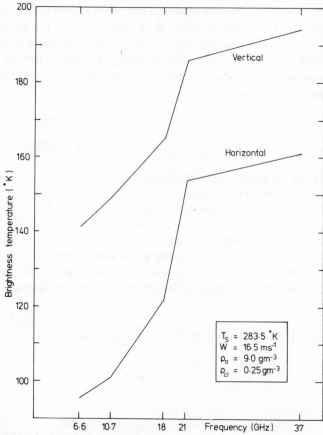

Fig. 30.3 – A typical microwave spectrum generated by the environmental model, with vertical and horizontal polarisations separated. The angle of incidence assumed is 52°, and the values of other geophysical parameters are indicated. Note that there is no rain in this example.

30.3.3.5 *Results*

A pair of spectra, for vertical and horizontal polarisations, which is typical of the results using this model is shown in Fig. 30.3. Many such spectra have been generated, for two main purposes. First, to investigate the variation in the spectra due to changes in the geophysical variables; this showed that a non-linear algorithm was required, and led to the development of the iterative algorithm. Secondly, to generate synthetic sets of microwave measurements to test the retrieval algorithm.

30.4 THE RETRIEVAL ALGORITHM

The overall scheme of the algorithm is shown in Fig. 30.4. It is based on the differential correction method used in orbit determination.

A guess is first made of the values of the parameters; in general these are the values obtained for the last cell. The environmental model then calculates the resulting brightness temperatures. These are compared with the measured brightness temperatures and residuals found, $\Delta\mathbf{T_B}$. If we assume that the residuals are small then we can write the following first-order equations:

$$\Delta T_{B1} = \frac{\partial T_{B1}}{\partial P_1}\,\Delta P_1 \; + \; \frac{\partial T_{B1}}{\partial P_2}\,\Delta P_2 \; + \ldots\ldots + \; \frac{\partial T_{B1}}{\partial P_5}\,\Delta P_5$$

$$\Delta T_{B10} = \frac{\partial T_{B10}}{\partial P_1}\,\Delta P_1 \; + \; \frac{\partial T_{B10}}{\partial P_2}\,\Delta P_2 \; + \ldots\ldots + \; \frac{\partial T_{B10}}{\partial P_5}\,\Delta P_5$$

$$(30.1)$$

where $\mathbf{P}$ is the parameter vector.

Before these may be solved for $\Delta\mathbf{P}$, we need to evaluate the partial derivatives, $(\partial\mathbf{T_B}/\partial\mathbf{P})$. There are 50 of these.

The parital derivatives may be evaluated numerically, since

$$\frac{\partial T_{Bi}}{\partial P_j} \cong T_{Bi}\,\frac{(P_1 . . P_j + \Delta P_j, . . , P_5) \; - T_{Bi} \; (P_1 . . , P_j, . . , P_5)}{\Delta P_j} \quad (30.2)$$

for $i = 1 \ldots 10$
$j = 1 \ldots 5$

Many runs of the environmental model are thus required, changing each parameter by a small amount in turn. The model has been arranged so that only those sections where a change occurs are recalculated, so that this procedure can be carried out as rapidly as possible (the problem of further increasing the speed of the model is discussed later).

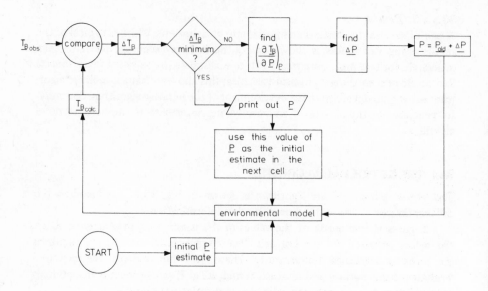

Fig. 30.4 – The overall scheme of the iterative algorithm.

The set of equations (30.1) is over-determined, and so the solutions for ΔP must be found by a 'least-squares' method. The solution is

$$\Delta P = (A^T . W . A)^{-1} A^T . W . \Delta T_B \tag{30.3}$$

where ΔP is the correction to the parameter estimate

   A is the matrix of partial derivatives of
      T_B as a function of P

   W is a weighting matrix for the channels; it is a
      10×10 diagonal matrix whose elements are the square
      of the confidence in each channel

   ΔT_B is the residual vector between the observed and
      calculated brightness temperatures.

The correction, ΔP, may now be applied to the previous estimate and the previous steps retraced.

   Each time the residual, ΔT_B, is found, the r.m.s. value is also calculated. This is used in the test for the final estimate. Obviously, the best estimate of P occurs when the r.m.s. residual is a minimum and it is assumed that here the r.m.s. residual changes only slightly from one iteration to the next. The test for minimum residual is therefore based on there being a small difference between successive r.m.s. residuals.

   Safeguards are included in the program to avoid a diverging or oscillating solution. These are:

a) A maximum of 20 iterations are performed for each set of brightness temperatures.

b) If the r.m.s. residual increases monotonically over 3 iterations the solution is assumed to be diverging and the parameters associated with the minimum residual are printed. Under these circumstances, the initial guess is altered and a second attempt to obtain a solution made. This second attempt normally converges.

c) The correction, ΔP, is not permitted to be greater than a preset value in any one step. This tends to damp out oscillation of the solution.

Error estimates are generated within the algorithm. In equation (30.3) the term $(\mathbf{A}^T \cdot \mathbf{W} \cdot \mathbf{A})^{-1}$ is called the covariance matrix, and the square roots of the diagonal elements of this give the errors in each environmental parameter.

Clearly many runs of the environmental model are required for each retrieval, and it is thus important to reduce the time required for each run as much as possible.

The greatest improvement in speed has been achieved by modifying the division of the atmosphere into layers. Absorption and emission from water vapour make the greatest contribution to the atmospheric component of the brightness temperature, so the thickness of each layer has been adjusted so that they all contain the same quantity of water vapour. By this means, the number of layers necessary has been considerably reduced, since the lower layers may be quite thin whilst higher layers are much thicker. The atmosphere has been modelled up to 8 km as 10 and 15 such layers, as illustrated in Fig. 30.5. A comparison has been made between the brightness temperatures calculated by

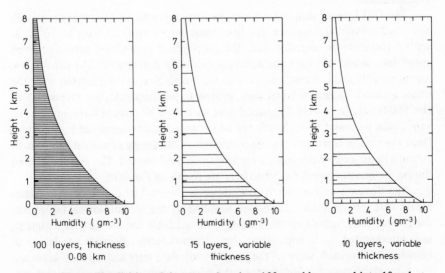

100 layers, thickness 0·08 km

15 layers, variable thickness

10 layers, variable thickness

Fig. 30.5 – The division of the atmosphere into 100 equal layers, and into 10 and 15 non-uniform layers.

this method and by the 100-layer model. The errors introduced by using the 15-layer model were found to be small (0.1°K in most channels; 0.2 and 0.3°K at 21 GHz), while reducing the number of layers to 10 adds only 0.1°K to the errors at 21 GHz.

The resulting increase in speed has been dramatic; the time required for each iteration (i.e. a run of the environmental model, plus evaluation of 50 partial derivatives, the other steps shown in Fig. 30.4 and the various checks and safeguards described) is about 1½ seconds of processing time on an IBM 370/158. Further increases in speed have been achieved by more efficient programming and the use of an optimising compiler.

30.5 EVALUATION OF THE ITERATIVE ALGORITHM

Testing of the iterative retrieval algorithm has been severely restricted by various instrument and data problems encountered by NASA. According to NASA's original schedule, fully calibrated sets of brightness temperatures, accumulated into standard 'cells', should have been available in 1979. In fact, such tapes never became available before the work described here was terminated, though such calibrated tapes are now being produced. Instrument difficulties also occurred. These have been the subject of thorough investigation by the Nimbus SMMR Experiment Team.

The principal instrumental problem has been diagnosed as a thermal variation, although that is not the full extent of hardware difficulties. In addition, some of the preprocessing algorithms developed by NASA have also proved problematical.

The CELL tapes supplied to BAe, in common with all the tapes then available, had unknown biases in the brightness temperatures, of up to 10°K. It should therefore be expected that the geophysical parameters extracted from these data would not have the accuracy originally anticipated. The plan to use measurements made during passes over the North Sea, in conjunction with the dense network of ground-truth sites, proved to be unrealistic; the beamwidth of the SMMR at the lowest frequency was such that an insignificant number of cells could be found in the North Sea which were not contaminated by radiation from the land in adjacent cells. Measurements made during passes over the North Atlantic, well away from land, were therefore used instead. The swaths covered during the measurements described here are shown in Fig. 30.6.

The decision to use North Atlantic passes instead of the North Sea passes meant that ground-truth information supplied by the Meteorological Office was used; these data include all the ground-truth available and are filtered, weighted and interpolated to a convenient grid. Unfortunately, because of delays in reaching this decision, some of the ground-truth data were lost, as they exceeded the normal lifetime for Meteorological Office records of this kind. In fact, the only data available for the whole North Atlantic area for the period for which tapes were available, were of wind speed. Synoptic charts of sea surface temperature, based on statistical climatological information, were also available for the appropriate days. These synoptic data are accurate to within a few degrees.

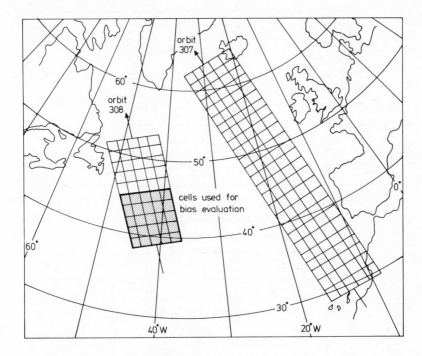

Fig. 30.6 – The SMMR swaths over the North Atlantic, which were used to assess biases, and to evaluate the algorithm.

We have made an attempt to determine the unknown biases in brightness temperature, for certain conditions. Measurements on the daylight side of the orbit only were considered and all the measurements were at (roughly) the same latitude. If the major biases are caused by thermal effects, then, it is argued, they should be roughly constant under these conditions, since the orbit is sun-synchronous. A grid of 25 cells in the North Atlantic was chosen, in an area where it was judged that the atmospheric conditions could be fairly accurately estimated. Then, using these estimates, and the available ground-truth, predictions of the brightness temperatures which should be observed by the SMMR were made using the environmental model. Comparisons with the actual brightness temperature observed allowed the mean biases to be determined. The values obtained are shown in Table. 30.1. This grid of 25 cells was not then used in any further comparisons (see Fig. 30.6).

Table 30.1

Empirical biases of the Nimbus–7 SMMR

Channel		Bias ($^\circ$K)
6.6 GHz	H	5.2
6.6 GHz	V	2.7
10.7 GHz	H	2.6
10.7 GHz	V	−0.8
18 GHz	H	−5.5
18 GHz	V	−15.5
21 GHz	H	12.7
21 GHz	V	9.9
37 GHz	H	5.9
37 GHz	V	1.0

(Biases are defined by

T_B (actual) = T_B (SMMR) − Bias)

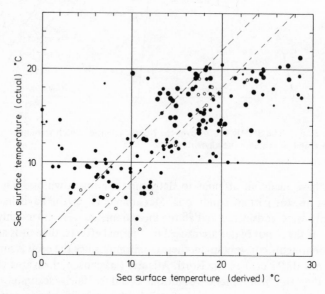

Final RMS residual

- $\bullet$ < 2.5° $\cdot$ > 7.5°
- $\bullet$ 2.5 – 7.5° $\circ$ cell near land

Fig. 30.7 − Comparison between the algorithm-derived sea surface temperature and the ground-truth values. The dashed lines indicate the magnitude of the estimated errors, as described in the text.

Runs of the iterative algorithm were then made using measurements of brightness temperature made in the North Atlantic area during orbits 307 and 308 of Nimbus-7, on 15 November 1978. A comparison of the predicted values of sea surface temperature and wind speed with the ground-truth values is shown in Fig. 30.7 and 30.8 respectively. The points are differentiated according to the r.m.s. residual associated with the parameter estimates, and points near land are also indicated. If the iterative algorithms were giving perfect results, the points would all lie along a line through the origin with slope 45°. This is clearly not the case, and yet, considering the magnitude of the error estimates, the poorly calibrated set of initial data and the crude way in which the biases were obtained, the results are surprisingly good.

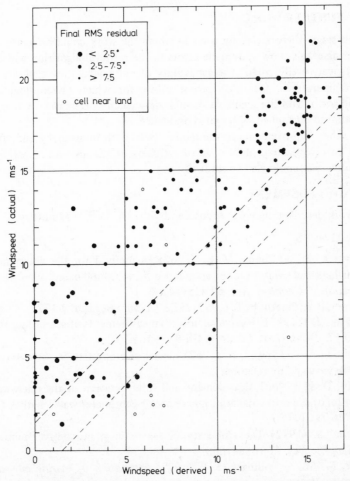

Fig. 30.8 – Comparison between the algorithm-derived wind speed and the ground-truth values.

The standard deviation of the values of parameter dervied from the algorithm is approximately:

Sea surface temperature   ±1°K

Wind speed   ±1 m/s

It should be noted that this does not reflect the overall error of the prediction since it does not include inaccuracies caused by biases arising from, for example, incorrect values of empirical biases bias errors in the model.

The errors in the 'actual' values are believed to be approximately

Sea surface temperature   ±2°K

Wind speed   ±1½ m/s

These are overall errors as judged by the Meteorological Office.

30.6 FURTHER WORK

At this stage, there are many ways in which the work could be continued. The scatter diagrams show systematic errors in the retrieval algorithm which should be relatively easily cured. The availability of more accurately calibrated brightness temperatures, for more recent orbits for which Meteorological Office ground-truth data are available, would allow such corrections to the retrieval algorithm to be applied with more confidence.

Futher improvements in the model itself, both in accuracy and efficiency, could be investigated. More efficient versions of the algorithm, perhaps using look-up tables, could also be studied.

ACKNOWLEDGEMENT

The work described here was sponsored by the UK Dept. of Industry.

REFERENCES

Barrett, A. H. & Chung, V. K. (1962) 'A method for the determination of high-altitude water-vapour abundance from ground-based microwave observation, *J. Geophys. Res.* **67** 4259–4266.

Gloersen, P. & Barath, F. T. (1977) *IEEE Oceanic Eng.* **OE 2** 172.

Hollinger, J. P. (1971) Passive microwave measurements of sea surface roughness *IEEE Trans. Geos. Electron.* **GE-9** (3) 165.

Saxton, J. A. & Lane, J. A. (1952) Electrical properties of sea water, *Wireless Engineer,* p.269 (October).

Staelin, D. H. (1966) Measurement and interpretation of the microwave spectrum of the terrestrial atmosphere near 1-centimeter wavelength *J. Geophys. Res.* **71**, 2875.

Stogryn, A. (1972) The emissivity of sea foam at microwave frequencies, *J. Geophy. Res.* **77** 1658–1666.

Wisler, M. M. & Hollinger, J. P. (1977) *Estimation of marine environmental parameters using microwave radiometric remote sensing systems* N.R.L. memo rept. 3661 (November 1977).

PART 6

Future European Programme

Chapter 31

Status and future plans for ERS-1

G. DUCHOSSOIS
European Space Agency, Paris (France)

31.1 INTRODUCTION

On the 28 October 1981, the Member States of the European Space Agency, joined by Norway and Canada, decided to initiate the first ESA Remote Sensing Satellite (ERS-1) Programme (Table 31.1). This first mission is oriented towards ice and ocean monitoring, the main mission objectives being of both scientific and economic nature and aiming at:

- increasing the scientific understanding of coastal zones and global ocean processes which, together with the monitoring of polar regions, will provide a major contribution to the World Climate Research Programme. ERS data used alone, or more commonly used in conjunction with complementary data from buoys, radio-sondes, research vessels, other near-surface platforms and other satellites in pre-arranged global or regional experiments, will enable significant advances to be made in physical oceanography, glaciology, and climatology;
- developing and promoting economic/commercial applications related to a better knowledge of ocean parameters and sea-state conditions.
  This is of importance in view of the increasing development of coastal and offshore activities and the adoption by countries of the 200 nautical mile economic zone. In addition, monitoring of sea-ice and icebergs will be of importance for industrial activities performed at high latitudes.

In the following sections an overview is given of the present status and future plans for ERS-1 due for launch in 1987 and preparing the way for a fully operational multi-satellite system in the 1990s.

31.2 ERS-1 STATUS

Since 1978 a number of studies have been conducted within European industry aimed at the preliminary definition of the ERS-1 programme including overall mission concept, preliminary instruments definition, and accommodation and data acquisition/processing/distribution strategy. A brief description is provided of the present overall ERS-1 system including the payload configuration, the launcher and satellite platform, the orbit, the ground segment, and the development plan.

Table 31.1
ERS-1 Programme

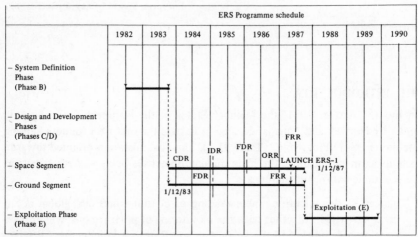

LEGEND
CDR Critical Design Review   FDR Final Design Review   FRR Flight Readiness Review
IDR Intermediate Design Review   QRR Qualification Result Review

Experimental/pre-operational mission

Experimental:

- Instrument validation
- Algorithm testing and tuning
- Ground segment performance validation

Pre-operational:

- Preparation of user community through pilot projects
- Demonstration of operational capability
- International coordination/cooperation

Main Characteristics:

Launch date: end 87
Total Mass: 2250 KG (PL: 950 kg; PFM: 1000 kg; Hydrazine: 300 kg)
Power: 2 Kw
Life duration: 2 years (target 3 years)
Launcher: Ariane 2 or 3

31.2.1 Payload description
The priority in the payload has been given to a comprehensive set of active microwave insturments able to observe as completely as possible the surface

wind and wave structure over the oceans. The set of instruments consists of a Wind Scatterometer to provide wind speeds to an accuracy of 10–20%, and wind directions to within about 20°, and a Radar Altimeter to provide significant wave heights to about 10% accuracy. Further, the SAR on SEASAT was successful in imaging waves, yielding, in most cases, the directions and wavelengths of the dominant waves and potentially the full two-dimensional wave energy spectrum.

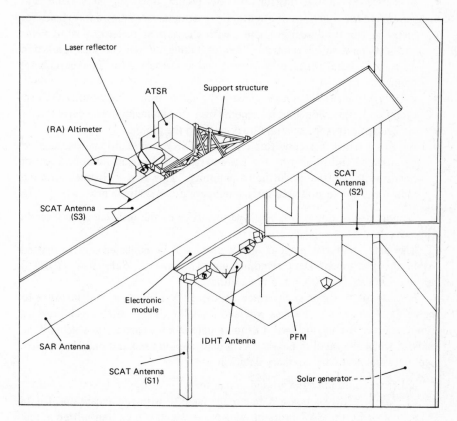

In the selected ERS–1 configuration, SAR small-scene wave images will be interleaved with scatterometer data of the same microwave wavelength in a global sampling scheme, thereby enabling the development of algorithms including the important coupling between the wind, the short back-scattering ripples, and the longer modulating waves. The simultaneous global sampling of wind and wave fields will furthermore enable the application of mutually supportive objective analysis schemes in which the wind and wave fields are estimated jointly through the application of a dynamic wave model.

For ERS-1, the Wind Scatterometer, Wave Scatterometer, and SAR will be combined as an Active Microwave Instrumentation (AMI). This approach leads to a reduction of the required mass, volume and cost by sharing common hardware.

To summarise, therefore, the ERS-1 baseline payload will consist of the following instruments:

- an active microwave instrumentation (AMI) operating in C-Band and combining the functions of a synthetic aperture radar (SAR), a wave scatterometer and a wind scatterometer, with the aim of measuring wind fields and the wave image spectrum and of taking *all-weather* high-resolution images of coastal zones, open oceans and ice areas and (on an experimental basis) over land;
- a Ku-band radar altimeter (RA), with the aim of measuring significant waveheight and of providing measurements over ice and major ocean currents;
- laser retroflectors for accurate tracking from the ground;
- the Along Track Scanning Radiometer (ATSR)†, an additional package, to be provided and funded by the UK and resulting from an announcement of opportunity to the scientific community. This is a 3-channel infra-red radiometer for accurate sea surface temperature measurements.

A short description of the main performance of the instruments is given in Appendix 1.

The data from the payload instruments will be collected and formatted within the Instrument Data Handling and Transmission Subsystem (IDHTS) before transmission at X-band to the ground.

Because of the global character of the ERS-1 mission, and in order to enable data acquisition in areas where the satellite is not within visibility of a ground station, the payload will include a data storage capability which will be read-out when the satellite passes over a dedicated ground station. Thus instrument data and necessary auxiliary data will be:

- either transmitted to the ground in real time,
- or stored in the on-board data storage facility, with the exception of data generated by the SAR in its imaging mode which will be transmitted in real time only,
- or both transmitted in real time to one station while recording for later playback to a dedicated ground station.

Representative sensor combinations to be considered include:

- AMI in its high-resolution imaging mode, either alone, or in combination with the Radar Altimeter. This combination will be mostly used over land, ice, and coastal areas within visibility of ground station visibility.

† A final decision to embark the ATSR will be taken at the end of Phase B

- AMI in its wave and wind measurement interleaved mode, in combination with the Radar Altimeter. This combination will be used over open ocean area. The data are recorded on-board and/or transmitted directly when in ground station visibility.
- ATSR for which it is assumed that it shall function continuously and in combination with all other instruments.

31.2.2 Launcher and satellite platform

ERS-1 will be launched from the Kourou Space Centre (French Guiana) by the Ariane 2 (or 3) launcher as a single passenger, owing to its dimensions.

It will make use of the multi-mission platform (PFM) developed in the framework of the French SPOT programme. Main characteristics of this platform are given in Appendix 2.

31.2.3 Orbit configuration

Considering the conflicting requirements:

- to monitor appropriately fast-changing features such as surface wind fields or sea-state;
- to achieve a global coverage of the Earth's surface with the sensors, the polar caps being accepted exceptions;
- to operate the sensors under constant geometrical conditions and, in some particular cases, under constant local time conditions to ease the data interpretation (AMI imaging mode for important applications);
- to simplify the technical requirements for the satellite thermal control and power generation;

and the diverse characteristics of the sensors selected for ERS-1 in terms of field of view and modes of operation, and the variety of their potential applications, it is difficult to achieve a compromise between coverage and repetition rate which is fully adequate for all sensors and for all potential applications; this difficulty will be overcome by the orbit manoeuvre capability to change the orbit repetition pattern and to test various coverage/repetition rate profiles on an experimental basis.

The nominal orbit is sun-synchronous and circular at an altitude of 777 km, providing a 3-day repetitive ground track pattern. The stability of the ground tracks will be better than ± 5 km with ± 1 km as a goal. The local time of Equator crossing is not frozen, and its optimisation is still the subject of further consideration by the users. However, a local time between 9.30 and 10.30 (descending node) is a realistic assumption. This local time will be maintained within ± 15 minutes.

Main characteristics of the ERS-1 nominal orbit are given in Appendix 3.

31.2.4 Ground segment concept
31.2.4.1 *Mission requirements*
Benefits to be expected from a mission such as ERS–1 will result essentially from the possibility of generating improved short-term and medium-term forecasts of weather and ocean conditions on a local or global basis. All-weather high-resolution imaging from the AMI in the imaging mode will be used to monitor ice coverage in polar or near polar regions or for surface pollution monitoring in combination with optical data from other sources, e.g. LANDSAT, SPOT. As far as ground segment design is concerned, it appears that some products will have to be generated within a few hours for use in operational systems.

Outside these operational users, there are many scientific activities for which longer delivery times are acceptable. This leads to a natural separation of the users' community into two classes: a 'real (or quasi-real) time' class for which the ERS–1 data will have to be processed and delivered within 3 to 6 hours, and a 'non-real time' class for which delivery times from one day to several weeks are adequate.

31.2.4.2 *Instrument duty cycles*
The following duty cycles have been used for the preliminary definition of the ground segment concept.

- HIGH BIT RATE MODE (AMI IMAGING MODE)

  Because of the high bit rate (100 mbps) on-board recording cannot take place and, therefore, data acquisition is only possible from areas in the coverage of available stations. The SAR will operate in a duty cycle of about 10%.

- LOW BIT RATE MODE

  During the low bit rate mode, the following sensors operate with a duty cycle of 40 to 100%:

  Altimeter: $\sim$ 10 kbps
  Wind Scatterometer: $\sim$ 1 kbps
  AMI wave mode: $\sim$ 300 kbps
  ATSR: $\sim$ 100 kbps

31.2.4.3 *Coverage*
ERS–1 is intended to allow worldwide coverage, compatible with the duty cycles of its payload instruments. The Agency will, accordingly, ensure real-time transmission from the satellite and effect any necessary coordination with national facilities for data acquisition.

Furthermore, the on-board tape recorders will give access to data from any part of the world via the playback data acquisition facility, provided that the global amount of data recorded per recording period will have an upper limit of around 5 Gbits.

31.2.4.4 *Deliverable products & their quality*
Level A   Raw Data

Raw data as received from the satellite and recorded on HDDT in satellite format

Level B   Quick annotated raw data/system-corrected data
RT/DT

Raw sensor data, time ordered, time-tagged, earth located. Internal calibration tables are attached but not applied. Geometric auxiliary information will include earth location data, making use of the latest available orbits and attitude information, i.e. the corrections defined in the auxiliary products can be applied in order to generate system-corrected data.

Level C   Quick-Look data
RT/DT

Data, in digital or photographic form, for payload monitoring, browse facility support and, if requested, fast delivery to users. Depending on the sensor, the spatial resolution and radiometric quality could be lower than that of the final product.

Level D   Quick thematic products
RT/DT

Geophysical data, from low bit rate sensors, after application of instrument transfer functions and removal of environmental effects. They will be extracted mainly from Level B data. They include, amongst others, wave spectra from SAR when this instrument is operated in its reduced mode.

RT = real time (normally hours)

DT = delayed time (one day to weeks)

31.2.4.5 *Ground segment concept description*
The ERS–1 ground segment will include

- The mission management and control facilities located at ESOC, and combined in the MMCC (Mission Management & Control Centre),
- One station, located in Kiruna for coverage reasons, for:
  - acquisition of all payload data (DAF: Data Acquisition Facility)
  - telemetry, telecommand and tracking and for provision of near-real-time products (RTPF: Real-Time Processing Facility) of a global nature

For the distribution of the near-real-time data to the users, two possibilities will be investigated:

- Transmission via land lines to regional hubs (WMO solution)
- transmission via satellites such as SIRIO–2/MDD or ECS/ESS.

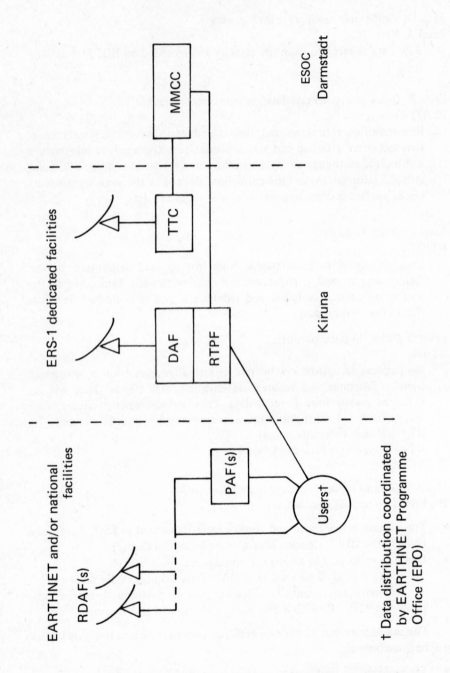

The other facilities of regional interest, RDAF(s) Real-Time Data Acquisition Facilities, and PAF(s) Processing and Archiving Facilities, which are not necessarily ERS-1 dedicated, but participate in the end-to-end data system, should be the elements of the EARTHNET programme or national facilities coordinated by the EARTHNET programme office.

31.2.5 Development plan

On 15 April 1982, most of the ESA Member States, plus Norway and Canada, had already officially stated their decision to finance the detailed system definition phase (Phase B) of the ERS-1 programme within industry. The final decision to proceed with phase C/D (hardware development) will be taken at the end of 1983 at the end of the Phase B.

The development schedule for the ERS-1 is expected to be as follows:

(i) Carrying out of ERS-1 Phase B studies (detailed system and sub-system design) by industry from mid 1982 to end 1983.
(ii) Approval and start of ERS-1 Phase C/D at end of 1983 (development and manufacture of the spacecraft/instruments, plus the setting up of an adequate ground segment).
(iii) Launch of ERS-1 by Ariane at the end of 1987. Lifetime of ERS-1 is expected to be 3 years.

It is worth mentioning that in order to prepare the necessary support technology, the ESA Member States approved in early 1979 the Remote Sensing Preparatory Programme (RSPP) aimed at:

- initiating the development of critical technologies both for optical and microwave payload elements and on-board and ground data management; and
- performing experiment campaigns with sensors embarked on aircraft e.g. campaign SAR 580.

31.3 FUTURE PLANS FOR ERS-1

31.3.1 Experimental/pre-operational system

It is essential when designing the ERS-1 mission to involve the user community in the development of the requirements, the development of the system, and the evaluation of the performance of the system. Two main user communities will make use of ERS-1 data, one scientifically oriented and the other applications oriented, each having different requirements and characteristics. The Agency has made a special effort to involve these communities in the definition of the mission requirements, and intends to continue this effort through the development and exploitation phases of ERS-1, considered as both an experimental and pre-operational system.

It will be experimental since, as the first ESA mission, it will have to demonstrate that the concept and technology for both the space and ground segments are right for the applications envisaged, and that the users are ready and able to use the data generated. This will require a number of activities before and after launch which are very important for the success of the mission, such as:

- simulation and optimisation of sensor performances (airborne testing),
- development and testing of algorithms and models,
- setting up and testing of data products,
- definition of distribution networks to meet user requirements,
- development of pilot-projects (small-scale) and demonstration missions (large-scale),
- continuation of research and development to optimise the use and value of the data provided by the satellite system.

On the other hand, ERS-1 will have to demonstrate, for some appropriate applications and on a limited scale, an operational capability. This will, of course, require that data or products be delivered in quasi-real-time to corresponding existing operational services or end users.

To this effect, the Agency has set up, or is setting up, teams of experts, called Instruments and Data Teams, to advise ESA on all user-related/scientific/ technical aspects of each instrument, and, for the overall system, on the operation aspects, data handling/dissemination, and the promotion aspects of the programme.

Scientific and application-oriented experts from the Participating States will contribute to these teams throughout the development and exploitation phases of the ERS-1 programme.

31.3.2 Preparation of follow-on operational missions

It is expected that, with ERS-1, a gradual transfer of applications from experimental to operational users will take place, preparing the future users for later operational satellite systems. As a first step, although not yet approved, it is planned to propose to the Participating States in ERS-1 to consider the launch of a second flight unit nearly identical to ERS-1 2 or 3 years after the launch of ERS-1, i.e. in 1989-1990, thus providing the user community with 5 to 6 years of continuous data.

In parallel, the Agency is participating in meetings of remote sensing satellite operators, and will host the second meeting of these operators, currently including Canada, Japan, India, USA, France, and ESA. This offers the possibility for coordination of missions of the same nature (i.e. land-oriented or ocean-oriented) with the aim of improving services to users in terms of measurement frequency, standardisation of products etc.

Lastly, the ultimate goal would be to set up or to contribute to an operational multi-satellite system for global rather than regional monitoring.

31.4 CONCLUSION

Within Europe there is a particular interest in a satellite monitoring programme over such important economic areas as the oceans, the coastal zones, and ice caps. Equally, there exists an active European scientific community which recognises the potential value of modern spaceborne sensors to longer-term studies of ocean processes and their relationships to climatological problems.

In conclusion, it is clear that if Europe wants to be present on the world scene with operational remote sensing satellite systems in the 1990s, it is necessary to develop now an experimental/pre-operational satellite programme with a first launch around 1987. All necessary capabilities, whether industrial, scientific or economic, exist in Europe, and the positive decision which has been recently taken by ESA Member States plus Canada and Norway to develop the ERS-1 programme should allow Europe not only to be present in the world competition but also to provide a contribution to future worldwide systems aiming at global monitoring of coastal zones and oceans.

APPENDIX 31.1

PRELIMINARY INSTRUMENT PERFORMANCES

AMI Imaging Mode (SAR)
Spatial resolution
$\quad$ 100 m X 100 m or 30 m X 30 m
Swath
$\quad$ 80 km minimum
Radiometric resolution
$\quad$ 1 dB for 100 X 100 m at $\sigma_0 = -18$ dB
$\quad$ 2.5 dB for 30 X 30 m at $\sigma_0 = -18$ dB
Incidence angle (on the ground) $\cong 23°$ at mid swath
Frequency
$\quad$ 5.3 GHz in HH polarisation
Mean RF power
$\quad$ < 400 W
Data rate
$\quad \cong 100$ Mb/s
Antenna size
$\quad \cong 10$ m X 1 m

AMI Wave Mode
Spectral samples
$\quad$ From <100 m to 1000 m in 12 log. steps,
$\quad$ resolution equal to step size

Angular samples
   Over $180°$ in azimuth in steps $<30°$, resolution equal to step size
Spatial samples
   Each 100 km along track looking at 5 km square
Sample accuracy
   $\pm 20\%$ spectral energy density of σ_0 spectrum (target figures)
Incidence angle
   $\cong 23°$ side-looking
Frequency
   5.3 GHz in HH polarisation
Mean RF power
   <100 W
Data rate
   $\cong 300$ kbps

AMI Wind Mode
Spatial resolution
   50 km
Wind speed
   4 to 24 m s^{-1}
Accuracy
   2 m s^{-1} or 10%
Wind direction accuracy
   $<20°$ (1σ); acceptable level of ambiguities TBD† (design goal: 0)
Swath
   min. 400 km one-sided ($25°$ to $55°$ incident)
Polarisation
   VV, 3 antenna beams
Frequency
   5.3 GHz
Antenna size
   3.6 m long, 0.3 m high, 2 off
   2.5 m long, 0.3 m high, 1 off
Data rate
   ~ 1 kbps

Radar Altimeter (RA)
Altitude measurement
   <10 cm (goal 5 cm) (1σ, 1 sec)
Significant wave height
   1 . . . . . 20 m ($\pm 10\%$ or 5 m)

†Under investigation with simulated data products

Backscatter coefficient measurement
   ± 1 dB (1σ)
Frequency
   13.5 GHz
Bandwidth
   300 MHz
Peak power
   500 W
Antenna size
   1 m ϕ
Data rate
   8.5 kbps

Along Track Scanning Radiometer (ATSR)‡
Spectral channels,
   3.7, 11 and 12 microns
Spatial resolution
   1 km $\times$ 1 km square
Radiometric resolution
   0.1° K
Absolute accuracy predicted
   0.5° K over 50 km $\times$ 50 km square in 80% cloud cover conditions
Swath width
   500 km
Power consumption
   48 W
Data rate
   110 kbps (after data reduction)
Mass
   30 kg

APPENDIX 31.2
   Main characteristics of the Multi-Mission Platform (PFM)

Compatibility with sun-synchronous, circular orbits of altitudes between 600 km
   and 1200 km
Compatibility with local time of satellite passes (ascending and descending node:
   08.00 to 16.00 hours)
Available power: 1.9 KW (beginning of life) provided by a solar generator
Power storage capacity: 4 $\times$ 23 Ampere hours

‡It is at present planned to complement the ATSR with a 2- or 4-channel microwave nadir
sounder.

Attitude control performance:
  Pointing towards Earth centre with yaw steering capability
  Stability on yaw: $1.1 \; 10^{-3}$ °/sec
  Pitch and roll: 7.10^{-4} °/sec
  Accuracy: $0.15°$ for all axes
Attitude measurement accuracy:
  Better than $0.15°$
  Use of Fine Attitude Measurement System (**FAMS**) with star sensor could improve above performance
Orbit control:
  Parallel and perpendicular to the orbit plane
  Maximum capability 580,000 Nsec (300 kg of hydrazine)
Satellite management:
  By on-board computer with 20 K words (16 bits) memory available for the payload management
Communications with ground:
  Telemetry and Telecommand via an S-band transponder compatible with ESA and NASA networks
  Transmitted power: up to 200 mW
  Data rate: 2 Kbits/sec.

APPENDIX 31.3
          Main characteristics of the ERS-1 Nominal Orbit

Semi-major axis a: 7046 km
Inclination i: $98.1°$
Eccentricity e: 10^{-3}
Nodal Period T: 98.18 min
Repeat cycle: 3 days
Orbital periods per day: $14\frac{2}{3}$

Index